Instructor's Guide / Test Bank for

MICROBIOLOGY
AN INTRODUCTION

SEVENTH EDITION

TORTORA • FUNKE • CASE

CHRISTINE L. CASE

Benjamin
Cummings

An Imprint of Addison Wesley Longman

San Francisco   Boston   New York
Capetown   Hong Kong   London   Madrid   Mexico City
Montreal   Munich   Paris   Singapore   Sydney   Tokyo   Toronto

ISBN 0-8053-7584-8

2 3 4 5 6 7 8 9 10—VG—05 04 03 02

Addison Wesley Longman, Inc.
1301 Sansome Street
San Francisco, California 94111

CONTENTS

CHAPTER TESTS   253

PREFACE

There is no substitute for an instructor's interaction with students, and a student's enthusiasm for learning is stimulated by your presence. This *Instructor's Guide for Microbiology: An Introduction,* Seventh Edition, provides some new ideas and reinforcement for teaching your course. If you are just beginning to teach microbiology, this guide can provide a framework for developing your course.

This guide is divided into three sections. The first section, *Introduction,* includes several alternative course outlines for use with *Microbiology: An Introduction,* Seventh Edition. For presentation of microbial diseases by etiology (taxonomic group), portal of entry, or method of transmission, sequences of topics and pertinent pages in the text are listed in this section.

The second section, *Chapter Notes,* includes six subsections: Learning Objectives, Chapter Summary, The Loop, New in This Edition, Answers to Study Questions, and Case History. *Learning Objectives* provides an overview of the chapter contents. They are the same as the objectives in the text. The scope of each chapter is highlighted in *Chapter Summary.* Cross-references to other chapters are listed in *The Loop.*

For the many users of the Sixth Edition, changes and additions new to this Seventh Edition are highlighted in the section *New in This Edition. Answers to Study Questions* in the text follows the Chapter Summary. These answers are brief but should be sufficient to provide you with insight regarding our intent in asking particular questions. Answers to the end-of-chapter Review and Multiple Choice questions are also available on the web site (http://www.microbiologyplace.com). Answers to the Critical Thinking and Clinical Applications questions will not be posted on this website, so you can use these as homework or test questions if you wish. The *Case History* in each chapter can be used for testing, class discussion, or homework. They can also be found online at http://www.microbiologyplace.com. The case histories require analysis and application of new information; additionally, many require quantitative analysis. In Microbiology for Allied Health at Skyline College, we make selected case histories available for extra credit. The students can choose one or two to turn in. Many are suitable for use as essay questions on tests; however, keep in mind that students will need time to think through the problems. "Microtriviology" (Chapter 11) can be used to encourage students to use reference materials such as *Bergey's Manual.*

The third section is *Chapter Tests.* Each chapter has objective test questions with answers provided in the left margin. The tests can be reproduced and used directly from the guide to test students' recall and understanding of material presented in the text. Essay questions or analytical problems for each chapter are also provided. These same test questions are also available in TestGen 3.0, a cross-platform CD-ROM, which you can request from your Benjamin Cummings representative.

Introduction

THE FIRST DAY

The first day of a semester is hectic. Introductory remarks on regulations, requirements, and grading are needed; roll must be taken; and students are adding and dropping classes. Generally, the first day is not a good day to present material on which students will be tested.

I begin my class by distributing the course syllabus and explaining it. Then I show 35-mm color slides for the remainder of the period. Since students are usually unfamiliar with microbiology, slides can introduce the subject with pictures of representative organisms, laboratory cultures, and environmental and industrial applications. A discussion of food and agriculture generates an awareness of the importance of microbiology.

Course content can emphasize general microbiology with examples and specific details from medical microbiology and biotechnology. First-day slides focus on ecology and applied microbiology. I sometimes give the students study questions that can be answered from the slide narration and Chapter 1.

EDUCATIONAL TECHNOLOGY

Transparencies

Acetate overhead transparencies of 366 full-color line drawings from the text are available free to instructors who adopt *Microbiology: An Introduction*, Seventh Edition.

Transparencies are a useful teaching tool because they accurately illustrate structures or events and eliminate the need to spend time carefully drawing on the chalkboard during a lecture. Moreover, transparencies can be used in a fully lit room so that the students can take notes.

The use of transparencies copied directly from the text will stimulate the students to use their text as a reference. Additionally, the students will not have to copy the entire transparency but can make notes in their text or take notes that refer to a specific figure in the text.

Slides

Eighty color slides are available free to instructors who adopt *Microbiology: An Introduction*, Seventh Edition. The slides present biological agents of disease and HIV-related topics with corresponding clinical photos.

Web Site

The web site includes the study outline, answers to selected study questions, new multiple choice test questions, updated feature boxes from the text, links to other web sites, and a monthly microbe identification exercise. Students can also email questions to the authors at the web site.

Interactive Student Tutorial CD

An Interactive Student Tutorial CD (v.2.0) is included with every textbook. The CD includes review questions for each chapter and interactive tutorials. Additionally, students can perform virtual experiments on enzyme activity, PCR, bacterial growth, and BOD

SUGGESTED USES FOR SPECIAL FEATURES IN MICROBIOLOGY: AN INTRODUCTION, SEVENTH EDITION

Learning Objectives

The objectives at the beginning of each major heading focus the student's attention on major concepts presented in the text. You may wish to modify the objectives into mastery objectives. To do this, identify the performance and conditions necessary for the student to show the desired competence.

For example, an additional sentence in the third objective on p. 7 of *Microbiology: An Introduction* ("Identify the contributions...") would define the test conditions. The students can be told whether whey will be expected to identify contributions from a list or write an essay on the historical background of microbiology including contributions made by five of these people.

Objective 2, p. 7 of *Microbiology: An Introduction* ("Compare the theories...") should tell the student what test conditions to expect. The student might anticipate writing an essay or making a list to show differences between these theories. Should the student expect to differentiate between these theories by providing supporting evidence for each theory? Sample questions are sometimes useful to clarify an objective.

Additions and deletions can be made to the lists of objectives to suit your needs.

Study Questions

Four levels of study questions are provided at the end of each chapter. The Review level allows students to test their recall of information. The Multiple Choice section includes questions that require recall and questions that require analysis. The Critical Thinking level provides problems that require knowledge and reasoning. Actual case histories are included in the Clinical Applications questions. Answers to the study questions are provided in this guide.

Study questions can be a basis for class discussion.

Further Reading

Further Reading is included on the web site. Efforts were made to provide references for each chapter that are a balance between general reading for breadth, advanced texts, and scientific papers. Many of the references used by the authors are listed. Most, if not all, of these references will be found in college libraries.

Learning to use references is an integral part of education. Too often, a lecture leaves the student with the notion that what they just heard is all there is to know. Periodically, I give library assignments. The students receive a sheet of questions that can be answered using selected references. Selections from Further Reading can be used this way.

A special reference section for Part Four of *Microbiology: An Introduction*, Seventh Edition, is provided. These are textbooks that are used in medical microbiology. I find that allied health students use these texts frequently for additional information on "unknown" reports or special topics on diseases.

MMWR, Clinical Problem-Solving, and Microbiolgy in the News boxes will be updated monthly on the microbiology web site.

Appendices

Pertinent topics in the Chapter Notes in this guide are cross-referenced to the Appendices in the text.

Mycology and Parasitology

In addition to bacteriology and virology, an overview of mycology and parasitology is provided in Chapter 12, and representative diseases are included in Chapters 21 through 26. A few examples from mycology and parasitololgy can provide students with an introduction to general biological principles as well as broaden their concept of disease-causing organisms.

The content of your course is determined by you and the other faculty involved in allied health programs. Most health personnel must have some familiarity with a wide range of disease-causing organisms. Your state public health department publishes reference material on diseases that may occur in your geographic area. When discussing mycology and parasitology, I find it useful to refer to organisms that local clinicians have encountered.

Scheduling Topics

The following outline is suggested for a one-semester course. It is based on 45 fifty-minute lectures.

Topic	Number of Lectures
Welcome and First-Day Business	1
The Microbial World and You	1
Chemical Principles	3
Observing Microorganisms Through a Microscope	1
Functional Anatomy of Prokaryotic and Eukaryotic Cells	3
Microbial Metabolism	3
Microbial Growth	1
The Control of Microbial Growth	2
Microbial Genetics	3
Biotechnology and Recombinant DNA	1
Classification of Microorganisms	0.5
Bacteria and Archaea	1
Fungi	1
Protozoa and Algae	1
Multicellular Parasites	1
Viruses	2
Principles of Disease and Epidemiology	0.5
Microbial Mechanisms of Pathogenicity	0.5
Nonspecific Defenses of the Host	1
Specific Defenses of the Host: The Immune Response	3
Disorders Associated with the Immune System	2
Practical Applications of Immunology	0.5
Antimicrobial Drugs	1
Microbial Diseases of the Skin and Eyes	1
Nosocomial Infections	1
Microbial Diseases of the Nervous System	1
Microbial Diseases of the Cardiovascular and Lymphatic Systems	1
Microbial Diseases of the Respiratory System	2
Microbial Diseases of the Digestive System	2
Microbial Diseases of the Urinary and Reproductive Systems	1
Environmental Microbiology	1.5
Applied and Industrial Microbiology	0.5

The text is flexible and can be adapted to suit the schedule you prefer. On the following pages, selected topics are grouped to assist you in preparing your course outline.

Biotechnology

Biochemistry

Many instructors do not cover basic chemistry (Chapter 2) as a lecture topic because chemistry is a prerequisite to their microbiology courses. Some instructors feel that they can incorporate the necessary basic concepts of chemistry into metabolism and genetics. In either case, Chapter 2 can provide a review for the students.

The following sections deal with the biochemical process in living cells.

Control of Microbial Growth

Environmental Microbiology

Acquired Immunodeficiency Syndrome (AIDS)

Cancer

Alternative Course Outlines

Specific diseases and etiologies can be covered by systems, taxa, or methods of transmission. It is up to you to decide which diseases need to be covered for each group of students. An instructor may wish to emphasize bacterial diseases but include a representative disease caused by a virus, fungus, protozoan, and helminth for comparison and breadth. Some nonbacterial agents are important causes of diseases worldwide. In a class in which all the students are in an allied health program, all the microbial diseases relevant to those students could be presented. For example, respiratory therapy students need to learn about diseases of the respiratory and circulatory systems. Although bacterial and viral diseases are the most common, protozoan and multicellular parasites will be encountered in clinical work. Additionally, liberal arts students often find examples from parasitology interesting.

Taxonomic Approach

For a listing of pathogens that enter through the skin/mucous membranes and parenteral route or by vectors, see the following section, Method of Transmission Approach.

Method of Transmission Approach

A discussion of the transmission of disease is on pages 417–419. Diseases acquired through the respiratory tract are usually transmitted by direct contact including droplet infection. Diseases acquired through the gastrointestinal tract are most often transmitted by indirect contact in food and water. These diseases are listed in the Portal of Entry Approach section of this guide.

Pathogens that enter through the skin/mucous membranes and parenteral route have the most varied methods of transmission and are listed below.

Diseases Acquired by Direct Contact Through the Skin/Mucous Membranes
Bacterial

Animal Reservoirs

This is a list of diseases acquired from animals by direct contact, indirect contact, or arthropod vectors. See p. 416.

ASM Curriculum Recommendations

ASM recommends the following core curriculum guidelines for all introductory microbiology courses. Microbiology: An Introduction, Seventh Edition and Laboraotry Experiments in Microbiology, Sixth Edition, support all of the ASM's curriculum recommendations. These core curriculum gudelines are meant to support the development of learning objectives that can be met within the introductory microbiology courses. The asterisks in the list below denote those themes and concepts considered essential to the laboratory content.

	Chapter(s) in Microbiology: An Introduction, Seventh Edition	Exercise(s) in Laboratory Experiments in Microbiology, Sixth Edition
Theme 1: Microbial cell biology*		
1. Information flow within a cell	8	27–28
2. Regulation of cellular activities	8	
3. Cellular structure and function*	4	3–7
4. Growth and division*	6, 28	20
5. Cell energy metabolism*	5	13–17
Theme 2: Microbial genetics*		27–31
1. Inheritance of genetic information	8	
2. Causes, consequences, and uses of mutations*	8–9	
3. Exchange and acquisition of genetic information	8	
Theme 3: Interactions and impact of microorganisms and humans*		
1. Host defense mechanisms	16–19	41–44
2. Microbial pathogenicity mechanisms*	15, 21–26	39–40, 45–49
3. Disease transmission	14	39–40
4. Antibiotics and chemotherapy*	7, 20	24–25
5. Genetic engineering	9	28, 30
6. Biotechnology	9, 28	30, 54, 56
Theme 4: Interactions and impact of microorganisms in the environment*		51–56
1. Adaptation and natural selection	8	
2. Symbiosis	27	
3. Microbial recycling of resources	27	
4. Microbes transforming the environment	6, 27–28	
Theme 5: Integrating themes*		
1. Microbial evolution	10	
2. Microbial diversity*	11–13, 27	32–38, 55

TRANSPARENCY MASTER

MENU À LA LABORATOIRE

Soup
Miso (soybeans arranged by *Aspergillus* and *Saccharomyces*)

Salad
Olives prepared by *Leuconostoc*
fleshy fungi (*Agaricus*) grown on thoroughbred
manure and seasoned by *Acetobacter* (vinegar)

Entrées
S. cerevisiae and (by request) *Lactobacillus* will prepare
rye, pumpernickel, and sourdough

Hawaiian Single-Cell Protein
A delightful casserole of sewage-fed cyanobacteria. Flavored with poi
(lactic-acid bacteria work their magic on taro root)

Thai Noodles
Noodles "proteinized" with *Torulopsis* yeast and flavored with fish sauce
made by a team of moderately halophilic *Bacillus* and coyneforms

Beef Bonanza
Tender slices of *Methylophilus*-fed beef marinated in soy sauce
(produced by a symphony of microbes)

Carne Macha
An assortment of sausages from *Pediococcus* and *Penicillium italicum*

Vegetable
Cabbage fermented to pH 3.5 by *L. plantarum*. Beans by *Bacillus natto*.

Desserts
* Chocolate prepared by *Kluyveromyces* and lactic acid bacteria.
Chef *Leu CoNostoc* will smother it in dextran, an α-1,6-glucose polymer
* Assorted cheeses
Streptococcus and *Lactobacillus*, assisted by *P. roquefortii* and *P. camenberti*

Drinks
Alcohol served by the sweet   Nonalcoholic beverages
fungus *Saccharomyces*   *Lactobacillus'* buttermilk
Beer   *Saccharomyces'* root beer
Wine   *Erwinia's* coffee

Chapter 1 *The Microbial World and You*

Learning Objectives

1. List several ways in which microbes affect our lives.
2. Recognize the system of scientific nomenclature that uses genus and specific epithet names.
3. Differentiate among the major groups of organisms studied in microbiology.
4. Explain the importance of observations made by Hooke and van Leeuwenhoek.
5. Compare the theories of spontaneous generation and biogenesis.
6. Identify the contributions to microbiology made by Needham, Spallanzani, Virchow, and Pasteur.
7. Identify the importance of Koch's postulates.
8. Explain how Pasteur's work influenced Lister and Koch.
9. Identify the importance of Jenner's work.
10. Identify the contributions to microbiology made by Ehrlich, Fleming, and Dubos.
11. Define bacteriology, mycology, parasitology, immunology, and virology.
12. Explain the importance of recombinant DNA technology.
13. List at least four beneficial activities of microorganisms.
14. List two examples of biotechnology that use genetic engineering and two examples that do not.
15. Define normal microbiota and resistance.
16. Define and describe several infectious diseases.
17. Define emerging infectious disease.

NEW IN THIS EDITION

- Classification of organisms into Bacteria, Archaea, and Eukarya domains.
- Emerging infectious disease discussion includes bovine spongiform encephalopathy, *E. coli* O157:H7, Ebola hemorrhagic fever, *Hantavirus* pulmonary syndrome, and cryptosporidiosis.

CHAPTER SUMMARY

Microbes in Our Lives

1. Living things too small to be seen with the unaided eye are called microorganisms.
2. Microorganisms are important in the maintenance of an ecological balance on Earth.
3. Some microorganisms live in humans and other animals and are needed to maintain the animal's health.
4. Some microorganisms are used to produce foods and chemicals.
5. Some microorganisms cause disease.

Naming and Classifying Microorganisms

1. In a nomenclature system designed by Carolus Linnaeus (1735), each living organism is assigned two names.
2. The two names consist of a genus and a specific epithet, both of which are underlined or italicized.

Types of Microorganisms

Bacteria

1. Bacteria are unicellular organisms. Because they have no nucleus, the cells are described as prokaryotic.
2. The three major basic shapes of bacteria are bacillus, coccus, and spiral.
3. Most bacteria have a peptidoglycan cell wall; they divide by binary fission; and they may possess flagella.
4. Bacteria can use a wide range of chemical substances for their nutrition.

Archaea

1. Archaea have prokaryotic cells; they lack peptidoglycan in their cell walls.
2. Archaea include methanogens, halophiles, and extreme thermophiles.

Fungi

1. Fungi (mushroom, molds, and yeasts) have eukaryotic cells (with a true nucleus). Most fungi are multicellular.
2. Fungi obtain nutrients by absorbing organic material from their environment.

Protozoa

1. Protozoa are unicellular eukaryotes.
2. Protozoa obtain nourishment by absorption or ingestion through specialized structures.

Algae

1. Algae are unicellular or multicellular eukaryotes that obtain nourishment by photosynthesis.
2. Algae produce oxygen and carbohydrates that are used by other organisms.

Viruses

1. Viruses are noncellular entities that are parasites of cells.
2. Viruses consist of a nucleic acid core (DNA or RNA) surrounded by a protein coat. An envelope may surround the coat.

Multicellular Animal Parasites

1. The principal groups of multicellular animal parasites are flatworms and roundworms, collectively called helminths.
2. The microscopic stages in the life cycle of helminths are identified by traditional microbiological procedures.

Classification of Microorganisms

1. All organisms are classified into Bacteria, Archaea, and Eukarya. Eukarya includes Protists, Fungi, Plants, and Animals.

A Brief History of Microbiology

The First Observations

1. Robert Hooke observed that plant material was composed of "little boxes"; he introduced the term *cell* (1665).

2. Hooke's observations laid the groundwork for development of the cell theory, the concept that all living things are composed of cells.

3. Antoni van Leeuwenhoek, using a simple microscope, was the first to observe microorganisms (1673).

The Debate over Spontaneous Generation

1. Until the mid-1880s, many people believed in spontaneous generation, the idea that living organisms could arise from nonliving matter.

2. Francesco Redi demonstrated that maggots appear on decaying meat only when flies are able to lay eggs on the meat (1668).

3. John Needham claimed that microorganisms could arise spontaneously from heated nutrient broth (1745).

4. Lazzaro Spallanzani repeated Needham's experiments and suggested that Needham's results were due to microorganisms in the air entering his broth (1765).

5. Rudolf Virchow introduced the concept of biogenesis: Living cells can arise only from preexisting cells (1858).

6. Louis Pasteur demonstrated that microorganisms are in the air everywhere and offered proof of biogenesis (1861).

7. Pasteur's discoveries led to the development of aseptic techniques used in laboratory and medical procedures to prevent contamination by microorganisms that are in the air.

The Golden Age of Microbiology

1. Rapid advances in the science of microbiology were made between 1857 and 1914.

Fermentation and Pasteurization

1. Pasteur found that yeast ferments sugars to alcohol and that bacteria can oxidize the alcohol to acetic acid.

2. A heating process called pasteurization is used to kill bacteria in some alcoholic beverages and milk.

The Germ Theory of Disease

1. Agostino Bassi (1835) and Pasteur (1865) showed a causal relationship between microorganisms and disease.

2. Joseph Lister introduced the use of a disinfectant to clean surgical dressings in order to control infections in humans (1860s).

3. Robert Koch proved that microorganisms caused disease. He used a sequence of procedures called Koch's postulates (1876), which are used today to prove that a particular microorganism causes a particular disease.

Vaccination

1. In a vaccination, immunity (resistance to a particular disease) is conferred by inoculation with a vaccine.
2. In 1798, Edward Jenner demonstrated that inoculation with cowpox material provides humans with immunity from smallpox.
3. About 1880, Pasteur discovered that avirulent bacteria could be used as a vaccine for fowl cholera; he coined the word *vaccine*.
4. Modern vaccines are prepared from living avirulent microorganisms or killed pathogens, from isolated components of pathogens, and by recombinant DNA techniques.

The Birth of Modern Chemotherapy: Dreams of a "Magic Bullet"

1. Chemotherapy is the chemical treatment of a disease.
2. Two types of chemotherapeutic agents are synthetic drugs (chemically prepared in the laboratory) and antibiotics (substances produced naturally by bacteria and fungi that inhibit the growth of other microorganisms).
3. Paul Ehrlich introduced an arsenic-containing chemical called salvarsan to treat syphilis (1910).
4. Alexander Fleming observed that the mold (fungus) *Penicillium* inhibited the growth of a bacterial culture. He named the active ingredient penicillin (1928).
5. Penicillin has been used clinically as an antibiotic since the 1940s.
6. In 1939, René Dubos discovered two antibiotics produced by the bacterium *Bacillus*.
7. Researchers are tackling the problem of drug-resistant microbes.

Modern Developments in Microbiology

1. Bacteriology is the study of bacteria, mycology is the study of fungi, and parasitology is the study of parasitic protozoa and worms.
2. Microbiologists are using genomics, the study of all of an organism's genes, to classify bacteria, fungi, and protozoa.
3. The study of AIDS, analysis of interferon action, and the development of new vaccines are among the current research interests in immunology.
4. New techniques in molecular biology and electron microscopy have provided tools for advancement of our knowledge of virology.
5. The development of recombinant DNA technology has helped advance all areas of microbiology.

Microbes and Human Welfare

1. Microorganisms degrade dead plants and animals and recycle chemical elements to be used by living plants and animals.
2. Bacteria are used to decompose organic matter in sewage.
3. Bioremediation processes use bacteria to clean up toxic wastes.

4. Bacteria that cause diseases in insects are being used as biological controls of insect pests. Biological controls are specific for the pest and do not harm the environment.

5. Using microbes to make products such as foods and chemicals is called biotechnology.

6. Using recombinant DNA, bacteria can produce substances such as proteins, vaccines, and enzymes.

7. In gene therapy, viruses are used to carry replacements for defective or missing genes into human cells.

8. Genetic engineering is used in agriculture to protect plants from frost and insects and to improve the shelf life of produce.

Microbes and Human Disease

1. Everyone has microorganisms in and on the body; these make up the normal microbiota or flora.

2. The disease-producing properties of a species of microbe and the host's resistance are important factors in determining whether a person will contract a disease.

3. An infectious disease is one in which pathogens invade a susceptible host.

4. An emerging infectious disease (EID) is a new or changing disease, showing an increase in incidence in the recent past or a potential to increase in the near future.

Contributions to the field of microbiology by the following individuals are noted in this chapter.

Oswald Avery	Robert Koch	Louis Pasteur
Augostino Bassi	Rebecca Lancefield	Francesco Redi
George Beadle	Laurent Lavoisier	Ignaz Semmelweis
Martinus Beijerink	Joshua Lederberg	Lazzaro Spallanzani
Francis Crick	Antoni van Leeuwenhoek	Wendell Stanley
René Dubos	Carolus Linnaeus	Edward Tatum
Paul Ehrlich	Joseph Lister	Rudolf Virchow
Alexander Fleming	Colin MacLeod	James Watson
Robert Hooke	Maclyn McCarty	Chaim Weizmann
Dmitri Iwanowsky	Jacques Monod	Sergei Winogradsky
François Jacob	John Needham	Carl Woese
Edward Jenner		

THE LOOP

The organisms studied in microbiology are defined. Topics introduced in the overview of microbiology can be covered in more depth by reading the following sections:

Bioremediation   pp. 17, 752–753
Classification   Chapter 10
Emerging infectious diseases   pp. 423–425
Industrial microbiology/biotechnology   Chapters 9 and 28
Koch's postulates   pp. 410–412
Vaccines   pp. 500–506

Answers

Review

1. The observations of flies coming out of manure and maggots coming out of dead animals, and the appearance of microorganisms in liquids after a day or two, led people to believe that living organisms arose from nonliving matter.

2. Pasteur's S-neck flasks allowed air to get into the beef broth, but the curves of the S trapped bacteria before they could enter the broth.

3. a. Certain microorganisms cause diseases in insects. Microorganisms that kill insects can be effective biological control agents because they are specific for the pest and do not persist in the environment.

   b. Carbon, oxygen, nitrogen, sulfur, and phosphorus are required for all living organisms. Microorganisms convert these elements into forms that are useful for other organisms. Many bacteria decompose material and release carbon dioxide into the atmosphere for plants to use. Some bacteria can take nitrogen from the atmosphere and convert it into a form that can be used by plants and other microorganisms.

   c. Normal microbiota are microorganisms that are found in and on the human body. They do not usually cause disease, and can be beneficial.

   d. Organic matter in sewage is decomposed by bacteria into carbon dioxide, nitrates, phosphates, sulfate, and other inorganic compounds in a wastewater treatment plant.

   e. Recombinant DNA techniques have resulted in insertion of the gene for insulin production into bacteria. These bacteria can produce human insulin inexpensively.

   f. Microorganisms can be used as vaccines. Some microbes can be genetically engineered to produce components of vaccines.

4. **Matching**

a, c	Studies biodegradation of toxic wastes.
h	Studies the causative agent of *Hantavirus* pulmonary syndrome.
a, d, f	Studies the production of human proteins by bacteria.
b	Studies the symptoms of AIDS.
e	Studies the production of toxin by *E. coli*.
c	Studies the life cycle of *Cryptosporidium*.
c	Develops gene therapy for a disease.
f	Studies the fungus *Candida albicans*.

5. **Matching**

l	Avery, MacLeod and McCarty
o	Beadle and Tatum
p	Berg
b	Dubos
r	Ehrlich
d	Fleming
j	Hooke
k	Iwanowski
c	Jacob and Monod

a	Jenner
m	Koch
s	Lancefield
e	Lederberg and Tatum
h	Lister
f	Pasteur
g	Stanley
i	van Leeuwenhoek
n	Virchow
q	Weizmann

6. *Erwinia carotovora* is the correct way to write this scientific name. Scientific names can be derived from the names of scientists. In this case, *Erwinia* is derived from Erwin F. Smith, an American plant pathologist. Scientific names also can describe the organism, its habitat, or its niche. *E. carotovora* is a pathogen of carrots (*vora* = "eat").

7. **Matching**

d	Algae
g	Archaea
c	Bacteria
b	Fungi
f	Helminths
e	Protozoa
a	Viruses

8. a. *B. thuringiensis* is sold as a biological insecticide.

   b. *Saccharomyces* is the yeast sold for making bread, wine, and beer.

Critical Thinking

1. Pasteur showed that life comes from preexisting life. The microorganisms that produced chemical and physical changes in beef broth and wine came from a few cells that entered the liquids from dust, containers, or the air. After showing that microorganisms could both grow on and change organic matter, Pasteur and others began to suspect that diseases were the result of microorganisms growing on living organic matter.

2. Semmelweis had observed an increased incidence of fever when medical students worked in obstetrics, as compared to the incidence during the students' summer break. The medical students were carrying bacteria from the autopsy room. Lister observed that compound bone fractures could result in death while recovery from simple fractures was without incident.

3. There are many! Check the dairy section for fermented products such as sour cream, yogurt, and cheese. Protein supplements often are yeasts. Bread, wine, and beer are products of yeasts and some bacteria. Sauerkraut is cabbage that has been fermented by lactobacilli. Vinegar is produced by bacterial growth on ethyl alcohol (wine). Xanthan, a thickener in many foods, is made by *Xanthomonas* bacteria.

4. Factors contributing to infectious disease include: mutations in existing organisms, spread of diseases to new areas, ecological disturbances such as deforestation, lack of immunization, pesticide resistance, and antibiotic resistance.

Clinical Applications

1. a. Treatment with penicillin suggests a bacterial cause because only bacterial diseases are treatable with this drug. The summer onset also suggested an infectious disease perhaps related to an outdoor activity such as swimming or contact with mosquitoes or ticks.
   b. Lyme disease.
   c. The tick vector is more active during these months. Additionally, people spend more time outdoors and potentially in contact with ticks during these months.

2. Pasteur showed that microbes were omnipresent and were responsible for "diseases" (i.e., spoilage) of food; Lister reasoned that these microbes might be responsible for diseases of people. Neither Lister nor Pasteur proved that microbes caused diseases. Koch provided a repeatable proof to demonstrate that a microbe causes a disease.

Case History: Are Ulcers an Infectious Disease?

Background

In 1981, the following information came to the attention of Barry Marshall, a gastroenterologist at the Royal Perth Hospital in Australia. Household members of ulcer patients do not develop antibodies against *Helicobacter*. However, clinical staff involved in obtaining biopsy samples from ulcer patients develop antibodies against *Helicobacter*. If acid-suppressive therapy is combined with antibiotics, ulcers usually do not recur. Marshall concluded that ulcers are an infectious disease.

Questions

What caused Marshall to reach his conclusion? What additional proof would be needed?

The Solution

The presence of antibodies against *Helicobacter* is evidence of current or prior infection by the organism. Exchange of bacteria of the intestinal and skin microbiota, which is normal among household members, does not transmit *Helicobacter*, but direct contact with stomach contents does. Marshall collected the additional proof by demonstrating Koch's postulates. Healthy volunteers were inoculated with *Helicobacter*; they developed symptoms of the disease; and the *Helicobacter* was recovered from them.

Chapter 2   *Chemical Principles*

Learning Objectives

1. Discuss the structure of an atom and its relation to the chemical properties of elements.
2. Define ionic bond, covalent bond, hydrogen bond, molecular weights, and mole.
3. Diagram three basic types of chemical reactions.
4. Identify the role of enzymes in chemical reactions.
5. List several properties of water that are important to living systems.
6. Define acid, base, salt, and pH.
7. Distinguish between organic and inorganic compounds.
8. Identify the building blocks of carbohydrates, simple lipids, phospholipids, proteins, and nucleic acids.
9. Identify the role of ATP in cellular activities.

NEW IN THIS EDITION

- Enhanced discussion of enzymes.
- New art program uses color consistently to provide a visual aid for understanding concepts.

CHAPTER SUMMARY

Introduction

1. The science of the interaction between atoms and molecules is called chemistry.
2. The metabolic activities of microorganisms involve complex chemical reactions.
3. Nutrients are broken down by microbes to obtain energy and to make new cells.

The Structure of Atoms

1. Atoms are the smallest units of chemical elements that enter into chemical reactions.
2. Atoms consist of a nucleus, which contains protons and neutrons, and electrons that move around the nucleus.
3. The atomic number is the number of protons in the nucleus; the total number of protons and neutrons is the atomic weight.

Chemical Elements

1. Atoms with the same number of protons and the same chemical behavior are classified as the same chemical element.

2.   Chemical elements are designated by abbreviations called chemical symbols.

3.   About 26 elements are commonly found in living cells.

4.   Atoms that have the same atomic number (are of the same element) but different atomic weights are called isotopes.

Electronic Configurations

1.   In an atom, electrons are arranged around the nucleus in electron shells.

2.   Each shell can hold a characteristic maximum number of electrons.

3.   The chemical properties of an atom are largely due to the number of electrons in its outermost shell.

How Atoms Form Molecules

Chemical Bonds

1.   Molecules are made up of two or more atoms; molecules consisting of at least two different kinds of atoms are called compounds.

2.   Atoms form molecules in order to fill their outermost electron shells.

3.   Attractive forces that bind the atomic nuclei of two atoms together are called chemical bonds.

4.   The combining capacity of an atom—the number of chemical bonds the atom can form with other atoms—is its valence.

Ionic Bonds

1.   A positively or negatively charged atom or group of atoms is called an ion.

2.   A chemical attraction between ions of opposite charge is called an ionic bond.

3.   To form an ionic bond, one ion is an electron donor and the other ion is an electron acceptor.

Covalent Bonds

1.   In a covalent bond, atoms share pairs of electrons.

2.   Covalent bonds are stronger than ionic bonds and are far more common in organisms.

Hydrogen Bonds

1.   A hydrogen bond exists when a hydrogen atom covalently bonded to one oxygen or nitrogen atom is attracted to another oxygen or nitrogen atom.

2.   Hydrogen bonds form weak links between different molecules or between parts of the same large molecule.

Molecular Weight and Moles

1.   The molecular weight is the sum of the atomic weights of all the atoms in a molecule.

2.   A mole of an atom, ion, or molecule is equal to its atomic or molecular weight expressed in grams.

Chemical Reactions

1.   Chemical reactions are the making or breaking of chemical bonds between atoms.

Energy in Chemical Reactions

1.   A change of energy occurs during chemical reactions.

2. Endergonic reactions require energy; exergonic reactions release energy.
3. In a synthesis reaction, atoms, ions, or molecules are combined to form a larger molecule.
4. In a decomposition reaction, a larger molecule is broken down into its component molecules, ions, or atoms.
5. In an exchange reaction, two molecules are decomposed, and their subunits are used to synthesize two new molecules.
6. The products of reversible reactions can readily revert back to form the original reactants.

How Chemical Reactions Occur
1. For a chemical reaction to take place, the reactants must collide with each other.
2. The minimum collision energy that can produce a chemical reaction is called its activation energy.
3. Specialized proteins called enzymes accelerate chemical reactions in living systems by lowering the activation energy.

Important Biological Molecules

Inorganic Compounds
1. Inorganic compounds are usually small, ionically bonded molecules.
2. Water and many common acids, bases, and salts are examples of inorganic compounds.

Water
1. Water is the most abundant substance in cells.
2. Because water is a polar molecule, it is an excellent solvent.
3. Water is a reactant in many of the decomposition reactions of digestion.
4. Water is an excellent temperature buffer.

Acids, Bases, and Salts
1. An acid dissociates into H^+ ions and anions.
2. A base dissociates into OH^- ions and cations.
3. A salt dissociates into negative and positive ions, neither of which is H^+ or OH^-.

Acid-Base Balance
1. The term pH refers to the concentration of H^+ in a solution.
2. A solution with a pH of 7 is neutral; a pH below 7 indicates acidity; a pH above 7 indicates alkalinity.
3. A pH buffer, which stabilizes the pH inside a cell, can be used in culture media.

Organic Compounds
1. Organic compounds always contain carbon and hydrogen.
2. Carbon atoms form up to four bonds with other atoms.
3. Organic compounds are mostly or entirely covalently bonded, and many of them are large molecules.

Structure and Chemistry

1. A chain of carbon atoms forms a carbon skeleton.
2. Functional groups of atoms are responsible for most of the properties of organic molecules.
3. The letter R may be used to denote the remainder of an organic molecule.
4. Frequently encountered classes of molecules are R—OH (alcohols), R—COOH (organic acids), H₂N—R—COOH (amino acids).
5. Small organic molecules may combine into very large molecules called macromolecules.
6. Monomers usually bond together by dehydration synthesis or condensation reactions that form water and a polymer.
7. Organic molecules may be broken down by hydrolysis, a reaction involving the splitting of water molecules.

Carbohydrates

1. Carbohydrates are compounds consisting of atoms of carbon, hydrogen, and oxygen, with hydrogen and oxygen in a 2:1 ratio.
2. Carbohydrates include sugars and starches.
3. Carbohydrates can be classified as monosaccharides, disaccharides, and polysaccharides.
4. Monosaccharides contain from three to seven carbon atoms.
5. Isomers are two molecules with the same chemical formula but different structures and properties—for example, glucose ($C_6H_{12}O_6$) and fructose ($C_6H_{12}O_6$).
6. Monosaccharides may form disaccharides and polysaccharides by dehydration synthesis.

Lipids

1. Lipids are a diverse group of compounds distinguished by their insolubility in water.
2. Simple lipids (fats) consist of a molecule of glycerol and three molecules of fatty acids.
3. A saturated lipid has no double bonds between carbon atoms in the fatty acids; an unsaturated lipid has one or more double bonds. Saturated lipids have higher melting points than unsaturated lipids.
4. Phospholipids are complex lipids consisting of glycerol, two fatty acids, and a phosphate group.
5. Steroids have carbon ring structures; sterols have a functional hydroxyl group.

Proteins

1. Amino acids are the building blocks of proteins.
2. Amino acids consist of carbon, hydrogen, oxygen, nitrogen, and sometimes sulfur.
3. Twenty amino acids occur naturally.
4. By linking amino acids, peptide bonds (formed by dehydration synthesis) allow the formation of polypeptide chains.
5. Proteins have four levels of structure: primary (sequence of amino acids), secondary (regular coils or pleats), tertiary (overall three-dimensional structure of a polypeptide), and quaternary (two or more polypeptide chains).
6. Conjugated proteins consist of amino acids combined with other organic or inorganic compounds.

Nucleic Acids

1. Nucleic acids—DNA and RNA—are macromolecules consisting of repeating nucleotides.

2. A nucleotide is composed of a pentose, a phosphate group, and a nitrogenous base. A nucleoside is composed of a pentose and a nitrogenous base.

3. A DNA nucleotide consists of deoxyribose (a pentose) and one of the following nitrogenous bases: thymine or cytosine (pyrimidines), or adenine or guanine (purines).

4. DNA consists of two strands of nucleotides wound in a double helix. The strands are held together by hydrogen bonds between purine and pyrimidine nucleotides: AT and GC.

5. An RNA nucleotide consists of ribose (a pentose) and one of the following nitrogenous bases: cytosine, guanine, adenine, or uracil.

6. Genes consist of sequences of nucleotides.

Adenosine Triphosphate (ATP)

1. ATP stores chemical energy for various cellular activities.

2. When the bond to ATP's terminal phosphate group is broken, energy is released.

3. The energy from decomposition reactions is used to regenerate ATP from ADP and phosphate.

THE LOOP

1. Have students study Chapter 2 and use the Study Questions as a self-test.

2. Have students study Chapter 2 and take a pretest for Chapter 5. Pretests can be administered individually during office hours, open laboratories, or study sessions. Fifteen questions from the Chapter 2 test bank can be given, and a student can show mastery with a score of at least 9 points. If a student does not achieve mastery, he or she can study and take a second chapter test.

3. Students with some chemistry but less than one year of college chemistry may find it useful to have the last half of this chapter, "Important Biological Molecules" (pp. 36–51), used as an introduction to Chapter 5, "Microbial Metabolism."

Answers

Review

1. Atoms with the same atomic number and chemical behavior are classified as chemical elements.
2. Refer to Figure 2.1.
3. ^{14}C and ^{12}C are isotopes of carbon. ^{12}C has 6 neutrons in its nucleus and ^{14}C has 8 neutrons.
4. Hydrogen bonds.
5. a. Ionic
   b. Single covalent bond
   c. Double covalent bonds
   d. Hydrogen bond
6. 10^4 or 10,000 times.
7.

Element	Atomic Weight	×	Number of Atoms	=	Total Weight of That Element
C	12	×	6	=	72
H	1	×	12	=	12
O	16	×	6	=	96

The molecular weight of $C_6H_{12}O_6$ is 180 grams.
8. a. Synthesis reaction, condensation, or dehydration
   b. Decomposition reaction, digestion, or hydrolysis
   c. Exchange reaction
   d. Reversible reaction
9. The enzyme lowers the activation energy required for the reaction, and therefore speeds up this decomposition reaction.
10. a. Lipid
    b. Protein
    c. Carbohydrate
    d. Nucleic acid

11.

a. Acetic acid

b. Ethyl alcohol

c. Acetaldehyde

d. Ethanolamine

e. Diethylether

12. a. Amino acids
    b. Right to left
    c. Left to right

13. Breaking of bonds between <u>phosphorus and oxygen</u>. These are <u>covalent</u> bonds.

14. The entire protein shows tertiary structure. No quaternary structure.

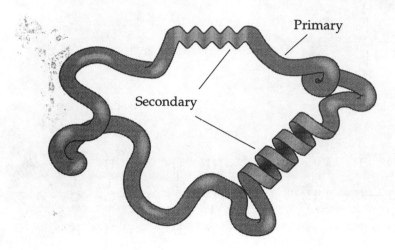

15.

Critical Thinking

1. a. Synthesis reaction.

   b. H_2CO_3 is an acid.

2. ATP and DNA have 5-carbon sugars. ATP has ribose, and DNA has deoxyribose; ATP and DNA contain the purine, adenine.

3. In order to maintain the proper fluidity, the percentage of unsaturated lipids decreases at the higher temperature.

4. These animals have cellulose-degrading bacteria in specialized structures in their digestive tracts.

Clinical Applications

1. The enzymes will degrade organic molecules on the clothing. Any stains caused by these molecules will be removed when the molecules are digested.

2. *T. ferrooxidans* can oxidize sulfur ("thio") as well as iron ("ferro"). The oxidation of sulfide in pyrite produces sulfuric acid, which dissolves the limestone. Gypsum forms in a subsequent exchange reaction.

$$2S^{2-} + 3O_2 + 2H_2O \longrightarrow 2SO_4^{2-} + 4H^+$$
$$2CaCO_3 + 4H^+ + 2SO_4^{2-} \longrightarrow 2CaSO_4 + 2H^+ + 2HCO_3^-$$

3. Since L-isomers are more common in nature, most cells, such as phagocytes, will be able to degrade the L-isomers. D-isomers will be resistant to metabolism by most cells.

4. Amphotericin B would not work against most bacteria because they lack sterols. Fungi have sterols and are generally susceptible to amphotericin B. Human cells have sterols.

Case History: Kesterson National Wildlife Refuge

Background

Kesterson National Wildlife Refuge, California. In the San Joaquin Valley, irrigation water wasn't draining properly, and crops were dying in the water-logged soil. In 1981, a drainage system was built that would channel irrigation runoff into shallow ponds called Kesterson Reservoir. In addition to receiving field runoff, the new reservoir was to be a waterfowl habitat. In 1983, an unusually large number of dead birds was found, indicating that something was wrong with the water.

Selenium from the soil (in the form of selenate, SeO_4^{2-}) was dissolving in the irrigation water and being carried to Kesterson Reservoir, where it stayed. The concentration of selenium in Kesterson rose to 29 times higher than that which was considered safe.

For now, to prevent the killing of more birds, Kesterson Reservoir is being filled with soil. But the selenium remains.

A number of bacteria, including *Bacillus*, *Acinetobacter*, and *Pseudomonas*, can convert selenate (SeO_4^{2-}) to nontoxic elemental selenium. The bacteria do this for their own survival—to prevent the accumulation of toxic levels in their cells.

Question

What is the chemical reaction that shows how the bacteria make Se^0, using hydrogen sulfide (H_2S) and selenate?

The Solution

$$SeO_4^{2-} + 4H_2S \longrightarrow Se^0 + 4H_2O + 4S^0$$

Chapter 3   *Observing Microorganisms Through a Microscope*

Learning Objectives

1. List the metric units of measurement used for microorganisms, and their metric equivalents.
2. Diagram the path of light through a compound microscope.
3. Define total magnification and resolution.
4. Identify a use for darkfield, phase-contrast, DIC, and fluorescence microscopy, and compare each with brightfield illumination.
5. Explain how electron microscopy differs from light microscopy.
6. Identify one use for the TEM, SEM, and scanned-probe microscope.
7. Differentiate between an acidic dye and a basic dye.
8. Compare simple, differential, and special stains.
9. List the steps in preparing a Gram stain, and describe the appearance of gram-positive and gram-negative cells after each step.
10. Compare and contrast the Gram stain and the acid-fast stain.
11. Explain why each of the following is used: capsule stain, endospore stain, flagella stain.

NEW IN THIS EDITION

- Scanned-probe microscopes are discussed.
- New photographs illustrate microscopic techniques.

CHAPTER SUMMARY

Units of Measurement

1. The standard unit of length is the meter (m).
2. Microorganisms are measured in micrometers, μm (10^{-6} m) and nanometers, nm (10^{-9} m).

Microscopy: The Instruments

1. A simple microscope consists of one lens; a compound microscope has multiple lenses.

Light Microscopy
Compound Light Microscopy

1. The most common microscope used in microbiology is the compound light microscope (LM).

2. We calculate the total magnification of an object by multiplying the magnification of the objective lens by the magnification of the ocular lens.
3. The compound light microscope uses visible light.
4. The maximum resolving power (ability to distinguish two points) of a compound light microscope is 0.2 μm; maximum magnification is 2000$\times$.
5. Specimens are stained to increase the difference between the refractive indexes of the specimen and the medium.
6. Immersion oil is used with the oil immersion lens to reduce light loss between the slide and the lens.
7. Brightfield illumination is used for stained smears.
8. Unstained cells are more productively observed using darkfield, phase-contrast, or DIC microscopy.

Darkfield Microscopy

1. The darkfield microscope shows a light silhouette of an organism against a dark background.
2. It is most useful to detect the presence of extremely small organisms.

Phase-Contrast Microscopy

1. The phase-contrast microscope uses a special condenser to enhance differences in the refractive indexes of the cell's parts and its surroundings.
2. It allows the detailed observation of living organisms.

Differential Interference Contrast (DIC) Microscopy

1. The DIC microscope provides a colored, three-dimensional image of the object being observed.
2. It allows detailed observations of living cells.

Fluorescence Microscopy

1. In fluorescence microscopy, specimens are first stained with fluorochromes and then viewed through a compound microscope by using an ultraviolet (or near-ultraviolet) light source.
2. The microorganisms appear as bright objects against a dark background.
3. Fluorescence microscopy is used primarily in a diagnostic procedure called fluorescent antibody technique.

Confocal Microscopy

1. In confocal microscopy, a specimen is stained with a fluorescent dye and illuminated one plane at a time.
2. Using a computer to process the images, two- and three-dimensional images of cells can be produced.

Electron Microscopy

1. A beam of electrons, instead of light, is used with an electron microscope.
2. Electromagnets, instead of glass lenses, control focus, illumination, and magnification.

3. Thin sections of organisms can be seen in an electron micrograph produced using a transmission electron microscope (TEM). Magnification: 10,000× to 100,000×. Resolving power: 2.5 nm.

4. Three-dimensional views of the surfaces of whole microorganisms can be obtained with a scanning electron microscope (SEM). Magnification: 1000× to 10,000×. Resolving power: 20 nm.

Scanned-Probe Microscopy

1. Scanning tunneling microscopy (STM) and atomic force microscopy (AFM) produce three-dimensional images of the surface of a molecule.

Preparation of Specimens for Light Microscopy

Preparing Smears for Staining

1. Staining means coloring a microorganism with a dye to make some structures more visible.

2. Fixing uses heat or alcohol to kill and attach microorganisms to a slide.

3. A smear is a thin film of material used for microscopic examination.

4. Bacteria are negatively charged, and the colored positive ion of a basic dye will stain bacterial cells.

5. The colored negative ion of an acidic dye will stain the background of a bacterial smear; a negative stain is produced.

Simple Stains

1. A simple stain is an aqueous or alcohol solution of a single basic dye.

2. It is used to make cellular shapes and arrangements visible.

3. A mordant may be used to improve bonding between the stain and the specimen.

Differential Stains

1. Differential stains, such as the Gram stain and acid-fast stain, divide bacteria into groups according to their reactions to the stains.

2. The Gram stain procedure uses a purple stain (crystal violet), iodine as a mordant, an alcohol decolorizer, and a red counterstain.

3. Gram-positive bacteria retain the purple stain after the decolorization step; gram-negative bacteria do not and thus appear pink from the counterstain.

4. Acid-fast microbes, such as members of the genera *Mycobacterium* and *Nocardia*, retain carbolfuchsin after acid-alcohol decolorization and appear red; non–acid-fast microbes take up the methylene blue counterstain and appear blue.

Special Stains

1. Stains such as the endospore stain and flagella stain color only certain parts of bacteria.

2. Negative staining is used to make microbial capsules visible.

THE LOOP

Chapter 3 should provide a good reference for laboratory exercises on microscopy and staining. The test questions can be used as laboratory quizzes.

Answers

Review

1. $1 \mu m = 10^{-6} m$
    $1 nm = 10^{-9} m$
    $1 \mu m = 10^3 nm$

2. a. Ocular lens
    b. Objective lens
    c. Diaphragm
    d. Condensor
    e. Illuminator

3. Ocular lens magnification ×   oil immersion lens magnification   =   total magnification of specimen
    10×   ×   100×   =   1000×

4. a. Compound light microscope
    b. Darkfield microscope
    c. Phase-contrast microscope
    d. Fluorescence microscope
    e. Electron microscope
    f. Differential interference contrast microscope

5. …that <u>a beam of electrons</u> focused by <u>magnets</u>…on <u>a television-like screen or photographic plate</u>.

6.
Type of Microscope	Maximum Magnification	Resolution
Compound light	2,000×	0.2 μm
Electron	100,000×	0.0025 μm

7. Bacterial cells have a slightly negative charge, and the colored positive ion of a basic dye is attracted to the negative charge of the cell. Acid dyes do not stain bacterial cells because the negatively charged colored ion is repelled by the like charge of the cell.

8. a. A simple stain is used to determine cell shape and arrangement.
    b. A differential stain is used to distinguish kinds of bacteria based on their reaction to the differential stain.
    c. A negative stain does not distort the cell and is used to determine cell shape, size, and the presence of a capsule.
    d. A flagella stain is used to determine the number and arrangement of flagella.

9. In a Gram stain, the mordant combines with the basic dye to form a complex that will not wash out of gram-positive cells. In a flagella stain, the mordant accumulates on the flagella so that they can be seen with a light microscope.

10. A counterstain stains the colorless non–acid-fast cells so that they are easily seen through a microscope.

11. In the Gram stain, the decolorizer removes the color from gram-negative cells. In the acid-fast stain, the decolorizer removes the color from non–acid-fast cells.

12. Endospore: safranin is the *counterstain*.
    Gram: safranin is the *counterstain*.

13.

Steps	Appearance after this step of	
	Gram-positive cells	Gram-negative cells
Crystal violet	Purple	Purple
Iodine	Purple	Purple
Alcohol-acetone	Purple	Colorless
Safranin	Purple	Red

Critical Thinking

1. The counterstain safranin can be omitted. Gram-positive bacteria will appear purple, and gram-negative bacteria will be colorless.

2. You would be able to discern two objects separated by the four distances given because each is equal to or greater than the resolving power of the microscope.

3. The high lipid content of acid-fast cell walls makes them impermeable to most stains. If the primary stain penetrates, the Gram stain decolorizer will not decolorize the cell. Therefore, acid-fast bacteria would be gram-positive if they could be Gram stained.

4. Inclusions as well as endospores may not stain in a Gram stain. The endospore stain will identify the unstained structure as an endospore.

Clinical Applications

1. Ehrlich observed that mycobacteria could not be decolorized with acid-alcohol, so he reasoned that an acidic disinfectant would not be able to penetrate the cell wall.

2. *N. gonorrhoeae* bacteria are gram-negative (red) diplococci, often found in the large human cells (phagocytes).

3. These are called clue cells. The large red cells are human mucosal cells; gram-positive bacteria on the surface of the human cells.

4. The presence of acid-fast rods suggests the elephant had a mycobacterial infection. Subsequent cultures verified the elephant had tuberculosis.

Case History: Electron Microscopy

Background

Samples prepared for transmission electron microscopy are embedded in an epoxy resin and sliced into ultrathin (100 nm) sections. The sections are usually stained with a heavy metal such as lead to enhance contrast. The sections are then examined with a transmission electron microscope. The photographs of these thin sections are put in order according to their position in the living cell and used to determine the shape of the original sample.

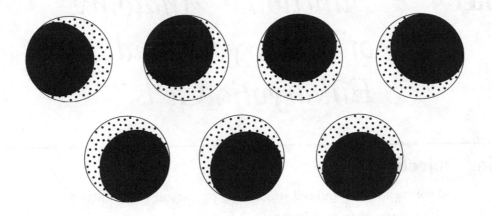

Question
Determine the appearance of the original, intact cell.

The Solution
A spirochete, see Figure 4.9. The solid area is the cross section of the cell body, and the "dots" are cross sections of axial filaments. The changing positions of the cell and axial filaments are due to the spiraling of the axial filaments around the cell.

Chapter 4 — Functional Anatomy of Prokaryotic and Eukaryotic Cells

Learning Objectives

1. Compare and contrast the overall cell structure of prokaryotes and eukaryotes.
2. Identify the three basic shapes of bacteria.
3. Describe the structure and function of the glycocalyx, flagella, axial filaments, fimbriae, and pili.
4. Compare and contrast the cell walls of gram-positive bacteria, gram-negative bacteria, archaea, and mycoplasmas.
5. Differentiate between protoplast and spheroplast.
6. Describe the structure, chemistry, and functions of the prokaryotic plasma membrane.
7. Define simple diffusion, osmosis, facilitated diffusion, active transport, and group translocation.
8. Identify the functions of the nuclear area, ribosomes, and inclusions.
9. Describe the functions of endospores, sporulation, and endospore germination.
10. Differentiate between prokaryotic and eukaryotic flagella.
11. Compare and contrast prokaryotic and eukaryotic cell walls and glycocalyxes.
12. Compare and contrast prokaryotic and eukaryotic plasma membranes.
13. Compare and contrast prokaryotic and eukaryotic cytoplasms.
14. Define organelle.
15. Describe the functions of the nucleus, endoplasmic reticulum, ribosomes, Golgi complex, lysosomes, vacuoles, mitochondria, chloroplasts, peroxisomes, and centrosomes.
16. Discuss evidence that supports the endosymbiotic theory of eukaryotic evolution.

NEW IN THIS EDITION

- Revised discussion of eukaryotic cells to include peroxisomes and centrosomes.
- New art illustrates eukaryotic cells.

CHAPTER SUMMARY

Comparing of Prokaryotic and Eukaryotic Cells: An Overview

1. Prokaryotic and eukaryotic cells are similar in their chemical composition and chemical reactions.
2. Prokaryotic cells lack membrane-enclosed organelles (including a nucleus).

3. Peptidoglycan is found in prokaryotic cell walls but not in eukaryotic cell walls.
4. Eukaryotic cells have a membrane-bound nucleus and other organelles.

The Prokaryotic Cell

1. Bacteria are unicellular, and most of them multiply by binary fission.
2. Bacterial species are differentiated by morphology, chemical composition, nutritional requirements, biochemical activities, and source of energy.

The Size, Shape, and Arrangement of Bacterial Cells

1. Most bacteria are from 0.20 to 2.0 µm in diameter and from 2 to 8 µm in length.
2. The three basic bacterial shapes are coccus (spheres), bacillus (rods), and spiral (twisted).
3. Pleomorphic bacteria can assume several shapes.

Structures External to the Cell Wall

Glycocalyx

1. The glycocalyx (capsule, slime layer, or extracellular polysaccharide) is a gelatinous polysaccharide and/or polypeptide covering.
2. Capsules may protect pathogens from phagocytosis.
3. Capsules allow adherence to surfaces, prevent desiccation, and may provide nutrients.

Flagella

1. Flagella are relatively long filamentous appendages consisting of a filament, hook, and basal body.
2. Prokaryotic flagella rotate to push the cell.
3. Motile bacteria exhibit taxis; positive taxis is movement toward an attractant, and negative taxis is movement away from a repellent.
4. Flagellar (II) protein functions as an antigen.

Axial Filaments

1. Spiral cells that move by means of an axial filament (endoflagellum) are called spirochetes.
2. Axial filaments are similar to flagella, except that they wrap around the cell.

Fimbriae and Pili

1. Fimbriae and pili are short, thin appendages.
2. Fimbriae help cells adhere to surfaces.
3. Pili join cells for the transfer of DNA from one cell to another.

The Cell Wall

Composition and Characteristics

1. The cell wall surrounds the plasma membrane and protects the cell from changes in water pressure.

2. The bacterial cell wall consists of peptidoglycan, a polymer consisting of NAG and NAM and short chains of amino acids.
3. Penicillin interferes with peptidoglycan synthesis.
4. Gram-positive cell walls consist of many layers of peptidoglycan and also contain teichoic acids.
5. Gram-negative bacteria have a lipoprotein–lipopolysaccharide–phospholipid outer membrane surrounding a thin peptidoglycan layer.
6. The outer membrane protects the cell from phagocytosis and from penicillin, lysozyme, and other chemicals.
7. Porins are proteins that permit small molecules to pass through the outer membrane; specific channel proteins allow other molecules to move through the outer membrane.
8. The lipopolysaccharide component of the outer membrane consists of sugars that function as antigens and lipid A, which is an endotoxin.

Cell Walls and the Gram Stain Mechanism

1. The crystal violet–iodine complex combines with peptidoglycan.
2. The decolorizer removes the lipid outer membrane of gram-negative bacteria and washes out the crystal violet.

Atypical Cell Walls

1. *Mycoplasma* is a bacterial genus that naturally lacks cell walls.
2. Archaea have pseudomurein; they lack peptidoglycan.
3. L forms are mutant bacteria with defective cell walls.

Damage to the Cell Wall

1. In the presence of lysozyme, gram-positive cell walls are destroyed, and the remaining cellular contents are referred to as a protoplast.
2. In the presence of lysozyme, gram-negative cell walls are not completely destroyed, and the remaining cellular contents are referred to as a spheroplast.
3. Protoplasts and spheroplasts are subject to osmotic lysis.
4. Antibiotics such as penicillin interfere with cell wall synthesis.

Structures Internal to the Cell Wall

The Plasma (Cytoplasmic) Membrane

1. The plasma membrane encloses the cytoplasm and is a phospholipid bilayer with peripheral and integral protein (the fluid mosaic model).
2. The plasma membrane is selectively permeable.
3. Plasma membranes carry enzymes for metabolic reactions, such as nutrient breakdown, energy production, and photosynthesis.
4. Mesosomes, irregular infoldings of the plasma membrane, are artifacts.
5. Plasma membranes can be destroyed by alcohols and polymyxins.

Movement of Materials Across Membranes

1. Movement across the membrane may be by passive processes, in which materials move from areas of higher to lower concentration and no energy is expended by the cell.
2. In simple diffusion, molecules and ions move until equilibrium is reached.
3. In facilitated diffusion, substances are transported by transporter proteins across membranes from areas of high to low concentration.
4. Osmosis is the movement of water from areas of high to low concentration across a selectively permeable membrane until equilibrium is reached.
5. In active transport, materials move from areas of low to high concentration by transporter proteins, and the cell must expend energy.
6. In group translocation, energy is expended to modify chemicals and transport them across the membrane.

Cytoplasm

1. Cytoplasm is the fluid component inside the plasma membrane.
2. The cytoplasm is mostly water, with inorganic and organic molecules, DNA, ribosomes, and inclusions.

The Nuclear Area

1. The nuclear area contains the DNA of the bacterial chromosome. Bacteria can also contain plasmids, which are extrachromosomal DNA circles.

Ribosomes

1. The cytoplasm of a prokaryote contains numerous 70S ribosomes; ribosomes consist of rRNA and protein.
2. Protein synthesis occurs at ribosomes; it can be inhibited by certain antibiotics.

Inclusions

1. Inclusions are reserve deposits found in prokaryotic and eukaryotic cells.
2. Among the inclusions found in bacteria are metachromatic granules (inorganic phosphate), polysaccharide granules (usually glycogen or starch), lipid inclusions, sulfur granules, carboxysomes (ribulose 1,5-diphosphate carboxylase), magnetosomes (Fe_3O_4), and gas vacuoles.

Endospores

1. Endospores are resting structures formed by some bacteria for survival during adverse environmental conditions.
2. The process of endospore formation is called sporulation; the return of an endospore to its vegetative state is called germination.

The Eukaryotic Cell

Flagella and Cilia

1. Flagella are few and long in relation to cell size; cilia are numerous and short.

2. Flagella and cilia are used for motility, and cilia also move substances along the surface of the cells.
3. Both flagella and cilia consist of an arrangement of nine pairs and two single microtubules.

The Cell Wall and Glycocalyx

1. The cell walls of many algae and some fungi contain cellulose.
2. The main material of fungal cell walls is chitin.
3. Yeast cell walls consist of glucan and mannan.
4. Animal cells are surrounded by a glycocalyx that strengthens the cell and provides a means of attachment to other cells.

The Plasma (Cytoplasmic) Membrane

1. Like the prokaryotic plasma membrane, the eukaryotic plasma membrane is a phospholipid bilayer containing proteins.
2. Eukaryotic plasma membranes contain carbohydrates attached to the proteins and sterols not found in prokaryotic cells (except *Mycoplasma* bacteria).
3. Eukaryotic cells can move materials across the plasma membrane by the passive processes used by prokaryotes, in addition to active transport and endocytosis (phagocytosis and pinocytosis).

Cytoplasm

1. The cytoplasm of eukaryotic cells includes everything inside the plasma membrane and external to the nucleus.
2. The chemical characteristics of the cytoplasm of eukaryotic cells resemble those of the cytoplasm of prokaryotic cells.
3. Eukaryotic cytoplasm has a cytoskeleton and exhibits cytoplasmic streaming.

Organelles

1. Organelles are specialized membrane-enclosed structures in the cytoplasm of eukaryotic cells.
2. The nucleus, which contains DNA in the form of chromosomes, is the most characteristic eukaryotic organelle.
3. The nuclear envelope is connected to a system of parallel membranes in the cytoplasm called the endoplasmic reticulum (ER).
4. The ER provides a surface for chemical reactions, serves as a transporting network, and stores synthesized molecules. Protein synthesis and transport occur on rough ER; lipid synthesis occurs on smooth ER.
5. 80S ribosomes are found in the cytoplasm or attached to the rough ER.
6. The Golgi complex consists of flattened sacs called cisterns. It functions in membrane formation and protein secretion.
7. Lysosomes are formed from Golgi complexes. They store powerful digestive enzymes.
8. Vacuoles are membrane-enclosed cavities derived from the Golgi complex or endocytosis usually found in plant cells that store various substances, help bring food into the cell, increase cell size, and provide rigidity to leaves and stems.
9. Mitochondria are the primary sites of ATP production. They contain small 70S ribosomes and DNA, and they multiply by binary fission.

10. Chloroplasts contain chlorophyll and enzymes for photosynthesis. Like mitochondria, they contain 70S ribosomes and DNA and multiply by binary fission.

11. A variety of organic compounds are oxidized in peroxisomes. Catalase in peroxisomes destroys H_2O_2.

12. The centrosome consists of the pericentriolar area and centrioles. Centrioles are 9 triplet microtubules involved in the formation of mitotic and flagellar microtubules.

The Evolution of Eukaryotes

1. According to the endosymbiotic theory, eukaryotic cells evolved from symbiotic prokaryotes living inside other prokaryotic cells.

THE LOOP

Methods of action of antibiotics from Chapter 20 can be included here to illustrate differences between prokaryotic and eukaryotic cells, as well as provide clinical applications to cell structure.

Answers

Review

1.

a.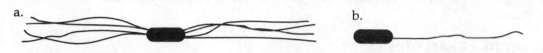

b.

c.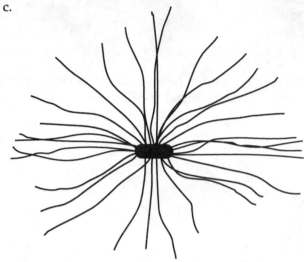

2. Endospore formation is called <u>sporogenesis</u>. It is initiated by <u>certain adverse environmental conditions</u>. Formation of a new cell from an endospore is called <u>germination</u>. This process is triggered by <u>favorable growth conditions</u>.

3.

a.    d.

b.   e.

c.   f.

4. Matching

<u>d</u> Cell wall

<u>f</u> Endospore

<u>a</u> Fimbriae

<u>c</u> Flagella

<u>a, e</u> Glycocalyx

<u>i</u> Pili

<u>b, h</u> Plasma membrane

<u>g</u> Ribosomes

5. An endospore is called a resting structure because it is a method of one cell "resting," or surviving, as opposed to growing and reproducing. The protective endospore wall allows a bacterium to withstand adverse conditions in the environment.

6. a. Both allow materials to cross the plasma membrane from a high concentration to a low concentration without expending energy. Facilitated diffusion requires carrier proteins.

   b. Both require enzymes to move materials across the plasma membrane. In active transport, energy is expended.

   c. Both move materials across the plasma membrane with an expenditure of energy. In group translocation, the substrate is changed after it crosses the membrane.

7. Mycoplasmas do not have cell walls.

8. a. Diagram (a) refers to a gram-positive bacterium because the lipopolysaccharide–phospholipid–lipoprotein layer is absent.

   b. The gram-negative bacterium initially retains the violet stain, but it is released when the outer membrane is dissolved by the decolorizing agent. After the dye–iodine complex enters, it becomes trapped by the peptidoglycan of gram-positive cells.

   c. The outer layer of the gram-negative cells prevents penicillin from entering the cells.

   d. Essential molecules diffuse through the gram-positive wall. Porins and specific channel proteins in the gram-negative outer membrane allow passage of small water-soluble molecules.

   e. Gram-negative.

9. An extracellular enzyme (amylase) hydrolyzes starch into disaccharides (maltose) and monosaccharides (glucose). A carrier enzyme (maltase) hydrolyzes maltose and moves one glucose into the cell. Glucose can be transported by group translocation as glucose-6-phosphate.

10. Matching

    c   Centriole
    d   Chloroplasts
    g   Golgi complex
    a   Lysosomes
    f   Mitochondria
    b   Peroxisomes
    e   Rough ER

11. A mitochondrion is an example of an organelle that resembles a prokaryotic cell. The inner membrane of a mitochondrion is arranged in folds similar to mesosomes. ATP is generated on this membrane just as it is in prokaryotic plasma membranes. Mitochondria can reproduce by binary fission, and they contain circular DNA and 70S ribosomes.

12. Phagocytosis. Pinocytosis.

13. Erythromycin inhibits protein synthesis in a prokaryotic cell; it will inhibit protein synthesis in mitochondria and chloroplasts.

Critical Thinking

1. Eukaryotic cells must be large enough to hold a nucleus and a mitochondrion (the minimum number of organelles). Prokaryotic cells contain molecules needed to carry on metabolic activities, but do not contain membrane-enclosed organelles, which require extra space.

2. *Micromonas* has a nucleus, one mitochondrion, one chloroplast, one Golgi complex, and one flagellum.

3. Like bacteria, archaea lack organelles. However, archaea also lack peptidoglycan cell walls. A more complete list of differences is in Table 10.1.

4. The large size of the organism caused the misidentification. Electron microscopy would reveal that this is a prokaryotic cell; chemical analysis of the cell wall would reveal peptidoglycan.

5. Water would passively leave the cell in a hypertonic environment. If a cell pumps K$^+$ in, water will follow, thus preventing plasmolysis.

Clinical Applications

1. Cell death released cell wall fragments. The gram-negative cell wall is responsible for the symptoms of septic shock.

2. The endospores allow survival in the presence of oxygen and during heating.

3. *Enterobacter, Pseudomonas,* and *Klebsiella* are gram-negative. Their cell walls contain lipid A endotoxin.

4. The bacteria were adhering to the inside of the pipes as a biofilm. Fimbriae and the glycocalyx allow the bacteria to adhere.

5. Bacterial endospores allow these bacteria to survive in products on store shelves. *B. thuringiensis* is sold as an insecticide, and *B. subtilis* as a fungicide.

Learning with Technology

	Morphology	Arrangement	Gram reaction
Micrococcus luteus	Coccus	Tetrads	+
Staphylococcus epidermidis	Coccus	Clusters	+
Enterococcus faecalis	Coccus	Chains	+
Enterobacter aerogenes	Rod	Singles	−
Proteus mirabilis	Rod	Singles	−
Bacillus subtilis	Rod	Chains	+

Case History: Coxiella burnetii

Background

The life cycle of *Coxiella burnetii* wasn't described until 1981, although the bacterium had been recognized more than 40 years earlier. Observations made by many researchers were finally assembled to show that this bacterium has a more complex life cycle than most. See if you can propose a life cycle for this bacterium from the information provided.

Coccoid and bacillary forms of *Coxiella burnetii* were first described in 1938. Subsequently, other researchers described round particles that passed through bacteriological filters (0.45 μm) and were capable of infecting guinea pig cells.

In 1981, electron microscopy studies of *Coxiella* revealed a large cell variant (LCV) and a small cell variant (SCV). The LCV has inner and outer membranes separated by a periplasm containing little peptidoglycan. The SCV lacks a periplasm and has a large peptidoglycan layer. LCVs develop a dense area in the periplasm at one end of the cell when nutrients are depleted or the pH increases. This area contains DNA and ribosomes.

In one study, suspensions of *C. burnetii* were put in distilled water, exposed to sonication (high-frequency vibration used to disrupt cells), and incubated at 45°C for 3 hr. Only SCVs were present after this treatment. *Coxiella* undergo binary fission in a host cell phagolysozyme. LCVs metabolize and divide more rapidly than SCVs.

Questions

1. Propose a life cycle for *Coxiella*.

2. Why do *Coxiella* show variable Gram stain results—that is, they may stain gram-positive or gram-negative? Should they be classified as gram-positive or gram-negative?

3. What disease does *C. burnetii* cause? Why can this disease be transmitted by airborne routes while other (closely related) rickettsia require insects and ticks for transmission to humans?

The Solution

1.

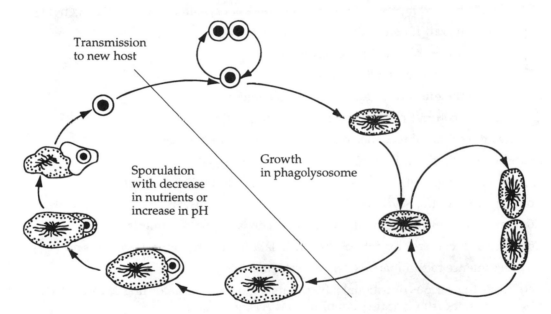

2. LCVs will stain gram-negative and SCVs, gram-positive. *Coxiella* is classified as gram-negative because the ultrastructure and chemical composition of the wall are gram-negative.

3. Q fever; SCVs (spores) allow this organism to survive outside of a host.

Chapter 5   *Microbial Metabolism*

Learning Objectives

1. Define metabolism, and describe the fundamental differences between anabolism and catabolism.
2. Identify the role of ATP as an intermediate between catabolism and anabolism.
3. Identify the components of an enzyme.
4. Describe the mechanism of enzymatic action.
5. List the factors that influence enzymatic activity.
6. Define ribozyme.
7. Explain what is meant by oxidation–reduction.
8. List and provide examples of three types of phosphorylation reactions that generate ATP.
9. Explain the overall function of biochemical pathways.
10. Describe the chemical reactions of glycolysis.
11. Explain the products of the Krebs cycle.
12. Describe the chemiosmotic model for ATP generation.
13. Compare and contrast aerobic and anaerobic respiration.
14. Describe the chemical reactions of, and list some products of, fermentation.
15. Describe how lipids and proteins are prepared for glycolysis.
16. Provide an example of the use of biochemical tests to identify bacteria.
17. Compare and contrast cyclic and noncyclic photophosphorylation.
18. Compare and contrast the light and dark reactions of photosynthesis.
19. Compare and contrast oxidative phosphorylation and photophosphorylation.
20. Write a sentence to summarize energy production in cells.
21. Categorize the various nutritional patterns among organisms according to carbon source and mechanisms of carbohydrate catabolism and ATP generation.
22. Describe the major types of anabolism and their relationship to catabolism.
23. Define amphibolic pathways.

NEW IN THIS EDITION

- Inset diagrams to illustrate orientation of metabolic pathways have been revised.

CHAPTER SUMMARY

Catabolic and Anabolic Reactions

1. The sum of all chemical reactions within a living organism is known as metabolism.

2. Catabolism refers to chemical reactions that result in the breakdown of more complex organic molecules into simpler substances. Catabolic reactions usually release energy.

3. Anabolism refers to chemical reactions in which simpler substances are combined to form more complex molecules. Anabolic reactions usually require energy.

4. The energy of catabolic reactions is used to drive anabolic reactions.

5. The energy for chemical reactions is stored in ATP.

Enzymes

1. Enzymes are proteins, produced by living cells, that catalyze chemical reactions by lowering the activation energy.

2. Enzymes are generally globular proteins with characteristic three-dimensional shapes.

3. Enzymes are efficient, can operate at relatively low temperatures, and are subject to various cellular controls.

Naming Enzymes

1. Enzyme names usually end in *-ase*.

2. The six classes of enzymes are defined on the basis of the types of reactions they catalyze.

Enzyme Components

1. Most enzymes are holoenzymes, consisting of a protein portion (apoenzyme) and a nonprotein portion (cofactor).

2. The cofactor can be a metal ion (iron, copper, magnesium, manganese, zinc, calcium, or cobalt) or
    a complex organic molecule known as a coenzyme (NAD, NADP, FMN, FAD, and coenzyme A).

Mechanism of Enzymatic Action

1. When an enzyme and substrate combine, the substrate is transformed, and the enzyme is recovered.

2. Enzymes are characterized by specificity, which is a function of their active sites.

Factors Influencing Enzymatic Activity

1. At high temperatures, enzymes undergo denaturation and lose their catalytic properties; at low temperatures, the reaction rate decreases.

2. The pH at which enzymatic activity is maximal is known as the optimum pH.

3. Within limits, enzymatic activity increases as substrate concentration increases.

4. Competitive inhibitors compete with the normal substrate for the active site of the enzyme. Noncompetitive inhibitors act on other parts of the apoenzyme or on the cofactor and decrease the enzyme's ability to combine with the normal substrate.

Feedback Inhibition

1. Feedback inhibition occurs when the end product of a pathway inhibits an enzyme's activity in the pathway.

Ribozymes

1. Ribozymes are enzymatic RNA molecules that cut and splice RNA in eukaryotic cells.

Energy Production

Oxidation–Reduction Reactions

1. Oxidation is the removal of one or more electrons from a substrate. Protons (H^+) are often removed with the electrons.
2. Reduction of a substrate refers to its gain of one or more electrons.
3. Each time a substance is oxidized, another is simultaneously reduced.
4. NAD^+ is the oxidized form; NADH is the reduced form.
5. Glucose is a reduced molecule; energy is released during a cell's oxidation of glucose.

The Generation of ATP

1. Energy released during certain metabolic reactions can be trapped to form ATP from ADP and Ⓟ (phosphate). Addition of Ⓟ to a molecule is called phosphorylation.
2. During substrate-level phosphorylation, a high-energy Ⓟ from an intermediate in catabolism is added to ADP.
3. During oxidative phosphorylation, energy is released as electrons are passed to a series of electron acceptors (an electron transport chain) and finally to O_2 or another inorganic compound.
4. During photophosphorylation, energy from light is trapped by chlorophyll, and electrons are passed through a series of electron acceptors. The electron transfer releases energy used for the synthesis of ATP.

Metabolic Pathways of Energy Production

1. Series of enzymatically catalyzed chemical reactions called biochemical pathways store energy in and release energy from organic molecules.

Carbohydrate Catabolism

1. Most of a cell's energy is produced from the oxidation of carbohydrates.
2. Glucose is the most commonly used carbohydrate.
3. The two major types of glucose catabolism are respiration, in which glucose is completely broken down, and fermentation, in which it is partially broken down.

Glycolysis

1. The most common pathway for the oxidation of glucose is glycolysis. Pyruvic acid is the end-product.
2. Two ATP and two NADH molecules are produced from one glucose molecule.

Alternatives to Glycolysis

1. The pentose phosphate pathway is used to metabolize five-carbon sugars; one ATP and 12 NADPH molecules are produced from one glucose molecule.
2. The Entner–Doudoroff pathway yields one ATP and two NADPH molecules from one glucose molecule.

Cellular Respiration Defined

1. During respiration, organic molecules are oxidized. Energy is generated from the electron transport chain.
2. In aerobic respiration, O_2 functions as the final electron acceptor.
3. In anaerobic respiration, the final electron acceptor is an inorganic molecule other than O_2.

Aerobic

The Krebs Cycle
1. Decarboxylation of pyruvic acid produces one CO_2 molecule and one acetyl group.
2. Two-carbon acetyl groups are oxidized in the Krebs cycle. Electrons are picked up by NAD^+ and FAD for the electron transport chain.
3. From one molecule of glucose, oxidation produces six molecules of NADH, two molecules of $FADH_2$, and two molecules of ATP.
4. Decarboxylation produces six molecules of CO_2.

The Electron Transport Chain
1. Electrons are brought to the electron transport chain by NADH.
2. The electron transport chain consists of carriers, including flavoproteins, cytochromes, and ubiquinones.

The Chemiosmotic Mechanism of ATP Generation
1. Protons being pumped across the membrane generate a proton motive force as electrons move through a series of acceptors or carriers.
2. Energy produced from movement of the protons back across the membrane is used by ATP synthase to make ATP from ADP and ℗ .
3. In eukaryotes, electron carriers are located on the inner mitochondrial membrane; in prokaryotes, electron carriers are in the plasma membrane.

A Summary of Aerobic Respiration

1. In aerobic prokaryotes, 38 ATP molecules can be produced from complete oxidation of a glucose molecule in glycolysis, the Krebs cycle, and the electron transport chain.
2. In eukaryotes, 36 ATP molecules are produced from complete oxidation of a glucose molecule.

Anaerobic Respiration

1. The final electron acceptors in anaerobic respiration include NO_3^-, SO_4^{2-}, and CO_3^{2-}.
2. The total ATP yield is less than aerobic respiration because only part of the Krebs cycle operates under anaerobic conditions.

Fermentation

1. Fermentation releases energy from sugars or other organic molecules by oxidation.
2. O_2 is not required in fermentation.
3. Two ATP molecules are produced by substrate-level phosphorylation.
4. Electrons removed from the substrate reduce NAD^+.
5. The final electron acceptor is an organic molecule.
6. In lactic acid fermentation, pyruvic acid is reduced by NADH to lactic acid.
7. In alcohol fermentation, acetaldehyde is reduced by NADH to produce ethanol.
8. Heterolactic fermenters can use the pentose phosphate pathway to produce lactic acid and ethanol.

Lipid Catabolism

1. Lipases hydrolyze lipids into glycerol and fatty acids.
2. Fatty acids and other hydrocarbons are catabolized by beta oxidation.
3. Catabolic products can be further broken down in glycolysis and the Krebs cycle.

Protein Catabolism

1. Before amino acids can be catabolized, they must be converted to various substances that enter the Krebs cycle.
2. Transamination, decarboxylation, and dehydrogenation reactions convert the amino acids to be catabolized.

Biochemical Tests

1. Bacteria and yeasts can be identified by detecting action of their enzymes.
2. Fermentation tests are used to determine whether an organism can ferment a carbohydrate to produce acid and gas.

Photosynthesis

1. Photosynthesis is the conversion of light energy from the sun into chemical energy; the chemical energy is used for carbon fixation.

The Light Reactions: Photophosphorylation

1. Chlorophyll *a* is used by green plants, algae, and cyanobacteria; it is found in thylakoid membranes.
2. Electrons from chlorophyll pass through an electron transport chain, from which ATP is produced by chemiosmosis.
3. In cyclic photophosphorylation, the electrons return to the chlorophyll.
4. In noncyclic photophosphorylation, the electrons are used to reduce NADP, and electrons are returned to chlorophyll from H_2O or H_2S.
5. When H_2O is oxidized by green plants, algae, and cyanobacteria, O_2 is produced.

The Dark Reactions: The Calvin–Benson Cycle

1. CO_2 is used to synthesize sugars.

Summary of Energy Production Mechanisms

1. Sunlight is converted to chemical energy in oxidation–reduction reactions carried on by phototrophs. Chemotrophs can use this chemical energy.
2. In oxidation–reduction reactions, energy is derived from the transfer of electrons.
3. To produce energy, a cell needs an electron donor (organic or inorganic), a system of electron carriers, and a final electron acceptor (organic or inorganic).

Metabolic Diversity Among Organisms

1. Photoautotrophs obtain energy by photophosphorylation and fix carbon from CO_2 via the Calvin–Benson cycle to synthesize organic compounds.
2. Cyanobacteria are oxygenic phototrophs. Green sulfur bacteria and purple sulfur bacteria are anoxygenic phototrophs.
3. Photoheterotrophs use light as an energy source and an organic compound for their carbon source or electron donor.
4. Chemoautotrophs use inorganic compounds as their energy source and carbon dioxide as their carbon source.
5. Chemoheterotrophs use complex organic molecules as their carbon and energy sources.

Metabolic Pathways of Energy Use

Polysaccharide Biosynthesis

1. Glycogen is formed from ADPG.
2. UDPNAc is the starting material for the biosynthesis of peptidoglycan.

Lipid Biosynthesis

1. Lipids are synthesized from fatty acids and glycerol.
2. Glycerol is derived from dihydroxyacetone phosphate, and fatty acids are built from acetyl CoA.

Amino Acid and Protein Biosynthesis

1. Amino acids are required for protein biosynthesis.
2. All amino acids can be synthesized either directly or indirectly from intermediates of carbohydrate metabolism, particularly from the Krebs cycle.

Purine and Pyrimidine Biosynthesis

1. The sugars composing nucleotides are derived from either the pentose phosphate pathway or the Entner–Doudoroff pathway.
2. Carbon and nitrogen atoms from certain amino acids form the backbones of the purines and pyrimidines.

The Integrations of Metabolism

1. Anabolic and catabolic reactions are integrated through a group of common intermediates.
2. Such integrated pathways are referred to as amphibolic pathways.

THE LOOP

Complete metabolic pathways are provided in Appendix C. The boxes in Chapter 2, Chapter 9, Chapter 11, Chapter 27, and Chapter 28 illustrate applications of microbial metabolism in bioremediation and industry. Chapters 27 and 28 can be included with study of Chapter 5 to provide students with applications of metabolism.

Answers

Review

1. Metabolism is the sum of all chemical reactions that occur within a living organism.

2. Catabolic reactions break down organic compounds and release energy, while anabolic reactions use the products of catabolism and energy to build cell material.

3.

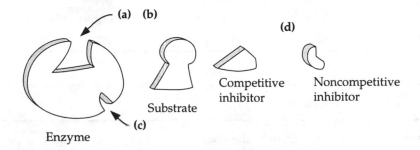

4. a. When the enzyme and substrate combine, the substrate molecule will be transformed.

    b. When the competitive inhibitor binds to the enzyme, the enzyme will not be able to bind with the substrate.

    c. When the noncompetitive inhibitor binds to the enzyme, the active site of the enzyme will be changed so the enzyme cannot bind with the substrate.

    d. The noncompetitive inhibitor.

5. (a) is the Calvin–Benson cycle, (b) is glycolysis, and (c) is the Krebs cycle.

6. Glycerol is catabolized by pathway (b) as dihydroxyacetone phosphate. Fatty acids by pathway (c) as acetyl groups.

7. In pathway (c) at α-ketoglutaric acid.

8. Glyceraldehyde-3-phosphate from the Calvin–Benson cycle enters glycolysis. Pyruvic acid from glycolysis is decarboxylated to produce acetyl for the Krebs cycle.

9. In (a), between glucose and glyceraldehyde-3-phosphate.

10. The conversion of pyruvic acid to acetyl, isocitric acid to α-ketoglutaric acid, and α-ketoglutaric acid to succinyl~CoA.

11. By pathway (c) as acetyl groups.

12.

	Uses	Produces
Calvin–Benson cycle	6 NADPH	
Glycolysis		2 NADH
Pyruvic acid ———> acetyl		1 NADH
Isocitric acid ———> α-ketoglutaric acid		1 NADH
α-ketoglutaric acid ———> Succinyl~CoA		1 NADH
Succinic acid ———> Fumaric acid		1 FADH$_2$
Malic acid ———> Oxaloacetic acid		1 NADH

13. Dihydroxyacetone phosphate; acetyl; oxaloacetic acid; α-ketoglutaric acid.

14. The optimum temperature for an enzyme is one that favors movement of molecules so the enzyme can "find" its substrate. Lower temperatures will decrease the rate of collisions and the rate of reactions. Increased temperatures will denature the enzyme.

15. Ethyl alcohol, lactic acid, butyl alcohol, acetone, and glycerol are some of the possible products. Refer to Table 5.4 and Figure 5.18b.

16.

Organism	Carbon Source	Energy Source
Photoautotroph	CO_2	Light
Photoheterotroph	Organic molecules	Light
Chemoautotroph	CO_2	Inorganic molecules
Chemoheterotroph	Organic molecules	Organic molecules

17.

ATP generated by	Reaction
Photophosphorylation	An electron, liberated from chlorophyll by light, is passed down an electron transport chain.
Oxidative phosphorylation	Cytochrome *c* passes two electrons to cytochrome *a*.
Substrate-level phosphorylation	

$$
\begin{array}{ccc}
CH_2 & & CH_3 \\
\| & & | \\
C - O \sim \textcircled{P} & \longrightarrow & C = O \\
| & & | \\
COOH & & COOH \\
\text{Phosphoenolpyruvic acid} & & \text{Pyruvic acid}
\end{array}
$$

Phosphoenolpyruvic acid   Pyruvic acid

18. a. Oxidation–reduction: A coupled reaction in which one substance is oxidized and one is reduced.

    b. The final electron acceptor in aerobic respiration is molecular oxygen; in anaerobic respiration, it is another inorganic molecule.

    c. In cyclic photophosphorylation, electrons are returned to chlorophyll. In noncyclic photophosphorylation, chlorophyll receives electrons from hydrogen atoms.

19. The pentose phosphate pathway produces pentoses for the synthesis of nucleic acids, precursors for the synthesis of glucose by photosynthesizing organisms, precursors in the synthesis of certain amino acids, and NADPH.

20. Oxidation

21. Reactions requiring ATP are coupled with reactions that produce ATP.

22.

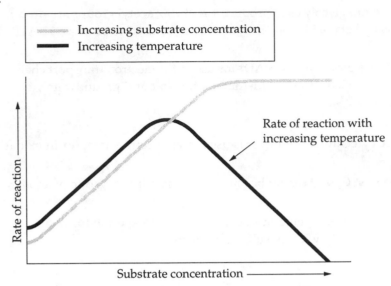

The reaction rate will increase until the enzymes are saturated.

Critical Thinking

1. Answers should include the following:
    Protons are pumped from one side of the membrane to the other; transfer of protons back across the membrane generates ATP.
    a. Outer portion is acidic and
    b. Has a positive electric charge.
    c. Energy-conserving sites are the three loci where H$^+$ molecules are pumped out.
    d. Kinetic energy is realized at ATP synthetase when protons cross the membrane.

2. *Streptococcus* is only capable of fermentation, which yields two molecules of ATP for each molecule of glucose consumed. Most of the energy that cells obtain from catabolism is from respiration.

3. NADH must be reoxidized so it can pick up hydrogen atoms again. NADH is usually reoxidized in respiration when electrons are transferred to the electron transport chain carrier molecules. In fermentation, NADH is reoxidized when electrons are transferred to pyruvic acid (reducing pyruvic acid to lactic acid, for example).

4. The rate at which an enzyme converts substrate to product is partly a function of initial concentration of substrate. The more substrate molecules available, the more frequently they access the active site of the enzyme. The reaction proceeds at a linear rate (black). When the concetration of substrate is high enough that all enzyme molecules have their active sites engaged, the reaction rate will remain constant. When the enzyme becomes saturated with competitive inhibitor, the reaction will completely stop (red).

5. Carbohydrate catabolism:
    a. Oxidation of glucose (glycolysis, Krebs, aerobic ETC).
    b. CO$_2$ fixation and oxidation of glucose.
    c. CO$_2$ fixation and oxidation of glucose (anaerobic ETC).

    Energy production. All three use chemiosmotic mechanisms. In addition to oxidative phosphorylation, *Spirulina* and *Ectothiorhodospira* use photophosphorylation.

6.   Glucose = 38 ATP.
     Butterfat = 217 ATP. Glycerol goes into glycolysis to produce 1 ATP. Six acetyl groups are produced from each of the 12-carbon chains by beta-oxidation. Each acetyl can be used to produce 12 ATPs in the Krebs cycle.

7.   Two electrons removed from As^{3+} are picked up by NAD^+ for use in the electron transport chain. *Thiobacillus* could be used to remove arsenic from industrial wastewater and groundwater.

Clinical Applications

1.   X factor is necessary to synthesize cytochromes. V factor is used as an electron acceptor in oxidation reactions.

2.   The drug ddC is missing an O atom on C_3 so it cannot be joined to the phosphate group of another nucleotide.

3.   Live bacteria or whole bacterial cells are not being injected. Enzymes are specific for their substrates, so streptokinase will react only with its substrate, fibrin.

Learning With Technology

1.

	E. coli	*E. Aerogenes*	*P. mirabilis*
Gram reaction and morphology	Gram-negative rods	Gram-negative rods	Gram-negative rods
Glucose fermentation	acid + gas	acid + gas	acid + gas
Lactose fermentation	acid + gas	acid + gas	no acid/gas
Urease production	–	–	+
Citrate utilization	–	+	–
Lysine decarboxylase	+	+	–
Phenlyalanine deaminase	–	–	+
MacConkey agar	Red	Red	White

2.   Glycolysis begins with glucose and most carbohydrates would have to be converted to glucose.

3.   Lysine decarboxylase removes a carboxyl (-COOH) group from lysine. Phenylalanine deaminase removes an amino ($-NH_2$) group from phenylalanine.

Case History: The Pasteur Effect

Background

In 1861, Louis Pasteur observed that when yeasts grow in a sugar and protein medium completely free of air, they ferment vigorously, and for every gram of yeast that forms, 60 to 80 grams of sugar disappear. If the experiment is carried out in the presence of air, for one gram of yeast that forms, only 4 to 10 grams of sugar are removed. The yeasts again ferment if transferred to a sugar-containing medium absent air.

When the experiment is repeated with a protein medium, the yeasts grow only in the presence of oxygen. Pasteur concluded that the yeasts can take oxygen from air, and in the absence of air, the yeasts take oxygen from the sugar.

Pasteur applied quantitative methods to his studies of fermentation and was the first to report on organisms that could live and reproduce in the absence of oxygen. His conclusion was, however, incorrect. These different behaviors of yeasts are known today as the Pasteur effect.

Questions
1. Explain the three yeast behaviors based on modern concepts of microbial metabolism.

2. What was incorrect about Pasteur's conclusion?

The Solution
1. The yeasts are able to grow anaerobically if a fermentable sugar is available. Fermentation consumes more sugar than respiration to produce the same amount of energy.

2. Pasteur assumed yeasts were taking oxygen from the sugar.

Chapter 6 *Microbial Growth*

NEW IN THIS EDITION

- A discussion of hyperthermophiles is included.
- A new diagram illustrates the relationships between hydrothermal vent worms and their symbiotic bacteria.

CHAPTER SUMMARY

The Requirements for Growth

1. The growth of a population is an increase in the number of cells or in mass.
2. The requirements for microbial growth are both physical and chemical.

Physical Requirements

1. On the basis of growth range of temperature, microbes are classified as psychrophiles (cold-loving), mesophiles (moderate-temperature–loving), and thermophiles (heat-loving).

2. The minimum growth temperature is the lowest temperature at which a species will grow; the optimum growth temperature is the temperature at which it grows best; and the maximum growth temperature is the highest temperature at which growth is possible.
3. Most bacteria grow best at a pH value between 6.5 and 7.5.
4. In a hypertonic solution, microbes undergo plasmolysis; halophiles can tolerate high salt concentrations.

Chemical Requirements

1. All organisms require a carbon source; chemoheterotrophs use an organic molecule, and autotrophs typically use carbon dioxide.
2. Nitrogen is needed for protein and nucleic acid synthesis. Nitrogen can be obtained from decomposition of proteins or from NH_4^+ or NH_3^-; a few bacteria are capable of nitrogen (N_2) fixation.
3. On the basis of oxygen requirements, organisms are classified as obligate aerobes, facultative anaerobes, obligate anaerobes, aerotolerant anaerobes, and microaerophiles.
4. Aerobes, facultative anaerobes, and aerotolerant anaerobes must have the enzymes superoxide dismutase ($2O_2^{2-} \cdot + 2H^+ \longrightarrow O_2 + H_2O_2$) and either catalase ($2H_2O_2 \longrightarrow 2H_2O + O_2$) or peroxidase ($H_2O_2 + 2H^+ \longrightarrow 2H_2O$).
5. Other chemicals required for microbial growth include sulfur, phosphorus, trace elements, and, for some microorganisms, organic growth factors.

Culture Media

1. A culture medium is any material prepared for the growth of bacteria in a laboratory.
2. Microbes that grow and multiply in or on a culture medium are known as a culture.
3. Agar is a common solidifying agent for a culture medium.

Chemically Defined Media

1. A chemically defined medium is one in which the exact chemical composition is known.

Complex Media

1. A complex medium is one in which the exact chemical composition is not known.

Anaerobic Growth Media and Methods

1. Reducing media chemically remove molecular oxygen (O_2) that might interfere with the growth of anaerobes.
2. Petri plates can be incubated in an anaerobic jar or anaerobic chamber.

Special Culture Techniques

1. Some parasitic and fastidious bacteria must be cultured in living animals or in cell cultures.
2. CO_2 incubators or candle jars are used to grow bacteria requiring an increased CO_2 concentration.

Selective and Differential Media

1. By inhibiting unwanted organisms with salts, dyes, or other chemicals, selective media allow growth of only the desired microbes.
2. Differential media are used to distinguish among different organisms.

Enrichment Culture

1. An enrichment culture is used to encourage the growth of a particular microorganism in a mixed culture.

Obtaining Pure Cultures

1. A colony is a visible mass of microbial cells that theoretically arose from one cell.
2. Pure cultures are usually obtained by the streak plate method.

Preserving Bacterial Cultures

1. Microbes can be preserved for long periods of time by deep-freezing or lyophilization (freeze-drying).

Growth of Bacterial Cultures

Bacterial Division

1. The normal reproductive method of bacteria is binary fission, in which a single cell divides into two identical cells.
2. Some bacteria reproduce by budding, aerial spore formation, or fragmentation.

Generation Time

1. The time required for a cell to divide or a population to double is known as the generation time.

Logarithmic Representation of Bacterial Populations

1. Bacterial division occurs according to a logarithmic progression (2 cells, 4 cells, 8 cells, etc.).

Phases of Growth

1. During the lag phase, there is little or no change in the number of cells, but metabolic activity is high.
2. During the log phase, the bacteria multiply at the fastest rate possible under the conditions provided.
3. During the stationary phase, there is an equilibrium between cell division and death.
4. During the death phase, the number of deaths exceeds the number of new cells formed.

Direct Measurement of Microbial Growth

1. A standard plate count reflects the number of viable microbes and assumes that each bacterium grows into a single colony; plate counts are reported as number of colony-forming units (CFU).
2. A plate count may be done either by the pour plate method or the spread plate method.

3. In filtration, bacteria are retained on the surface of a membrane filter and then transferred to a culture medium to grow and subsequently be counted.

4. The most probable number (MPN) method can be used for microbes that will grow in a liquid medium; it is a statistical estimation.

5. In a direct microscopic count, the microbes in a measured volume of a bacterial suspension are counted with the use of a specially designed slide.

Estimating Bacterial Numbers by Indirect Methods

1 A spectrophotometer is used to determine turbidity by measuring the amount of light that passes through a suspension of cells.

2. An indirect way of estimating bacterial numbers is measuring the metabolic activity of the population, for example, acid production or oxygen consumption.

3. For filamentous organisms such as fungi, measuring dry weight is a convenient method of growth measurement.

THE LOOP

Appendix D, "Exponents, Exponential Notation, Logarithms, and Generation Time," is useful here. Bioenhancement with N and P is described in the box in Chapter 2, p. 35. The use of MPN in water quality testing is discussed on pp. 176–177.

Answers

Review

1. In binary fission, the cell elongates and the chromosome replicates. Next, the nuclear material is evenly divided. The plasma membrane invaginates toward the center of the cell. The cell wall thickens and grows inward between the membrane invaginations; two new cells result.

2. Refer to Figure 6.14. The period of no cell division is called lag phase. During lag phase, the bacteria are synthesizing enzymes that are necessary for growth. In log phase, the cells are dividing at the maximum rate under the conditions provided. The number of cells dividing equals the number of cells dying in stationary phase. When the number of deaths exceeds the number of divisions, death phase is observed.

3. Carbon (C) is required for synthesis of molecules that make up a living cell. Carbon-containing compounds also are required as an energy source for heterotrophs.

4. Most bacteria grow best between pH 6.5 and 7.5.

5. The addition of salt or sugar to foods increases the osmotic pressure for microorganisms on the food. The resulting hypertonic environment causes plasmolysis of the microbial cells.

6. a. Catalyzes the breakdown of H_2O_2 to O_2 and H_2O.

    b. H_2O_2; peroxide ion is O_2^{2-}.

    c. Catalyzes the breakdown of H_2O_2;

    $$NADH + H^+ + H_2O_2 \xrightarrow{\text{Peroxidase}} NAD^+ + 2H_2O$$

    d. $O_2^- \cdot$; this diatom has one unpaired electron.

    e. Converts superoxide to O_2 and H_2O_2;

    $$2O_2^- \cdot + 2H^+ \xrightarrow{\text{Superoxide dismutase}} O_2 + H_2O_2$$

    The enzymes are important in protecting the cell from the strong oxidizing agents, peroxide and superoxide, that form during respiration.

7. Both environments prevent molecular oxygen from reaching the bacterial cells. In reducing media, thioglycolate combines with dissolved oxygen, thereby removing it from the medium. In an anaerobic incubator, air is replaced with an atmosphere of CO_2 (and N_2). *Clostridium* is an obligate anaerobe that lacks superoxide dismutase and catalase. Consequently, the accumulation of superoxides and peroxides will kill the cell in an aerobic environment.

8. Direct methods are those in which the microorganisms are seen and counted. Direct methods are direct count, standard plate count, filtration, and most probable number.

9. The growth rate of bacteria slows down with decreasing temperatures. Mesophilic bacteria will grow slowly at refrigeration temperatures and will remain dormant in a freezer. Bacteria will not spoil food quickly in a refrigerator.

10. Number of cells $\times$ $2^{n \text{ generations}}$ = Total number of cells

    6 $\times$ 2^7 = 768

11. Petroleum can meet the carbon and energy requirements for an oil-degrading bacterium; however, nitrogen and phosphate are usually not available in large quantities. Nitrogen and phosphate are essential for making proteins, phospholipids, nucleic acids, and ATP.

12. A chemically defined medium is one in which the exact chemical composition is known. A complex medium is one in which the exact chemical composition is not known.

13.

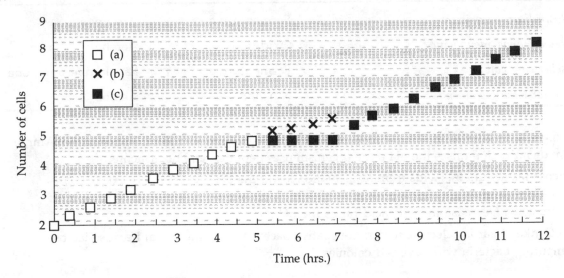

Critical Thinking

1. a. At x, the bacteria began a second lag phase during which they synthesized enzymes required to use the second carbon source.

    b. The first substrate provided the better growth conditions. The slope of the line is steeper, indicating that the bacteria grew faster.

2. In the presence of oxygen, H_2O_2 forms in *Clostridium*. The H_2O_2 accumulates in these catalase-negative cells and kills them. H_2O_2 does not form in *Streptococcus*.

3. Glucose provides a fermentable carbohydrate for chemoheterotrophs. Glucose is the carbon and energy source. Other macronutrients including nitrogen are provided in inorganic compounds in the "minimal salts."

4. a. A
    b. B
    c. A
    d. A
    e. A

Clinical Applications

1. 1.68×10^8. They are the progeny of the original 10.

2. At least 53°C; 60°C was recommended after this study. The bacteria get in the food during preparation and those buried inside do not get hot enough to be killed.

3. Product 2 decreased bacterial numbers by 89% compared to a 17% decrease for both Products 1 and 3. All the bacteria probably did not grow. Only those that could grow aerobically on nutrient agar were counted in the experiment.

Learning with Technology

1.

Organism	O-F	
Clostridium butyricum	Fermentative	Anaerobe
Pseudomonas mendocina	Oxidative	Aerobe
Escherichia coli	Fermentative	Facultative anaerobe
Arthrobacter globiformis	Neither	Aerobe

2. Acids from fermentation and the Krebs cycle lower the pH if the organism uses glucose. Ammonia from deamination raises the pH if the organism uses peptides, and not glucose.

3. Glucose fermentation: Acids from fermentation lower the pH.
    Urea agar: Ammonia raises the pH.
    Citrate agar: Decrease in citric acid raises the pH.

4. MacConkey agar is selective for gram-negative bacteria and differential because lactose fermenting bacteria will have red colonies.

Case History: Determining the Effectiveness of a Food Preservative

Background

In order to determine whether a newly synthesized chemical might be a useful food preservative, the chemical was tested for its ability to inhibit bacterial growth.

Control. 500 ml of cottage cheese was inoculated with 2 ml of a 24-hr culture of *Pseudomonas aeruginosa* and incubated at 25°C. Five hours after inoculation, a standard plate count showed there were 200 bacterial cells/ml in the cottage cheese. After 29 hours at 25°C, there were 1,000,000 cells/ml in the cottage cheese.

Experiment. 500 ml of cottage cheese containing the preservative was inoculated with 2 ml of a 24-hr culture of *P. aeruginosa*. After 6 hours of incubation at 25°C, a standard plate count was performed. There were 700 bacterial cells/ml in the cottage cheese. After 38 hours, there were 61,000,000 bacterial cells/ml in the cottage cheese.

Number	Log
1	0.00
2	0.30
5	0.70
6	0.78
24	1.38
32	1.51
200	2.30
700	2.85
1.00×10^6	6.00
6.10×10^6	6.79
6.10×10^7	7.79

Questions

1. Why were plate counts used instead of direct microscopic counts or turbidity measurements?

2. How did the control cottage cheese and the experiment cottage cheese differ? Was this a fair test?

3. Determine the effectiveness of the new food preservative.

4. Does this type of test determine bacteriostatic or bactericidal activity?

The Solution

1. The particles in cottage cheese would interfere with direct counts and turbidity.

2. The new chemical was added to the experiment and was lacking in the control. Yes, this is a fair test.

3. The two tests had the same generation time, proving the new food preservative was *not* effective.

4. Both. The answer is "bactericidal" when the number of bacteria declines and "bacteriostatic" if the number of bacteria stays the same.

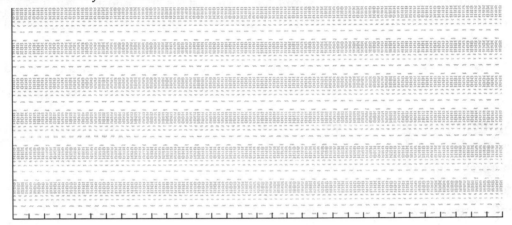

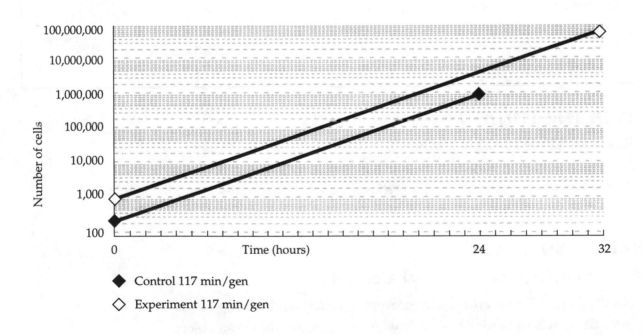

◆ Control 117 min/gen

◇ Experiment 117 min/gen

Source: King, A. D., H. G. Bayne, L. Jurd, and C. L. Case. "Antimicrobial properties of natural phenols and related compounds," *Antimicrobial Agents and Chemotherapy* 1:263–267, March, 1972.

Chapter 7 — The Control of Microbial Growth

Learning Objectives

1. Define the following key terms related to microbial control: sterilization, disinfection, antisepsis, degerming, sanitization, biocide, germicide, bacteriostasis and asepsis.
2. Describe the patterns of microbial death caused by treatments with microbial control agents.
3. Describe the effects of microbial control agents on cellular structures.
4. Compare the effectiveness of moist heat (boiling, autoclaving, pasteurization) and dry heat.
5. Describe how filtration, low temperature, desiccation, and osmotic pressure suppress microbial growth.
6. Explain how radiation kills cells.
7. List the factors related to effective disinfection.
8. Interpret results of use-dilution and filter paper tests.
9. Identify the methods of action and preferred uses of chemical disinfectants.
10. Differentiate between halogens used as antiseptics and as disinfectants.
11. Identify the appropriate uses for surface-active agents.
12. List the advantages of glutaraldehyde over other chemical disinfectants.
13. Identify the method of sterilizing plastic labware.
14. Explain how microbial growth is affected by the type of microbe and the environmental conditions.

NEW IN THIS EDITION

- Biocides and bisphenols are included.
- Revised discussions of conditions that influence microbial control are included.
- A new Clinical Problem-Solving box shows the link between resistance to chemical disinfectants and hospital-acquired infections.

CHAPTER SUMMARY
The Terminology of Microbial Control

1. Control of microbial growth can prevent infections and food spoilage.
2. Sterilization is the process of destroying all microbial life on an object.
3. Commercial sterilization is heat treatment of canned foods to destroy *C. botulinum* endospores.
4. Disinfection is the process of reducing or inhibiting microbial growth on a nonliving surface.
5. Antisepsis is the process of reducing or inhibiting microorganisms on living tissue.
6. The suffix *-cide* means to kill; the suffix *-stat* means to inhibit.
7. Bacterial contamination is called sepsis.

The Rate of Microbial Death

1. Bacterial populations subjected to heat or antimicrobial chemicals usually die at a constant rate.
2. Such a death curve, when plotted logarithmically, shows this constant death rate as a straight line.
3. Time to kill a microbial population is proportional to the number of microbes.
4. Microbial species and life cycle phases (i.e., endospores) have different susceptibilities to physical and chemical controls.
5. Organic matter may interfere with heat treatments and chemical control agents.
6. Longer exposure to lower heat can produce the same effect as shorter time at higher heat.

Actions of Microbial Control Agents

Alteration of Membrane Permeability

1. The susceptibility of the plasma membrane is due to its lipid and protein components.
2. Certain chemical control agents damage the plasma membrane by altering its permeability.

Damage to Proteins and Nucleic Acids

1. Some microbial control agents damage cellular proteins by breaking hydrogen and covalent bonds.
2. Other agents interfere with DNA and RNA replication and protein synthesis.

Physical Methods of Microbial Control

Heat

1. Heat is frequently used to eliminate microorganisms.
2. Moist heat kills microbes by denaturing enzymes.
3. Thermal death point (TDP) is the lowest temperature at which all bacteria in a liquid culture will be killed in 10 minutes.
4. Thermal death time (TDT) is the length of time required to kill all bacteria in a liquid culture at a given temperature.
5. Decimal reduction time (DRT) is the length of time in which 90% of a bacterial population will be killed at a given temperature.
6. Boiling (100°C) kills many vegetative cells and viruses within 10 minutes.
7. Autoclaving (steam under pressure) is the most effective method of moist heat sterilization. The steam must directly contact the material for it to be sterilized.
8. In pasteurization, a high temperature is used for a short time (72°C for 15 seconds) to destroy pathogens without altering the flavor of the food. Ultra-high-temperature (UHT) treatment (140°C for 3 sec.) is used to sterilize dairy products.
9. Methods of dry heat sterilization include direct flaming, incineration, and hot-air sterilization. Dry heat kills by oxidation.
10. Different methods that produce the same effect (reduction in microbial growth) are called equivalent treatments.

Filtration

1. Filtration is the passage of a liquid or gas through a filter with pores small enough to retain microbes.
2. Microbes can be removed from air by high-efficiency particulate air filters.
3. Membrane filters composed of nitrocellulose or cellulose acetate are commonly used to filter out bacteria, viruses, and even large proteins.

Low Temperature

1. The effectiveness of low temperatures depends on the particular microorganism and the intensity of the application.
2. Most microorganisms do not reproduce at ordinary refrigerator temperatures (0°–7°C).
3. Many microbes survive (but do not grow) at the subzero temperatures used to store foods.

Desiccation

1. In the absence of water, microorganisms cannot grow but can remain viable.
2. Viruses and endospores can resist desiccation.

Osmotic Pressure

1. Microorganisms in high concentrations of salts and sugars undergo plasmolysis.
2. Molds and yeasts are more capable than bacteria of growing in materials with low moisture or high osmotic pressure.

Radiation

1. The effects of radiation depend on its wavelength, intensity, and duration.
2. Ionizing radiation (gamma rays, X rays, and high-energy electron beams) has a high degree of penetration and exerts its effect primarily by ionizing water and forming highly reactive hydoxyl radicals.
3. Ultraviolet (UV) radiation, a form of nonionizing radiation, has a low degree of penetration and causes cell damage by making thymine dimers in DNA that interfere with DNA replication; the most effective germicidal wavelength is 260 nm.
4. Microwaves can kill microbes indirectly as materials get hot.

Chemical Methods of Microbial Control

1. Chemical agents are used on living tissue (as antiseptics) and on inanimate objects (as disinfectants).
2. Few chemical agents achieve sterility.

Principles of Effective Disinfection

1. Careful attention should be paid to the properties and concentration of the disinfectant to be used.
2. The presence of organic matter, degree of contact with microorganisms, and temperature should also be considered.

Evaluating a Disinfectant

1. In the use-dilution test, bacterial (*Salmonella choleraesuis, Staphylococcus aureus,* and *Pseudomonas aeruginosa*) survival in the manufacturer's recommended dilution of a disinfectant is determined.
2. Viruses, endospore-forming bacteria, mycobacteria, and fungi can also be used in the use-dilution test.
3. In the disk-diffusion method, a disk of filter paper is soaked with a chemical and placed on an inoculated agar plate; a clear zone of inhibition indicates effectiveness.

Types of Disinfectants

Phenol and Phenolics

1. Phenolics exert their action by injuring plasma membranes.

Bisphenols

1. Bisphenols such as triclosan (over the counter) and hexachlorophene (prescription) are widely used in household products.

Biguanides

1. Chlorhexidine damages plasma membranes of vegetative cells.

Halogens

1. Some halogens (iodine and chlorine) are used alone or as components of inorganic or organic solutions.
2. Iodine may combine with amino acids to inactivate enzymes and other cellular proteins.
3. Iodine is available as a tincture (in solution with alcohol) or as an iodophor (combined with an organic molecule).
4. The germicidal action of chlorine is based on the formation of hypochlorous acid when chlorine is added to water.
5. Chlorine is used as a disinfectant in gaseous form (Cl_2) or in the form of a compound, such as calcium hypochlorite, sodium hypochlorite, and chloramines.

Alcohols

1. Alcohols exert their action by denaturing proteins and dissolving lipids.
2. In tinctures, they enhance the effectiveness of other antimicrobial chemicals.
3. Aqueous ethanol (60–90%) and isopropanol are used as disinfectants.

Heavy Metals and Their Compounds

1. Silver, mercury, copper, and zinc are used as germicidals.
2. They exert their antimicrobial action through oligodynamic action. When heavy metal ions combine with sulfhydryl (—SH) groups, proteins are denatured.

Surface-Active Agents

1. Surface-active agents decrease the tension between molecules that lie on the surface of a liquid; soaps and detergents are examples.
2. Soaps have limited germicidal action but assist in the removal of microorganisms through scrubbing.
3. Acid-anionic detergents are used to clean dairy equipment.

Quaternary Ammonium Compounds (Quats)
1. Quats are cationic detergents attached to NH_4^+.
2. By disrupting plasma membranes, they allow cytoplasmic constituents to leak out of the cell.
3. Quats are most effective against gram-positive bacteria.

Chemical Food Preservatives
1. SO_2, sorbic acid, benzoic acid, and propionic acid inhibit fungal metabolism and are used as food preservatives.
2. Nitrate and nitrite salts prevent germination of *Clostridium botulinum* endospores in meats.

Antibiotics
1. Nisin and natamycin are antibiotics used to preserve foods, especially cheese.

Aldehydes
1. Aldehydes such as formaldehyde and glutaraldehyde exert their antimicrobial effect by inactivating proteins.
2. They are among the most effective chemical disinfectants.

Gaseous Chemosterilizers
1. Ethylene oxide is the gas most frequently used for sterilization.
2. It penetrates most materials and kills all microorganisms by protein denaturation.

Peroxygens (Oxidizing Agents)
1. Ozone, peroxide, and peracetic acid are used as antimicrobial agents.
2. They exert their effect by oxidizing molecules inside cells.

Microbial Characteristics and Microbial Control
1. Gram-negative bacteria are generally more resistant than gram-positive bacteria to disinfectants and antiseptics.
2. Mycobacteria, endospores, and protozoan cysts and oocysts are very resistant to disinfectants and antiseptics.
3. Nonenveloped viruses are generally more resistant than enveloped viruses to disinfectants and antiseptics.

Environment
1. Organic matter (such as vomit and feces) frequently affects the actions of chemical control agents.
2. Disinfectant activity is enhanced by warm temperatures.

THE LOOP

The chemical agents covered in this chapter are antiseptics and disinfectants. Antibiotics and other chemotherapeutic antimicrobials are discussed in Chapter 20.

Answers

Review

1. a. Lysis.
    b. Altered permeability and leakage of cell contents.
    c. Destruction of enzymes and structural proteins such as those in the plasma membrane.
    d. Interference with protein synthesis and cell division.

2. Autoclave. Due to the high specific heat of water, moist heat is readily transferred to cells.

3. Most organisms that cause disease or rapid spoilage of food are destroyed by pasteurization.

4. Variables that affect determination of the thermal death point are
    a. The innate heat resistance of the strain of bacteria.
    b. The past history of the culture, whether it was freeze-dried, wetted, etc.
    c. The clumping of the cells during the test.
    d. The amount of water present.
    e. The organic matter present.
    f. Media and incubation temperature used to determine viability of the culture after heating.

5. a. Ionizing radiation can break DNA directly. However, due to the high water content of cells, the formation of free radicals (H· and OH·) that break DNA strands is likely to occur.
    b. Ultraviolet radiation damages DNA by the formation of thymine dimers.

6. Microorganisms tend to die at a constant rate over a period of time. The constant rate is indicated by the straight line after exposure to the bactericidal compound.

7.

Sterilization Method	Temp.	Time	Type	Preferred Use	Mechanism of Action
Autoclaving	121°C	15 min	Moist	Media, equipment	Protein denaturation
Hot air	170°C	2 hr	Dry	Glassware	Oxidation
Pasteurization	72°C	15 sec	Moist	Milk, alcoholic drinks	Protein denaturation

8. All three processes kill microorganisms; however, as moisture and/or temperatures are increased, less time is required to achieve the same result.

9. Salts and sugars create a hypertonic environment. Salts and sugars (as preservatives) do not directly affect cell structures or metabolism; instead, they alter the osmotic pressure. Jams and jellies are preserved with sugar; meats are usually preserved with salt. Molds are more capable of growth in high osmotic pressure than bacteria.

10. 1. Acts rapidly.
    2. Attacks all, or a wide range of, microbes.
    3. Is able to penetrate.
    4. Readily mixes with water.
    5. Is not hampered by organic matter.
    6. Stable.
    7. Does not stain or corrode.
    8. Nontoxic.
    9. Pleasant odor.
    10. Economical.
    11. Safe to transport.

11.

Method of Action	Standard Use
a. Disrupts plasma membrane	Skin surfaces
b. Inhibits protein function	Antiseptic
c. Oxidation	Disinfect water
d. Denatures proteins, destroys lipids	Skin surfaces
e. Oligodynamic	AgNO$_3$ to prevent gonococcal eye infections
f. Inactivation of proteins	Chemical sterilizer
g. Denatures proteins	Chemical sterilizer
h. Oxidation	Antiseptic

12. Disinfectant B is preferable because it can be diluted more and still be effective.

13. Quaternary ammonium compounds are most effective against gram-positive bacteria. Gram-negative bacteria that were stuck in cracks or around the drain of the tub would not have been washed away when the tub was cleaned. These gram-negative bacteria could survive the washing procedure. Some pseudomonads can grow on quats that have accumulated.

Critical Thinking

1. a. Z.
   b. No. A culture medium would have to be inoculated from the zone of inhibition to determine the presence of viable bacteria.
2. a. Acid-resistant cell wall.
   b. Metabolizes many organic molecules.
   c. Endospores.
3. a. Disinfectant B diluted with distilled water.
   b. Can't tell; the test was done on *Salmonella*.
4. Bactericidal effects of microwave radiation are due to heat.

Clinical Applications

1. a. Hot water does not achieve sterilization, and
   b. There were no check valves to prevent backflow. Also, adapters and other invasive items should be disposable whenever possible.
2. Iodophors are approved for antiseptic uses and not for disinfection.
3. *Serratia* from the environment entered the jar from the air, wash water, or hands. The effectiveness of the quat was reduced by the cotton, and quat-resistant *Serratia* were able to survive. The bacteria were introduced into the methylprednisolone by swabbing the top of the vial. The disinfectant needs to be changed.

Case History: The Effect of Closure Type on Preventing Microbial Contamination of Cosmetics

Bacteria were isolated and identified from containers of shampoo and skin lotion that were in normal use. The containers had different types of closures and the products had low antimicrobial contents. The shampoos had been in use for 3 weeks and the lotions for 2 weeks.

Data

Organism	Contamination incidence (%)				
	Shampoo		Skin lotion		
	Screw	Slit	Screw	Flip	Pump
Citrobacter	2	0	0	0	0
Enterobacter	4	4	2	0	0
Klebsiella	1	1	2	0	0
Pseudomonas	1	1	2	0	0
Serratia	2	0	1	0	0
Gram-neg. rod, nonfermenter	0	0	1	1	0
Gram-neg. rod, fermenter	1	0	0	0	0

Questions
1. How did the bacteria get into the products?
2. Which type of closure is the best? Why do you suppose this closure works best?
3. Why did the researchers test for these bacteria?

The Solution
1. From the hands of the users.
2. The pump prevents the user from dropping bacteria into the opening. The slit/flip closures are also good because the openings are small and easily re-closed.
3. These are all gram-negative rods. Gram-negative bacteria are resistant to the antimicrobial action of detergents, a primary ingredient in the shampoos.

Chapter 8   *Microbial Genetics*

NEW IN THIS EDITION

- Explanation of $5' \rightarrow 3'$ orientation is included.
- Chromosome mapping and genomics are described.
- DNA chips are introduced.
- Vertical and horizontal gene transmission are defined.

CHAPTER SUMMARY

Structure and Function of the Genetic Material

1. Genetics is the study of what genes are, how they carry information, how their information is expressed, and how they are replicated and passed to subsequent generations or other organisms.
2. DNA in cells exists as a double-stranded helix; the two strands are held together by hydrogen bonds between specific nitrogenous base pairs: A-T and C-G.
3. A gene is a segment of DNA, a sequence of nucleotides, that codes for a functional product, usually a protein.

4. When a gene is expressed, DNA is transcribed to produce RNA; mRNA is then translated into proteins.
5. The DNA in a cell is duplicated before the cell divides, so each daughter cell receives the same genetic information.

Genotype and Phenotype

1. Genotype is the genetic composition of an organism—its entire DNA.
2. Phenotype is the expression of the genes—the proteins of the cell and the properties they confer on the organism.

DNA and Chromosomes

1. The DNA in a chromosome exists as one long double helix associated with various proteins that regulate genetic activity.
2. Bacterial DNA is circular; the chromosome of *E. coli*, for example, contains about 4 million base pairs and is approximately 1000 times longer than the cell.
3. Genomics is the molecular characterization of genomes.
4. Information contained in the DNA is transcribed into RNA and translated into proteins.

DNA Replication

1. During DNA replication, the two strands of the double helix separate at the replication fork, and each strand is used as a template by DNA polymerases to synthesize two new strands of DNA according to the rules of nitrogenous base pairing.
2. The result of DNA replication is two new strands of DNA, each having a base sequence complementary to one of the original strands.
3. Because each double-stranded DNA molecule contains one original and one new strand, the replication process is called semiconservative.
4. DNA is synthesized in one chemical direction called $5' \rightarrow 3'$. At the replication fork, the leading strand is synthesized continuously and the lagging strand, discontinuously.
5. DNA polymerase proofreads new molecules of DNA and removes mismatched bases before continuing DNA synthesis.
6. Each daughter bacterium receives a chromosome identical to the parent's.

RNA and Protein Synthesis

1. During transcription, the enzyme RNA polymerase synthesizes a strand of RNA from one strand of double-stranded DNA, which serves as a template.
2. RNA is synthesized from nucleotides containing the bases A, C, G, and U, which pair with the bases of the DNA sense strand.
3. The starting point for transcription, where RNA polymerase binds to DNA, is the promoter site; the region of DNA that is the endpoint of transcription is the terminator site; RNA is synthesized in the $5' \rightarrow 3'$ direction.
4. Translation is the process in which the information in the nucleotide base sequence of mRNA is used to dictate the amino acid sequence of a protein.
5. The mRNA associates with ribosomes, which consist of rRNA and protein.
6. Three-base segments of mRNA that specify amino acids are called codons.
7. The genetic code refers to the relationship among the nucleotide base sequence of DNA, the corresponding codons of mRNA, and the amino acids for which the codons code.

8. The genetic code is degenerate; that is, most amino acids are coded for by more than one codon.

9. Of the 64 codons, 61 are sense codons (which code for amino acids), and 3 are nonsense codons (which do not code for amino acids and are stop signals for translation).

10. The start codon, AUG, codes for methionine.

11. Specific amino acids are attached to molecules of tRNA. Another portion of the tRNA has a base triplet called an anticodon.

12. The base pairing of codon and anticodon at the ribosome results in specific amino acids being brought to the site of protein synthesis.

13. The ribosome moves along the mRNA strand as amino acids are joined to form a growing polypeptide; mRNA is read in the $5' \rightarrow 3'$ direction.

14. Translation ends when the ribosome reaches a stop codon on the mRNA.

15. In prokaryotes, translation can begin before transcription is complete.

The Regulation of Gene Expression in Bacteria

1. Regulating protein synthesis at the gene level is energy-efficient because proteins are synthesized only as they are needed.

2. Constitutive enzymes are always present in a cell. Examples are genes for most of the enzymes in glycolysis.

3. For these gene regulatory mechanisms, the control is aimed at mRNA synthesis.

Repression and Induction

1. Repression controls the synthesis of one or several (repressible) enzymes.

2. When cells are exposed to a particular end product, the synthesis of enzymes related to that product decreases.

3. In the presence of certain chemicals (inducers), cells synthesize more enzymes. This process is called induction.

4. An example of induction is the production of β-galactosidase by *E. coli* in the presence of lactose, so lactose can be metabolized.

The Operon Model of Gene Expression

1. The formation of enzymes is determined by structural genes.

2. In bacteria, a group of coordinately regulated structural genes with related metabolic functions and the promoter and operator sites that control their transcription are called an operon.

3. In the operon model for an inducible system, a regulatory gene codes for the repressor protein.

4. When the inducer is absent, the repressor binds to the operator and no mRNA is synthesized.

5. When the inducer is present, it binds to the repressor so that it cannot bind to the operator; thus, mRNA is made and enzyme synthesis is induced.

6. In repressible systems, the repressor requires a corepressor in order to bind to the operator site; thus, the corepressor controls enzyme synthesis.

7. Transcription of structural genes for catabolic enzymes (such as β-galactosidase) is induced by the absence of glucose. cAMP and CAP must bind to a promoter in the presence of an alternative carbohydrate (such as lactose).

8. The presence of glucose inhibits metabolism of alternative carbon sources by catabolic repression.

Mutation: Change in the Genetic Material

1. A mutation is a change in the nitrogenous-base sequence of DNA; that change causes a change in the product coded for by the mutated gene.

2. Many mutations are neutral, some are disadvantageous, and others are beneficial.

Types of Mutations

1. A base substitution occurs when one base pair in DNA is replaced with a different base pair.

2. Alterations in DNA can result in missense mutations (which cause amino acid substitutions) or nonsense mutations (which create stop codons).

3. In a frameshift mutation, one or a few base pairs are deleted or added to DNA.

4. Mutagens are agents in the environment that cause permanent changes in DNA.

5. Spontaneous mutations occur without the presence of a mutagen.

Mutagens

1. Chemical mutagens include base-pair mutagens (for example, nitrous acid), nucleoside analogs (for example, 2-aminopurine and 5-bromouracil), and frameshift mutagens (for example, benzpyrene).

2. Ionizing radiation causes the formation of ions and free radicals that react with DNA; base substitutions or breakage of the sugar-phosphate backbone result.

3. Ultraviolet radiation is nonionizing; it causes bonding between adjacent thymines.

4. Damage to DNA caused by ultraviolet radiation can be repaired by enzymes that cut out and replace the damaged portion of DNA.

5. Photoreactivation enzymes repair thymine dimers in the presence of visible light.

Frequency of Mutation

1. Mutation rate is the probability that a gene will mutate when a cell divides; the rate is expressed as 10 to a negative power.

2. Mutations usually occur randomly along a chromosome.

3. A low rate of spontaneous mutations is beneficial in providing the genetic diversity needed for evolution.

Identifying Mutants

1. Mutants can be detected by selecting or testing for an altered phenotype.

2. Positive selection involves the selection of mutant cells and rejection of nonmutated cells.

3. Replica plating is used for negative selection—to detect, for example, auxotrophs that have nutritional requirements not possessed by the parent (nonmutated) cell.

Identifying Chemical Carcinogens

1. The Ames test is a relatively inexpensive and rapid test for identifying possible chemical carcinogens.
2. The test assumes that a mutant cell can revert to a normal cell in the presence of a mutagen and that many mutagens are carcinogens.
3. Histidine auxotrophs of *Salmonella* are exposed to an enzymatically treated potential carcinogen, and reversions to the nonmutant state are selected.

Genetic Transfer and Recombination

1. Genetic recombination, the rearrangement of genes from separate groups of genes, usually involves DNA from different organisms; it contributes to genetic diversity.
2. In crossing over, genes from two chromosomes are recombined into one chromosome containing some genes from each original chromosome.
3. Vertical gene transfer occurs during reproduction when genes are passed from an organism to its offspring.
4. Horizontal gene transfer in bacteria involves a portion of the cell's DNA being transferred from donor to recipient.
5. When some of the donor's DNA has been integrated into the recipient's DNA, the resultant cell is called a recombinant.

Transformation in Bacteria

1. During this process, genes are transferred from one bacterium to another as "naked" DNA in solution.
2. This process was first demonstrated in *Streptococcus pneumoniae*, and occurs naturally among a few genera of bacteria.

Conjugation in Bacteria

1. This process requires contact between living cells.
2. One type of genetic donor cell is an F+; recipient cells are F-. F cells contain plasmids called F factors; these are transferred to the F- cells during conjugation.
3. When the plasmid becomes incorporated into the chromosome, the cell is called an Hfr (high-frequency recombinant).
4. During conjugation, an Hfr can transfer chromosomal DNA to an F-. Usually, the Hfr chromosome breaks before it is fully transferred.

Transduction in Bacteria

1. In this process, DNA is passed from one bacterium to another in a bacteriophage and is then incorporated into the recipient's DNA.
2. In generalized transduction, any bacterial genes can be transferred.

Plasmids and Transposons

1. Plasmids are self-replicating circular molecules of DNA carrying genes that are not usually essential for survival of the cell.
2. There are several types of plasmids, including conjugative plasmids, dissimilation plasmids, plasmids carrying genes for toxins or bacteriocins, and resistance factors.

3. Transposons are small segments of DNA that can move from one region of a chromosome to another region of the same chromosome or to a different chromosome or a plasmid.

4. Transposons are found in the main chromosomes of organisms, in plasmids, and in the genetic material of viruses. They vary from simple (insertion sequences) to complex.

5. Complex transposons can carry any type of gene, including antibiotic-resistance genes, and are thus a natural mechanism for moving genes from one chromosome to another.

Genes and Evolution

1. Diversity is the precondition of evolution.

2. Genetic mutation and recombination provide a diversity of organisms, and the process of natural selection allows the growth of those best adapted for a given environment.

THE LOOP

Generalized transduction is covered in this chapter; specialized transduction is discussed in Chapter 13. Genetic engineering techniques (Chapter 9) and industrial microbiology (Chapter 28) can be covered with this chapter. Antibiotics that interfere with protein synthesis can be included (see Figure 20.4).

Answers

Review

1. DNA consists of a strand of alternating sugars (deoxyribose) and phosphate groups with a nitrogenous base attached to each sugar. The bases are adenine, thymine, cytosine, and guanine. DNA exists in a cell as two strands twisted together to form a double helix. The two strands are held together by hydrogen bonds between their nitrogenous bases. The bases are paired in a specific, complementary way: A-T and C-G. The information held in the sequence of nucleotides in DNA is the basis for synthesis of RNA and proteins in a cell.

2. a.

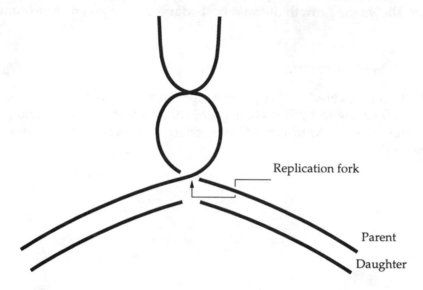

   b. DNA polymerases synthesize a complementary strand of DNA from a DNA template. RNA polymerase starts each fragment of the lagging strand with an RNA primer.
   c. Each new double-stranded DNA molecule contains one original strand and one new strand.

3. a. ATAT<u>TACTTT</u><u>GCATGGACT</u>.
   b. met-lys-arg-thr•(end).
   c. TATAATGAAACGTTCCTGA.
   d. No change.
   e. Cysteine substituted for arginine.
   f. Proline substituted for threonine (missense mutation).
   g. Frameshift mutation.
   h. Adjacent thymines might polymerize.
   i. ACT.

4. One end of the mRNA molecule becomes associated with a ribosome. Ribosomes are composed of rRNA and protein. The anticodon of a tRNA with its activated amino acid pairs with the mRNA codon at the ribosome.

5. A mutant is isolated by direct selection because it grows on a particular medium. The colonies on an antibiotic-containing medium can be identified as resistant to that antibiotic.

   A mutant is isolated by indirect selection because it does not grow on a particular medium. Replica plating could be employed to inoculate an antibiotic-containing medium. Colonies that did not grow on this medium can be isolated from the original plate and are antibiotic sensitive.

6. **Matching**
     b   A mutagen that is incorporated into DNA in place of a normal base.
     d   A mutagen that causes the formation of highly reactive ions.
     c   A mutagen that alters adenine so that it base-pairs with cytosine.
     a   A mutagen that causes insertions.
     e   A mutagen that causes the formation of pyrimidine dimers.

7. The basis for the Ames test is that a mutated cell can revert to a cell resembling the original, non-mutant cell by undergoing another mutation. The reversion rate of histidine auxotrophs of *Salmonella* in the presence of a mutagen will be higher than the spontaneous rate (in the absence of a mutagen).

8. Plasmids are small, self-replicating circles of DNA that are not associated with the chromosome. The F plasmid can be integrated into the chromosome. The F plasmid can be transferred from a donor to a recipient cell in conjugation. When the F plasmid becomes integrated into the chromosome, the cell is called an Hfr. During conjugation between an Hfr and an F cell, the chromosome of the Hfr cell, with its integrated F factor, replicates, and the new copy of the chromosome is transferred to the recipient cell.

9. a. ...a <u>repressor</u> protein must be bound tightly to the <u>operator</u> site...it will bind to the <u>repressor</u> so that <u>transcription</u> can occur.
    b. ...called a <u>corepressor,</u> causes the <u>repressor</u> to bind to the <u>operator</u>. Derepression is by removal of the <u>corepressor</u>, C in this case, when the <u>corepressor</u> is needed in the cell.
    c. None; constitutive enzymes are produced at certain necessary levels regardless of the amount of substrate or end-product.

10. Light repair; dark repair; proofreading by DNA polymerase.

11. a. The genetic makeup of an organism.
     b. The external manifestations of the genotype.
     c. Rearrangement of genes to form new combinations; in nature, this usually occurs between members of the same species; in vitro, recombinant DNA is made from genes of different species.

12. CTTTGA. Endospores and pigments offer protection against UV radiation. Additionally, repair mechanisms can remove and replace thymine polymers.

13. a. Culture 1 will remain the same. Culture 2 will convert to F+ but will have its original genotype.
     b. The donor and recipient cells' DNA can recombine to form combinations of A+B+C+ and A-B-C-. If the F plasmid also is transferred, the recipient cell may become F+.

14. Semiconservative replication ensures the offspring cell will have one correct strand of DNA. Any mutations that may have occurred during DNA replication have a greater chance of being correctly repaired.

15. Mutation and recombination provide genetic diversity. Environmental factors select for the survival of organisms through natural selection. Genetic diversity is necessary for the survival of some organisms through the processes of natural selection. Organisms that survive may undergo further genetic change, resulting in the evolution of the species.

Critical Thinking

1. Cancerous cells are growing faster than normal cells. Mutations have a greater effect when a cell is growing because it is synthesizing DNA and enzymes. The probability of a lethal mutation also is increased in rapidly growing cells.

2. The cell does not regulate the rate at which DNA is synthesized, but it regulates the rate at which replication forks on the chromosomes are initiated. The cell initiates multiple forks so that a daughter cell will inherit a complete chromosome plus additional portions from multiple replication forks. Chromosome replication begins during or immediately after division.

3. a. Mercuric ion.
    b. To detoxify it.
    c. Detoxifying mercuric ion will allow the cell to live where other organisms may not be able to.

Clinical Applications

1. a. Chloroquine interferes with transcription; erythromycin interferes with translation; acyclovir interferes with DNA replication.

   b. Erythromycin is specific for bacterial ribosomes.

   c. Acyclovir is effective against viruses because a viral enzyme tries to use it as a DNA nucleotide.

   d. Chloroquine and acyclovir will have the most effects on the host because they affect eukaryotic DNA. The effects of erythromycin on mitochondrial ribosomes are small for short-term use.

   e. Chloroquine is used against malaria. Malaria is caused by a eukaryote (protozoan).

   f. Acyclovir is used against *Herpesvirus* infections. Erythromycin affects bacterial (70S) ribosomes, not viruses or eukaryotes.

2. Sequence B is the most dissimilar and, therefore, probably not closely related to the others. The amino acid sequence reflects the RNA (genome) of the virus.

3. 28% of the nucleotides are different, however, they differ in only one of the seven amino acids. Mutations account for the difference. HHV-8 causes Kaposi's sarcoma.

Case History: Mapping a Bacterial Chromosome

Background

Conjugation mapping can be used to locate genes on a bacterial chromosome. In this example, the F+ strain is able to synthesize methionine, valine, leucine, and histidine. The F⁻ strain is unable to synthesize these amino acids; therefore, they must be supplied in the growth medium. Use the results from replica-plating to answer the questions.

Conjugation and Chromosome Mapping

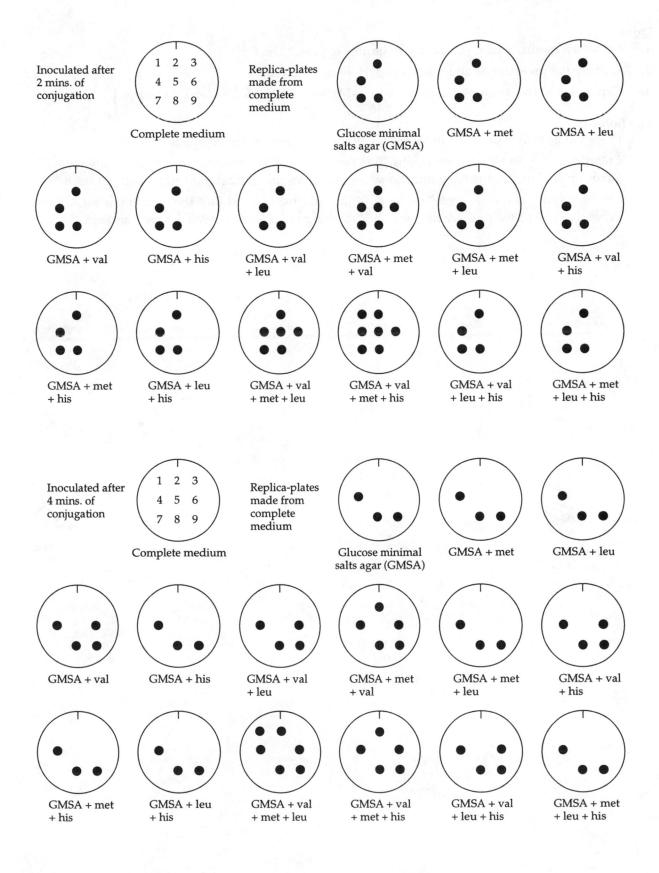

Questions

1. On which media, if any, should the F+ be able to grow? The F–?

2. Identify the prototrophs and auxotrophs. How can you tell?

3. Can you map the chromosome of this species? What further information do you need?

The Solution

1. F+ should be able to grow on all the media; F– can only grow on the complete medium.

2. Colonies 1, 3, 5, and 9 are prototrophs. These are the F+ cells. Colonies 2, 4, 6, 7, and 8 are the auxotrophs. Auxotrophs have nutritional requirements that the original strain did not have.

3. The map order is *leu*, *his*, *met*. You need additional time-trials to find the location of *val*. Other genes must be tested to determine whether anything occurs between *leu*, *his*, *met*, and *val*.

Chapter 9   *Biotechnology and Recombinant DNA*

Learning Objectives

1. Compare and contrast genetic engineering, recombinant DNA, and biotechnology.
2. Identify the roles of a clone and a vector in genetic engineering.
3. Compare selection and mutation.
4. Define restriction enzymes, and outline how they are used to make recombinant DNA.
5. List the four properties of vectors.
6. Describe the use of plasmid and viral vectors.
7. Outline the PCR and provide an example of its use.
8. Describe five ways of getting DNA into a cell.
9. Describe how a gene library is made.
10. Differentiate between cDNA and synthetic DNA.
11. Explain how each of the following are used to locate a clone: antibiotic-resistance genes, DNA probes, gene products.
12. List one advantage of engineering the following: *E. coli*, *Saccharomyces cerevisiae*, mammalian cells, plant cells.
13. List at least five applications of genetic engineering.
14. Diagram the Southern blot procedure and provide an example of its use.
15. Diagram DNA fingerprinting and provide an example of its use.
16. Outline genetic engineering with *Agrobacterium*.
17. List the advantages of, and problems associated with, the use of genetic engineering techniques.
18. Discuss some possible results of sequencing the human genome.

NEW IN THIS EDITION
- Selection and mutation techniques are discussed.
- Antisense DNA technology is introduced.

CHAPTER SUMMARY

Introduction to Biotechnology

1. Biotechnology is the use of microorganisms, cells, or cell components to make a product.

Recombinant DNA Technology

1. Closely related organisms can exchange genes in natural recombination.

2.   Genes can be transferred between unrelated species via laboratory manipulations called genetic engineering.

3.   Recombinant DNA refers to DNA that has been artifically manipulated to combine genes from two different sources.

An Overview of Recombinant DNA Procedures

1.   A desired gene is inserted into a DNA vector such as a plasmid or viral genome.

2.   The vector inserts the DNA into a new cell, which is grown to form a clone.

3.   Large quantities of the gene or the gene product can be harvested from the clone.

Tools of Biotechnology

Selection

1.   Microbes with desirable traits are selected for culturing by artificial selection.

Mutation

1.   Mutagens are used to cause mutations that might result in a microbe with desirable traits.

2.   Site-directed mutagenesis is used to change a specific codon in a gene.

Restriction Enzymes

1.   Prepackaged kits are available for many genetic-engineering techniques.

2.   A restriction enzyme recognizes and cuts only one particular nucleotide sequence in DNA.

3.   Some restriction enzymes produce sticky ends—short stretches of single-stranded DNA at the ends of the DNA fragments.

4.   Fragments of DNA produced by the same restriction enzyme will spontaneously join by hydrogen bonding. DNA ligase can covalently link the DNA backbones.

Vectors

1.   A plasmid containing a new gene can be inserted into a cell by transformation.

2.   Shuttle vectors are plasmids that can exist in several different species.

3.   A viral vector containing a new gene can be inserted into a cell by transduction.

Polymerase Chain Reaction

1.   The polymerase chain reaction (PCR) is used to make multiple copies of a desired piece of DNA enzymatically.

2.   The PCR can be used to increase the amounts of DNA in samples to detectable levels. This may allow sequencing of genes, diagnosis of genetic disease, or detection of viruses.

Inserting Foreign DNA into Cells

1.   Cells can take up naked DNA by transformation. Chemical treatments are used to make cells that do not naturally transform competent to take up DNA.

2.   Pores made in protoplasts and animal cells by electric current in the process of electroporation can provide entrance for new pieces of DNA.

3.   Protoplast fusion is the joining of cells whose cell walls have been removed.

4.   Foreign DNA can be introduced into plant cells by shooting DNA-coated particles into the cells.

5.   Foreign DNA can be injected into animal cells by using a fine glass micropipette.

Obtaining DNA

1. Gene libraries can be made by cutting up an entire genome with restriction enzymes and inserting the fragments into bacterial plasmids or phages.
2. cDNA made from mRNA by reverse transcription can be cloned in gene libraries.
3. Synthetic DNA can be made in vitro by a DNA-synthesis machine.

Selecting a Clone

1. Antibiotic resistance markers on plasmid vectors are used to identify cells containing the engineered vector by direct selection.
2. In blue/white screening, the vector contains the genes for amp^R and β-galactosidase.
3. The desired gene is inserted into the β-galactosidase-gene site, destroying the gene.
4. Clones containing the recombinant vector will be resistant to ampicillin and unable to hydrolyze X-gal (white colonies). Clones containing the vector without the new gene will be blue. Clones lacking the vector will not grow.
5. Clones containing foreign DNA can be tested for the desired gene product.
6. A short piece of labeled DNA called a DNA probe can be used to identify clones carrying the desired gene.

Making a Gene Product

1. *E. coli* is used to produce proteins by genetic engineering because it is easily grown and its genetics are well understood.
2. Efforts must be made to ensure that *E. coli*'s endotoxin does not contaminate a product intended for human use.
3. To recover the product, *E. coli* must by lysed or the gene must be linked to a gene that produces a naturally secreted protein.
4. Yeasts can be genetically engineered and are likely to secrete the gene product continuously.
5. Mammalian cells can be engineered to produce proteins such as hormones for medical use.
6. Plant cells can be engineered and used to produce plants with new properties.

Applications of Genetic Engineering

1. Cloned DNA is used to produce products, study the cloned DNA, and alter the phenotype of an organism.

Medical Therapy

1. Synthetic genes linked to the β-galactosidase gene in a plasmid vector were inserted into *E. coli*, allowing *E. coli* to produce and secrete the two polypeptides used to make human insulin.
2. Cells can be engineered to produce a pathogen's surface protein, which can be used as a subunit vaccine.
3. Animal viruses can be engineered to carry a gene for a pathogen's surface protein. When the virus is used as a vaccine, the host develops an immunity to the pathogen.

Scientific and Medical Applications

1. Recombinant DNA techniques can be used to increase understanding of DNA, for genetic fingerprinting, and for gene therapy.
2. DNA sequencing machines are used to determine the nucleotide base sequence in a gene.
3. Southern blotting can be used to locate a gene in a cell.
4. Genetic screening uses Southern blotting to look for mutations responsible for inherited disease in humans.
5. Gene therapy can be used to cure genetic diseases by replacing the defective or missing gene.
6. Southern blotting is used in DNA fingerprinting to compare DNA recovered from a crime scene with that of a suspect.
7. DNA probes can be used to identify a pathogen quickly in body tissue or food.

Agricultural Applications

1. Cells from plants with desirable characteristics can be cloned to produce many identical cells. These cells can then be used to produce whole plants from which seeds can be harvested.
2. Plant cells can be engineered by using the Ti plasmid vector. The tumor-producing T genes are replaced with desired genes, and the recombinant DNA is inserted into *Agrobacterium*. The bacterium naturally transforms its plant hosts.
3. *Rhizobium* has been engineered for enhanced nitrogen fixation.
4. *Pseudomonas* has been engineered to produce *Bacillus thuringiensis* toxin against insects.
5. Bovine growth hormone is being produced by *E. coli*.

Safety Issues and the Ethics of Genetic Engineering

1. Strict safety standards are used to avoid accidental release of genetically engineered microorganisms.
2. Some microbes used in genetic engineering have been altered so that they cannot survive outside of the laboratory.
3. Microorganisms that are intended for use in the environment may be engineered to contain suicide genes so that the organisms do not persist in the environment.
4. Genetic technology raises ethical questions such as (a) should employers and insurance companies have access to a person's genetic records, (b) will some people be targeted for either breeding or sterilization, and (c) will genetic counseling be available to everyone?
5. Genetic engineering techniques may provide new treatments for disease and new diagnostic tools.
6. Genetic engineering techniques are being used to map the human genome through the Human Genome Project.
7. This will provide tools for diagnosis and possible repair of genetic diseases.

THE LOOP

This chapter can follow Chapter 8, "Microbial Genetics," and can be covered with Chapter 27. Chapter 28 includes industrial applications of microorganisms. The boxes on pp. 35, 251, 328, and 786 provide additional examples of biotechnology.

Answers

Review

1. Recombinant DNA (rDNA) is DNA that is combined from different sources. In nature, rDNA results from conjugation, transduction, and tranformation. Genetic engineering is the artifical making of rDNA.

2. a. Both are DNA. cDNA is a segment of DNA made by RNA-dependent DNA polymerase. It is not necessarily a gene; a gene is a transcribable unit of DNA that codes for protein or RNA.

   b. Both are DNA. A restriction fragment is a segment of DNA produced when a restriction endonuclease hydrolyzes DNA. It is not usually a gene; a gene is a transcribable unit of DNA that codes for protein or RNA.

   c. Both are DNA. A DNA probe is a short, single-stranded piece of DNA. It is not a gene; a gene is a transcribable unit of DNA that codes for protein or RNA.

   d. Both are enzymes. DNA polymerase synthesizes DNA one nucleotide at a time using a DNA template; DNA ligase joins pieces (strands of nucleotides) together.

   e. Both are DNA. Recombinant DNA results from joining DNA from two different sources; cDNA results from copying a strand of RNA.

3. a. A desired gene can be spliced into a plasmid and inserted into a cell by transformation.

   b. A desired gene can be spliced into a viral genome and inserted into a cell by transduction.

   c. Antibiotic resistance genes are used as markers or labels on plasmids so that the cell containing the plasmid can be found by direct selection on an antibiotic-containing medium.

   d. A genetically engineered bacterium should be producing a new protein product. Radioactively labeled antibodies against a specific protein can be used to locate the bacterial colony producing the protein.

4. Restriction fragments from one source can be cloned in microbial cells to make a gene library. Synthetic DNA is made in a lab.

5. In protoplast fusion, two wall-less cells fuse together to combine their DNA. A variety of genotypes can result from this process. In b, c, and d, specific genes are inserted directly into the cell.

6. *BamH* I, *EcoR* I, and *Hind* III make sticky ends. Fragments of DNA produced with the same restriction enzyme will spontaneously anneal to each other at their sticky ends.

7. The gene can be spliced into a plasmid and inserted into a bacterial cell. As the cell grows, the number of plasmids will increase. The polymerase chain reaction can make copies of a gene using DNA polymerase in vitro.

8. In a eukaryotic cell, RNA polymerase copies DNA; RNA processing removes the introns, leaving the exons in the mRNA. cDNA can be made from the mRNA by reverse transcriptase.

9. See Tables 9.1 and 9.2.

10. You probably used a few plant cells in a Petri plate for your experiment. How will you select the plant cells that actually have the new Ti plasmid? You can grow these cells on plant-cell culture media with tetracycline. Only the cells with the new plasmid will grow.

Critical Thinking

1. *EcoR* I: 2 fragments
   Hind III: 2 fragments
   Both enzymes: 4 fragments
   The smallest fragment containing the *tet* gene is cut at one end by *Hind* III and at the other end by *EcoR* I.

2. Isolate cDNA (or synthesize DNA) for the desired gene from HIV. Insert the HIV gene into vaccinia virus's DNA. Infect the host cells with vaccinia virus. As the virus reproduces, it will cause production of the HIV protein.

3. "Normal" DNA polymerase is denatured by the heating step, so a technician would have to add DNA polymerase to the reaction vessel every 2 minutes. DNA polymerase from *Thermus* is not denatured by 90°C, so fresh DNA polymerase does not have to be added every 2 minutes.

4. Large colonies are ampicillin-resistant because they are growing. Smaller ampicillin-sensitive colonies may appear later after the antibiotic has been degraded by the ampicillin-resistant bacteria. White colonies have the new gene.

Clinical Applications

1. Sample 2 was positive for *V. cholerae* because the probe paired with DNA in lane 2. *V. cholerae* ingested by oysters is a source of disease for humans. The PCR doesn't require isolation and additional incubation for test results.

2. The vector and new gene fragments appear in the fifth lane; therefore, transformation did occur.

Learning with Technology

Xanthomonas frageriae

Making a Plasmid Model

Isolating a plasmid vector

Plasmids are self-replicating, extra-chromosomal molecules of DNA. Cut the three plasmid pieces on the solid lines and paste together to form a circular molecule of DNA. In DNA synthesis, the 5' end of one sugar is attached to the 3' end of the preceding sugar. This plasmid carries the (shaded) genes for resistance to kanamycin (*Km*) and resistance to ampicillin (*Ap*).

Isolating a gene

Genes can be isolated or, if small enough, synthesized. Cut out the gene on the solid lines.

Restriction enzyme digestion

Now, you must locate appropriate restriction sites. Can you find a restriction site for *Bam*H I? For *Hae* III? For *Eco*R I? Your scissors are the restriction enzyme *Eco*R I which cuts between the G and A in the sequence GAATTC on both strands of double-stranded DNA. The enzyme reads from the 5' end of DNA. Note that the two strands of DNA are complementary so you will have staggered (or *sticky*) ends after your cuts. Cut the plasmid and gene. Although enzymes work by trial-and-error to find their substrate, the carets (^) will help you locate the correct sequence.

Ligation

DNA ligase covalently joins pieces of DNA. Match the complementary bases of the staggered ends to paste the gene into the plasmid.

Voilà!

A recombinant plasmid.

Questions
1. Why were the *Bam*H I and *Hae* III sites not useful for this experiment?
2. How can you clone a plasmid by PCR? In a living cell?
3. How will you identify cells carrying the recombinant plasmid? Cells carrying the original plasmid? Cells without a plasmid?

Answers

1. Neither of these sites occurs in the plasmid and at the ends of the new gene. *Hae* III does not produce the sticky ends needed for splicing.

2. PCR can make copies of DNA using DNA polymerase and DNA primers. The plasmid can be inserted into a cell so it will replicate itself in the cell.

3.

	Nutrient agar	Nutrient agar + ampicillin	Nutrient agar + kanamycin
Cells with recombinant plasmid	Growth	Growth	No growth
Cells with original plasmid	Growth	Growth	Growth
Cells without plasmid	Growth	No growth	No growth

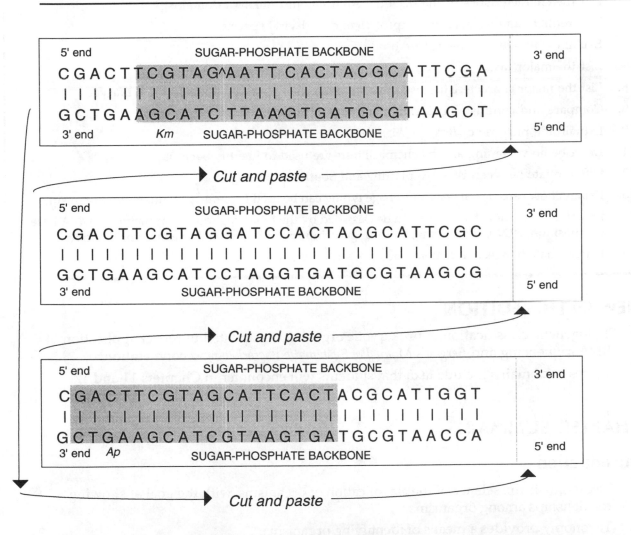

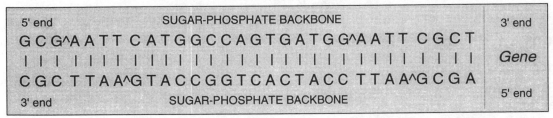

Chapter 10 Classification of Microorganisms

Learning Objectives

1. Define taxonomy, taxon, and phylogeny.
2. Discuss the limitations of a two-kingdom classification system.
3. Discuss the advantages of the three-domain system.
4. List the characteristics of the Bacteria, Archaea, and Eukarya domains.
5. Differentiate among eukaryotic, prokaryotic, and viral species.
6. Explain why scientific names are used.
7. List the major taxa.
8. List the major characteristics used to differentiate the four kingdoms in the Eukarya.
9. Compare and contrast classification and identification.
10. Explain the purpose of *Bergey's Manual*.
11. Describe how staining and biochemical tests are used to identify bacteria.
12. Differentiate between Western blotting and Southern blotting.
13. Explain how serological tests and phage typing can be used to identify an unknown organism.
14. Describe how a new bacterium can be classified by the following molecular methods: DNA base composition, rRNA sequencing, DNA fingerprinting, PCR, and nucleic acid hybridization.
15. Differentiate between a dichotomous key and a cladogram.

NEW IN THIS EDITION

- Phylogenetic classification of prokaryotic organisms is updated to reflect application of rRNA sequencing and *Bergey's Manual of Systematic Bacteriology*, second edition.
- Figures of domains include taxa that students will encounter in Chapters 11 and 12.

CHAPTER SUMMARY

Introduction

1. Taxonomy is the science of the classification of organisms, with the goal of showing relationships among organisms.
2. Taxonomy provides a means of identifying organisms.

The Study of Phylogenetic Relationships

1. Phylogeny refers to evolutionary history.
2. The taxonomic hierarchy shows evolutionary or phylogenetic relationships among organisms.

3. Living organisms were once divided into just plants and animals: bacteria, fungi, and algae were classified with plants; protozoa were classified as animals.
4. Bacteria were separated into the Kingdom Prokaryotae in 1969.

The Three Domains
1. Currently, living organisms are divided into three domains.
2. In this system, plants, animals, fungi, and protists belong to the Eukarya Domain.
3. Bacteria (with peptidoglycan) form a second domain.
4. Archaea (with unusual cell walls) are placed in the Archaea Domain.

A Phylogenetic Hierarchy
1. Organisms are grouped into taxa according to phylogenetic (from a common ancestor) relationships.
2. Some of the information for eukaryotic relationships is obtained from the fossil record.
3. Prokaryotic relationships are determined by molecular biology techniques.

Classification of Organisms

Scientific Nomenclature
1. According to scientific nomenclature, each organism is assigned two names (binomial nomenclature): a genus and a specific epithet, or species.
2. Rules for the assignment of names to bacteria are established by the International Committee on Systematic Bacteriology.
3. Rules for naming fungi and algae are published in the International Code for Botanical Nomenclature.
4. Rules for naming protozoa are found in the International Code of Zoological Nomenclature.

The Taxonomic Hierarchy
1. A eukaryotic species is a group of organisms that interbreeds but does not breed with individuals of another species.
2. Similar species are grouped into a genus; similar genera are grouped into a family; families into an order; orders, into a class; classes, into a division or phylum; and phyla, into a kingdom.

Classification of Prokaryotes
1. *Bergey's Manual of Systematic Bacteriology* is the standard reference on bacterial classification. In *Bergey's Manual*, bacteria are divided into four divisions.
2. A group of bacteria derived from a single cell is called a strain.
3. Closely related strains constitute a species.

Classification of Eukaryotes
1. Eukaryotic organisms may be classified into the Kingdoms Fungi, Plantae, or Animalia.
2. Protists are mostly unicellular organisms; these organisms are currently being assigned to kingdoms.
3. Fungi are absorptive heterotrophs that develop from spores.

4. Multicellular photoautotrophs are placed in the Kingdom Plantae.
5. Multicellular ingestive heterotrophs are classified as Animalia.

Classification of Viruses

1. Viruses are not placed in a kingdom. They are not composed of cells and cannot grow without a host cell.
2. A viral species is a population of viruses with similar characteristics that occupies a particular ecological niche.

Methods of Classifying and Identifying Microorganisms

1. *Bergey's Manual of Determinative Bacteriology* is the standard reference for laboratory identification of bacteria.
2. Morphological characteristics are useful in the identification of microorganisms, especially when aided by differential staining techniques.
3. The possession of various enzymes as determined by biochemical tests is used in the identification of microorganisms.
4. Serological tests, involving the reactions of microorganisms with specific antibodies, are useful in determining the identity of strains and species as well as relationships among organisms. ELISA and Western blot are examples of serological tests.
5. Phage typing is the identification of bacterial species and strains by the determination of their susceptibility to various phages.
6. The sequences of amino acids in proteins of related organisms are similar.
7. Related organisms have identical proteins; this characteristic can be ascertained by PAGE "fingerprints."
8. Flow cytometry measures physical and chemical characteristics of cells.
9. The percentage of G-C pairs in the nucleic acid of cells can be used in the classification of organisms.
10. The number and sizes of DNA fragments produced by restriction enzymes are used to determine genetic similarities.
11. The sequence of bases in rRNA can be used in the classification of organisms.
12. PCR can be used to detect small amounts of microbial DNA in a sample.
13. Single strands of DNA, or of DNA and RNA, from related organisms will hydrogen-bond to form a double-stranded molecule; this bonding is called nucleic acid hybridization.
14. Southern blotting and DNA probes are examples of hybridization techniques.
15. Dichotomous keys are used for identification of organisms. Cladograms show phylogenetic relationships between organisms.

THE LOOP

Chapter 10 can be assigned with other topics:

Answers

Review

1. Taxonomy is the science of classifying organisms to establish the relatedness between groups of organisms.

2. The three distinct chemical types of cells (see Table 10.1).

3. Living organisms cannot be grouped into two groups. For example, plant and animal is not acceptable because if fungi are grouped with plants, the definition of plants can't include *cellulose* and *photosynthesis*. If fungi are grouped with animals, the definition of animals can't include *no cell wall* and *ingestive*. The goal is to look for a "natural" scheme; that is, what criteria can be used to characterize all organisms.

4. Fungi: Unicellular or multicellular organisms that absorb organic nutrients; noncellulose cell walls; lack flagella.

    Plantae: Multicellular eukaryotes with tissue formation; cellulose cell walls; generally photosynthetic.

    Animalia: Multicellular eukaryotes with tissue formation; develop from an embryo (gastrula); lacking cell walls; ingest organic nutrients through a mouth of some kind.

5. a. Both are prokaryotic. They differ in composition of their cell walls, plasma membranes, and rRNAs.

    b. Both are bound by ester-linked plasma membranes. Eukarya have membrane-bound organelles.

    c. Both use methionine as the start signal. Eukarya have membrane-bound organelles and ester-linked membranes.

6. Binomial nomenclature is the system of assigning a genus and specific epithet to each organism.

7. Common names are not specific and can be misleading. According to the rules of scientific nomenclature, each organism has only one binomial.

8. The genus name must be written out so the reader knows what organism is being discussed, since the abbreviation for both of these species is *E. coli.*

9. Domain, kingdom, phylum, class, order, family, genus, species.

10. A eukaryotic species is a group of closely related organisms having limited geographical distribution that interbreeds but does not breed with other species. Species can be distinguished morphologically. Because of the distinct differences between eukaryotic organisms and bacteria, a bacterial species is defined as a population of cells with similar characteristics. A viral species is a population of viruses with similar characteristics that occupies a particular ecological niche.

11. (See Table 10.5)

    Used primarily for identification:
    morphological characteristics
    differential staining
    biochemical tests
    serology
    phage typing
    fatty acid profiles

Used primarily for taxonomic classification:
    flow cytometry
    DNA base composition
    DNA fingerprinting
    rRNA sequencing
    PCR
    nucleic acid hybridization

Data obtained from laboratory tests employing any (or all) of these twelve techniques can be assimilated using numerical taxonomy to provide information on classification.

12. Most microorganisms do not contain structures that are readily fossilized, making it difficult to obtain information on the evolution of microorganisms. Recent developments in molecular biology have provided techniques for determining evolutionary relationships amongst bacteria.

13. A and D appear to be most closely related because they have similar G-C moles %. No two are the same species.

Critical Thinking

1. A and D are most closely related.

2. Based on the nucleic acid composition, *Micrococcus* and *Staphylococcus* are not related. *Micrococcus* is in the phylum Actinobacteria; *Staphylococcus* is in the phylum Firmicutes.

3. DNA probe. Labeled DNA will hybridize with homologous DNA indicating (a) identity if the probe is known DNA and the homologous DNA is from an unknown bacterium, or (b) relatedness when the two organisms are known.

   PCR. The primer used in PCR will hybridize with homologous DNA so unrelated DNA will not be copied. Or, after making copies by PCR, a DNA probe can be used to locate specific DNA.

4. SF = "Streptococcus faecalis." SF broth is used to culture *Enterococcus faecalis*.

Clinical Applications

1. The patient had plague. The *Yersinia* (gram-negative rod) was missed in the first Gram stain. After culturing gram-negative rods, biochemical tests were not conclusive because *Yersinia* is biochemically inactive. Plague can be transmitted by the respiratory route so the patient's contacts were given prophylactic antibiotic treatment.

2. Incorrect identification could be due to mutation to sucrose+, misreading of the indicator in the sucrose fermentation test, or contamination by a sucrose+ organism.

3. One possible key is shown below. Alternate keys could be made starting with morphology or glucose fermentation.

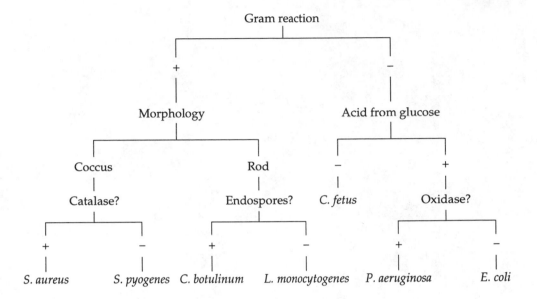

4.

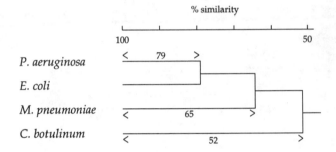

Learning with Technology

1. *Aeromonas hydrophila*
2. *Neisseria gonorrhoeae*

Designing a Taxonomic Key

Background

Classification is one of the fundamental concerns of science. Facts and objects must be arranged in an orderly fashion before unifying principles can be discovered and used as the basis for prediction. The purpose of taxonomy is to group objects according to degrees of relatedness. For example, all breeds of domestic dogs belong to one species; dogs differ from coyotes, but these two animals are similar enough to be included in the same genus. Wolves, dogs, and coyotes share enough numbers of characteristics to be included in one family. The dog family and cat family are grouped into the same order because of the numbers of teeth they have. Bacteria can also be grouped according to shared characteristics.

Data

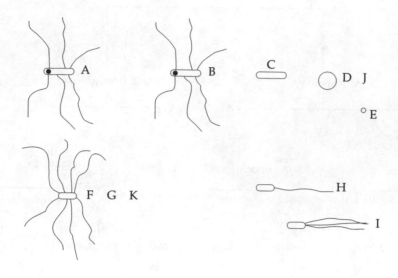

Figure	A	B	C	D	E	F	G	H	I	J	K
Catalase	–	+	–	+	+	+	+	+	+	+	+
G-C moles %	43	50	45	40	30	53	53	58	60	66	53
Gelatin hydrolyzed	+	+	+	–	–	+	+	+	–	–	–
Gram stain	+	+	+	+	–	–	–	–	–	+	–
H₂S produced	+	+	–	–	–	–	+	–	+	–	+
Indole produced	+	+	+	–	–	+	+	–	–	–	+
Lactose fermented	–	–	+	+	+	+	+	–	–	–	–
Metabolism	Anaerobe	Aerobe	Anaerobe	Facultative anaerobe	Facultative anaerobe	Facultative anaerobe	Facultative anaerobe	Aerobe	Aerobe	Aerobe	Facultativ anaerobe
Methyl red	+	–	+	+	+	+	–	–	–	–	–
Sterols required	–	–	–	–	–	–	–	–	+	–	–
Voges–Proskauer	+	+	–	–	–	–	+	–	–	–	+
Glucose fermented	Acid & gas	Acid only	Acid only	Acid only	Acid only	Acid & gas	Acid & gas	—	Acid only	—	Acid & ga

Question

1. Design a classification system grouping related species and, to the extent possible, show which groups are related.

The Solution

Students' schemes will vary; however, they should separate the following groups:

Wall-less   I

Gram-positive wall   A, B, C, D, J

Gram-negative wall   E, F, G, H, K

Chapter 11

The Prokaryotes: Domains Bacteria and Archaea

Learning Objectives

The Learning Objectives in this chapter will help students become familiar with these organisms and look for similarities and differences between organisms. In the textbook, an example is provided for Learning Objective 1 only.

1. Make a dichotomous key to distinguish among the α-proteobacteria described in this chapter.

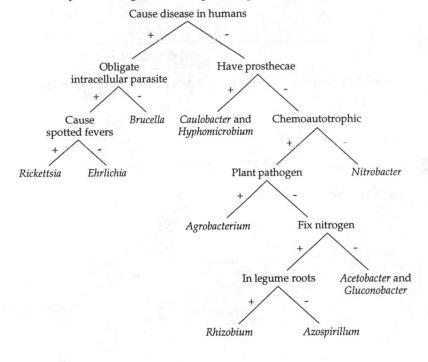

2. Make a dichotomous key to distinguish among the β-proteobacteria described in this chapter.

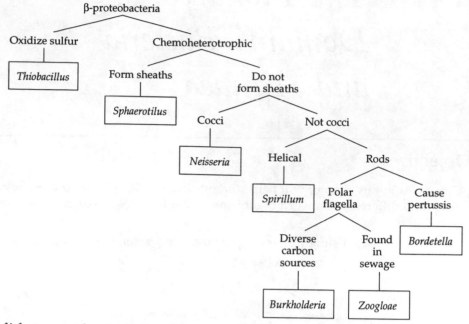

3. Make a dichotomous key to distinguish among the orders of γ-proteobacteria described in this chapter.

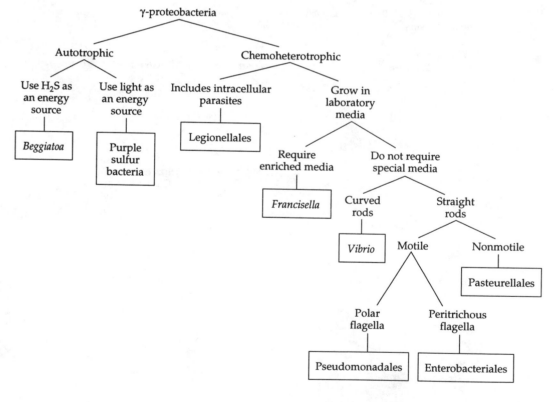

4. Make a dichotomous key to distinguish among the δ–proteobacteria described in this chapter.

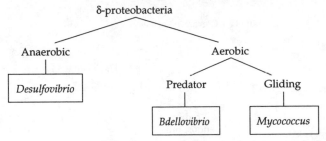

5. Make a dichotomous key to distinguish among the ε–proteobacteria described in this chapter.

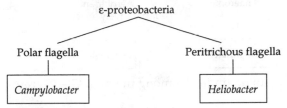

6. Make a dichotomous key to distinguish among the gram-negative nonproteobacteria bacteria described in this chapter.

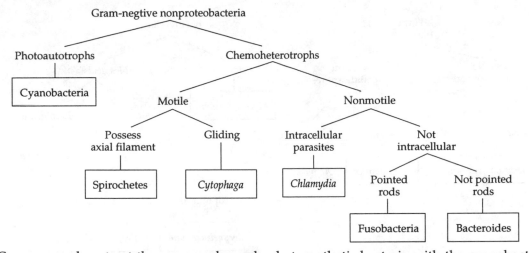

7. Compare and contrast the green and purple photosynthetic bacteria with the cyanobacteria.

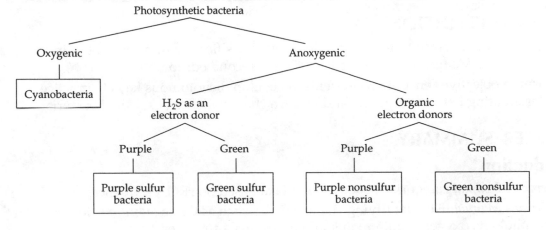

8. Make a dichotomous key to distinguish among the low G + C gram-positive bacteria described in this chapter.

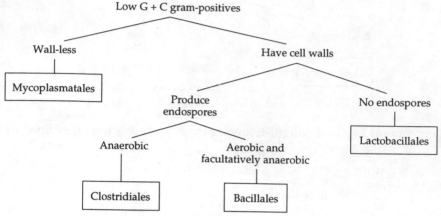

9. Make a dichotomous key to distinguish among the high G + C gram-positive bacteria described in this chapter.

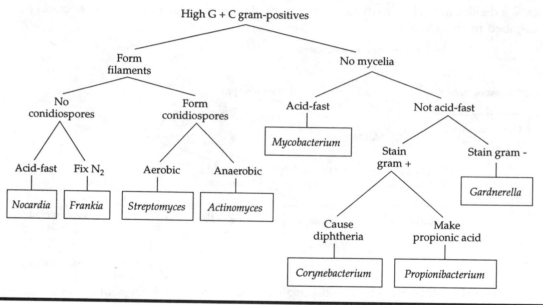

NEW IN THIS EDITION

*   Completely revised to reflect the phylogenetic classification of prokaryotes, in accordance with *Bergey's Manual of Systematic Bacteriology*, second edition.
*   Learning objectives encourage students to develop dichotomous keys to practice differentiating between bacteria and to learn characteristics of selected bacteria.

CHAPTER SUMMARY

Introduction

1. *Bergey's Manual* categorizes bacteria into taxa based on rRNA sequences.
2. *Bergey's Manual* lists identifying characteristics such as Gram stain reaction, cellular morphology, oxygen requirements, and nutritional properties.

Prokaryotic Groups

1. Prokaryotic organisms are classified into two domains: Archaea and Bacteria.

Domain Bacteria

1. Bacteria are essential to life on Earth.

The Proteobacteria

1. Members of the phylum Proteobacteria are gram-negative.
2. The α-proteobacteria include nitrogen-fixing bacteria, chemoautotrophs, and chomoheterotrophs.
3. The β-proteobacteria include chemoautotrophs and chemoheterotrophs.
4. Pseudomonadales, Legionellales, Vibrionales, Enterobacteriales, and Pasteurellales are in the γ-proteobacteria.
5. Purple and green photosynthetic bacteria are photoautotrophs that use light energy and CO_2 and do not produce O_2.
6. *Myxococcus and Bdellovibrio* in the δ-Proteobactria prey on other bacteria.
7. The ε-proteobacteria include *Campylobacter* and *Helicobacter*.

The Nonproteobacteria Gram-Negative Bacteria

1. Several phyla of gram-negative bacteria are not related phylogenetically to the Proteobacteria.
2. Cyanobacteria are photoautotrophs that use light energy and CO_2 and do produce O_2.
3. Chemoheterotrophic examples are *Chlamydia*, spirochetes, *Bacteroides*, and *Fusobacterium*.

The Gram-positive Bacteria

1. In *Bergey's Manual*, gram-positive bacteria are divided into those that have low G + C moles percent and those that have high G + C moles percent.
2. Low G + C gram-positive bacteria include common soil bacteria, the lactic acid bacteria, and several human pathogens.
3. High G + C gram-positive bacteria include mycobacteria, corynebacteria, and actino-mycetes.

Domain Archaea

1. Extreme halophiles, extreme thermophiles, and methanogens are included in the archaea.

Microbial Diversity

1. Few of the total number of different prokaryotes have been isolated and identified.
2. PCR can be used to uncover the presence of bacteria that can't be cultured in the laboratory.

THE LOOP   ○———▭

The Study Questions do not ask students to recall characteristics of specific groups of bacteria. At Skyline College, this memorization is required during a study of Part Four and appropriate laboratory work. At this point, students are shown how bacteria are classified and identified. Review question 1 provides a preliminary key for identification of unknowns in the laboratory, and names and characteristics of the large groups of bacteria.

Answers

Review

1.   I. Gram-positive
      A. Endospore-forming rod
          1. *Clostridium*
          2. *Bacillus*
      B. Nonendospore-forming
          1. Cells are rods
             a. *Streptomyces*
             b. *Mycobacterium*
          2. Cells are cocci
             a. *Streptococcus*
             b. *Staphylococcus*
  II. Gram-negative
      A. Cells are helical or curved
          1. *Treponema*
          2. *Spirillum*
      B. Cells are rods
          1. *Pseudomonas*
          2. *Escherichia*
  III. Lack of cell Walls
      A. *Mycoplasma*
      B. *Chlamydia*
  IV. Obligate intracellular parsites
      A. *Rickettsia*
      B. *Coxiella*

2.   a. Both are oxygenic photoautotrophs. Cyanobacteria are prokaryotes; algae are eukaryotes.

    b. Both are chemoheterotrophs capable of forming mycelia; some form conidia. Actinomycetes are prokaryotes; fungi are eukaryotes.

    c. Both are large rod-shaped bacteria. *Bacillus* forms endospores, *Lactobacillus* is a fermentative non-endospore-forming rod.

    d. Both are small rod-shaped bacteria. *Pseudomonas* has an oxidative metabolism; *Escherichia* is fermentative. *Pseudomonas* has polar flagella; *Escherichia* has peritrichous flagella.

    e. Both are helical bacteria. *Leptospira* (a spirochete) has an axial filament. *Spirillum* has flagella.

    f. Both are gram-negative, anaerobic bacteria. *Veillonella* are cocci, and *Bacteroides* are rods.

    g. Both are obligatory intracellular parasites.

    h. Both lack peptidoglycan cell walls. *Ureaplasma* are archaea; *Mycoplasma* are bacteria (see Table 10.2).

3. **Matching**

Nitrogen-fixing	(d)	*Frankia*
Anoxygenic	(i)	Purple bacteria
Oxygenic	(a)	Cyanobacteria
Oxidize NO_2^-	(h)	*Nitrobacter*
Reduce CO_2	(f)	Methanogens
Slime layer	(b)	*Cytophaga* and (g) Myxobacteria
Myxocysts	(g)	Myxobacteria
Anaerobic	(c)	*Desulfovibrio*
Thermophilic	(k)	*Sulfolobus*
Filaments	(j)	*Sphaerotilus*
Projections	(e)	*Hyphomicrobium*

Critical Thinking

1. See Appendix A.
    a. Firmicutes, Actinobacteria
    b. Proteobacteria, nonproteobacteria except *Chlamydia*
    c. Archaea
    d. Mollicutes, *Chlamydia*
2. *Chromatium* and *Escherichia* are both classified as Proteobacteria.
3. a. *Methanobacterium*
    b. *Bacillus*
    c. *Lactobacillus*
    d. *Pseudomonas*

Clinical Applications

1. *Neisseria (meningitidis)*
2. *Salmonella (enterica)*
3. *Listeria (monocytogenes)*

Learning With Technology

1.

	Streptococcus salivarius	*Streptococcus epidermidis*
Gram reaction	Gram-positive	Gram-positive
Morphology	Cocci	Cocci
Arrangement	Chains	Clusters
O-F reaction	Fermentative	Fermentative
Catalase reaction	Negative	Positive
Starch hydrolysis	Positive	Negative
Oxidase reaction	Negative	Negative

2.

	Pseudomonas stutzeri	*Providencia rettgeri*
Gram reaction and Morphology	Gram-negative rods	Gram-negative rods
Motility	Motile	Motile
O-F reaction	Oxidative	Fermentative
Glucose fermentation	Negative	Acid and gas
Catalase reaction	Positive	Positive
Oxidase reaction	Positive	Negative

3. *Neisseria meningitidis*

4. *Salmonella choleraesuis* susp. *arizonae* (now *S. enterica*)

5. *Listeria monocytogenes*

6. *Erwinia amylovora*

Microtriviology*

Students at Skyline College enjoy answering these trivia questions. Ten to twenty questions can be given to the students.

Trivia?! Certainly not a proper description of the science of microbiology! However, facts are sometimes encountered that are not common knowledge within the community of microbiologists; such obscure items could be described as microtrivia. Microtriviology is the study of these lesser known facts; the pursuit of microtriviology is intended to inform and entertain.

Using your text, laboratory references, and the library, answer the following questions.

1. What color are the colonies of *Chromobacterium* spp.?

2. Which is larger in diameter, the "average" *Chlamydia* or the vaccinia virus?

3. What is the general nature of the pigment characteristically produced by *Bacteriodes melaninogenicus*?

4. What illegitimate genus name have *Salmonella typhi*, *Shigella dysenteriae*, and *Actinobacillus equui* historically shared in synonymy?

5. When was the first edition of *Bergey's Manual of Determinative Bacteriology* published? What edition of *Bergey's Manual of Determinative Bacteriology* is currently in use?

6. Who originally described the bacterium now known as *Escherichia coli* (*Bacterium coli commune*) and isolated its type species?

7. What is the current classification of bacteria formerly termed members of the Bethesda-Ballerup group?

8. Are the actinomycetes considered bacteria or fungi?

9. After whom was the genus *Erwinia* named?

10. How did *Salmonella* get its name?

11. With what group of bacteria is Runyan associated?

12. What bacterium fixes nitrogen in root nodules of nonleguminous plants?

13. What is the habitat of *Cristispira*?

14. What are X and V factors?

15. What is the *Serratia marcescens* red pigment characteristically produced at room temperature by many strains?

*Philip A. Geis. "Microtriviology." A series of eight articles published in *SIM News*.

16. The two genera published simultaneously with *Legionella* that share its precedence for the appropriate bacteria are
    a. *Chainia* and *Wangiella*
    b. *Fluoribacter* and *Tatlockia*
    c. *Derxia* and *Gluconobacter*
    d. *Rickettsia* and *Bdellovibrio*

17. Name an endospore-forming coccus.

18. What two staining characteristics do vegetative cells of *Mycobacterium tuberculosis*, ascospores of *Sacchararomyces cerevisiae*, chlamydospores of *Candida albicans*, and endospores of *Bacillus subtilis* have in common?

19. Match the following *Salmonella* spp. with the term that most closely describes the origin of the strain name.

1.	*S. verity*	a.	The Mikado
2.	*S. gilbert*	b.	A racehorse
3.	*S. oysterbeds*	c.	Texas
4.	*S. manhattan*	d.	Oysters Rockefeller
5.	*S. jukestown*	e.	Truth
6.	*S. nashua*	f.	Chinese egg
7.	*S. tejas*	g.	New York
8.	*S. ank*	h.	Netherlands
		i.	A jukebox in Georgetown
		j.	Address unknown
		k.	Kansas

20. Match the following microorganisms with the term that best describes the biologic or geographic source of its original isolation.

1.	*Corynebacterium tritici*	a.	Central Africa
2.	*Azotobacter vinelandii*	b.	Wheat
3.	*Torulopsis apis*	c.	Beetle
4.	*Dermatophilus congolensis*	d.	Grapevine
5.	*Streptomyces bikiniensis*	e.	Bee
		f.	Swimming suits
		g.	From a shore
		h.	Vineland, New Jersey
		i.	Bikini atoll

Answers

1.	Violet	11.	*Mycobacteria*
2.	Vaccinia	12.	*Frankia*
3.	Heme derivative	13.	Crystalline styles of mollusks
4.	*Eberthella*	14.	X = heme derivatives; V = NAD+ or NADP+
5.	1923; 9th	15.	Prodigiosin
6.	Theodor Escherich	16.	B
7.	*Citrobacter*	17.	*Sporosarcina*
8.	Bacteria	18.	Gram-positive and acid-fast
9.	Erwin F. Smith	19.	1.e 2.a 3.d 4.k 5.i 6.b 7.c 8.j
10.	By Ligiéres for Daniel Salmon	20.	1.b 2.h 3.e 4.a 5.i

Chapter 12   The Eukaryotes: Fungi, Algae, Protozoa, and Helminths

Learning Objectives

1. List the defining characteristics of fungi.
2. Differentiate between sexual and asexual reproduction, and describe each of these processes in fungi.
3. List the defining characteristics of the three phyla of fungi described in this chapter.
4. Identify two beneficial and two harmful effects of fungi.
5. List the distinguishing characteristics of lichens, and describe their nutritional needs.
6. Describe the roles of the fungus and the alga in a lichen.
7. List the defining characteristics of algae.
8. List the outstanding characteristics of the five divisions of algae discussed in this chapter.
9. Compare and contrast cellular slime molds and plasmodial slime molds.
10. List the defining characteristics of protozoa.
11. Describe the outstanding characteristics of the seven phyla of protozoa discussed in this chapter, and give an example of each.
12. Differentiate between an intermediate host and a definitive host.
13. List the distinguishing characteristics of parasitic helminths.
14. Provide a rationale for the elaborate life cycles of parasitic worms.
15. List the characteristics of the three groups of parasitic helminths, and give an example of each.
16. Describe a parasitic infection in which humans serve as a definitive host, as an intermediate host, and as both.
17. Define arthropod vector.
18. Differentiate between a tick and a mosquito, and name a disease transmitted by each.

NEW IN THIS EDITION

- The phylogenetic classification of fungi and protozoa is updated.
- A new *MMWR* box on *Pfiesteria* and possible estuarine-associated syndrome is included.

CHAPTER SUMMARY

Fungi

1. The number of serious fungal infections is increasing.
2. The study of fungi is called mycology.
3. Fungi are aerobic or facultatively anaerobic chemoheterotrophs.
4. Most fungi are decomposers, and a few are parasites of plants and animals.

Characteristics of Fungi

1. A fungal thallus consists of filaments of cells called hyphae; a mass of hyphae is called a mycelium.
2. Yeasts are unicellular fungi. To reproduce, fission yeasts divide symmetrically, whereas budding yeasts divide asymmetrically.
3. Buds that do not separate from the mother cell form pseudohyphae.
4. Pathogenic dimorphic fungi are yeastlike at 37°C and moldlike at 25°C.
5. The following spores can be produced asexually: chlamydospores, sporangiospores, and conidiospores (including arthrospores and blastospores).
6. Fungi are classified according to the type of sexual spore that they form.
7. Sexual spores are usually produced in response to special circumstances, often changes in the environment.
8. Fungi can grow in acidic, low-moisture, aerobic environments.
9. They are able to metabolize complex carbohydrates.

Medically Important Phyla of Fungi

1. The Zygomycota have coenocytic hyphae and produce sporangiospores and zygospores.
2. The Ascomycota have septate hyphae and produce ascospores and frequently conidiospores.
3. Basidiomycota have septate hyphae and produce basidiospores; some produce conidiospores.
4. Teleomorphic fungi produce sexual and asexual spores; anamorphic fungi produce asexual spores only.

Fungal Diseases

1. Systemic mycoses are fungal infections deep within the body and affect many tissues and organs.
2. Subcutaneous mycoses are fungal infections beneath the skin.
3. Cutaneous mycoses affect keratin-containing tissues such as hair, nails, and skin.
4. Superficial mycoses are localized on hair shafts and superficial skin cells.
5. Opportunistic mycoses are caused by normal microbiota or fungi that are not usually pathogenic.
6. Opportunistic mycoses include mucormycosis, caused by some zygomycetes; aspergillosis, caused by *Aspergillus*; and candidiasis, caused by *Candida*.
7. Opportunistic mycoses can infect any tissues. However, they are usually systemic.

Economic Effects of Fungi

1. *Saccharomyces* and *Trichoderma* are used in the production of foods.
2. Fungi are used for biological control of pests.
3. Mold spoilage of fruits, grains, and vegetables is more common than bacterial spoilage of these products.
4. Many fungi cause diseases in plants (for example, in potatoes, chestnuts, and elms).

Lichens

1. A lichen is a symbiotic combination of an alga and a fungus.
2. The alga photosynthesizes, providing carbohydrates for the lichen; the fungus provides a holdfast.
3. Lichens colonize habitats that are unsuitable for either the alga or fungus alone.
4. Lichens may be classified on the basis of morphology as crustose, foliose, or fruticose.
5. Lichens are used for their pigments and as air quality indicators.

Algae

1. Algae are unicellular, filamentous, or multicellular (thallic).
2. Most algae live in aquatic environments.

Characteristics of Algae

1. All algae are eukaryotic photoautotrophs.
2. The thallus (or body) of multicellular algae usually consists of a stipe, a holdfast, and blades.
3. Algae reproduce asexually by cell division and fragmentation.
4. Many algae reproduce sexually.
5. Algae are photoautotrophs that produce oxygen.
6. Algae are classified as plants or protists according to their structures and pigments.

Selected Divisions of Algae

1. Brown algae (kelp) may be harvested for algin.
2. Red algae grow deeper in the ocean than other algae because their red pigments can absorb the blue light that penetrates to deeper levels.
3. Green algae have cellulose and chlorophyll *a* and *b* and store starch.
4. Diatoms are unicellular and have pectin and silica cell walls; some produce a neurotoxin.
5. Dinoflagellates produce neurotoxins that cause paralytic shellfish poisoning and ciguatera.

Roles of Algae in Nature

1. Algae are the primary producers in aquatic food chains.
2. Planktonic algae produce most of the molecular oxygen in the Earth's atmosphere.
3. Petroleum is the fossil remains of planktonic algae.
4. Unicellular algae are symbionts in such animals as *Tridacna*.

Slime Molds

1. Cellular slime molds resemble amoebas and ingest bacteria by phagocytosis.
2. Plasmodial slime molds consist of a multinucleated mass of protoplasm that engulfs organic debris and bacteria as it moves.

Protozoa

1. Protozoa are unicellular, eukaryotic chemoheterotrophs.
2. Protozoa are found in soil and water and as normal microbiota in animals.

Characteristics of Protozoa

1. The vegetative form is called a trophozoite.
2. Asexual reproduction is by fission, budding, or schizogony.
3. Sexual reproduction is by conjugation.
4. During protozoan conjugation, two haploid nuclei fuse to produce a zygote.
5. Some protozoa can produce a cyst for protection during adverse environmental conditions.
6. Protozoa have complex cells that may include a pellicle, a cytostome, and an anal pore.

Medically Important Phyla of Protozoa

1. Archaezoa lack mitochondria, and have flagella; they include *Trichomonas* and *Giardia*.
2. Microsporidia lack mitochondria and microtubules; microsporans cause diarrhea in AIDS patients.
3. Rhizopoda are amoeba; they include *Entamoeba* and *Acathamoeba*.
4. Apicomplexa have apical organelles for penetrating host tissue; they include *Plasmodium* and *Cryptosporidium*.
5. Ciliophora move by means of cilia; *Balantidium coli* is the only human parasitic ciliate.
6. Euglenozoa move by means of flagella, and lack sexual reproduction; they include *Trypanosoma*.

Helminths

1. Parasitic flatworms belong to the Phylum Platyhelminthes.
2. Parasitic roundworms belong to the Phylum Nematoda.

Characteristics of Helminths

1. Helminths are multicellular animals; a few are parasites of humans.
2. The anatomy and life cycles of parasitic helminths are modified for parasitism.
3. The adult stage of a parasitic helminth is found in the definitive host.
4. Each larval stage of a parasitic helminth requires an intermediate host.
5. Helminths can be monoecious or dioecious.

Platyhelminths

1. Flatworms are dorsoventrally flattened animals; parasitic flatworms may lack a digestive system.
2. Adult trematodes or flukes have an oral and ventral sucker with which they attach to and feed on host tissue.
3. Eggs of trematodes hatch into free-swimming miracidia that enter the first intermediate host; two generations of rediae develop in the first intermediate host; the rediae become cercariae that bore out of the first intermediate host and penetrate the second intermediate host; cercariae encyst as metacercariae in the second intermediate host; after they are ingested by the definitive host, the metacercariae develop into adults.

4. A cestode, or tapeworm, consists of a scolex (head) and proglottids.

5. Humans serve as the definitive host for the beef tapeworm, and cattle are the intermediate host.

6. Humans serve as the definitive host and can be an intermediate host for the pork tapeworm.

7. Humans serve as the intermediate host for *Echinococcus granulosus;* the definitive hosts are dogs, wolves, and foxes.

Nematodes

1. Members of the Phylum Nematoda are roundworms with a complete digestive system.

2. The nematodes that infect humans with their eggs are *Enterobius vermicularis* (pinworm) and *Ascaris lumbricoides.*

3. The nematodes that infect humans with their larvae are *Necator americanus* (hookworm), *Trichinella spiralis,* and anisakine worms.

Arthropods as Vectors

1. Joint-legged animals, including ticks and insects, belong to the Phylum Arthropoda.

2. Arthropods that carry diseases are called vectors.

3. Elimination of vectorborne diseases is best done by the control or eradication of the vectors.

THE LOOP

This chapter is divided so that you can select the units that meet the needs of your class.

	Study Questions
Fungi (pp. 331–342)	Review 1, 2, 3, 4; Critical Thinking 4; Clinical 2, 4, 5
Algae (pp. 344–349)	Review 7
Lichens (pp. 342–344)	Review 5, 6
Slime molds (pp. 354–356)	Review 8; Critical Thinking 2
Protozoa (pp. 349–354)	Review 9, 10, 11; Critical Thinking 5; Clinical 3
Helminths (pp. 356–363)	Review 12, 13, 14; Critical Thinking 1, 3; Clinical 1
Arthropods (pp. 363–365)	Review 15

Detailed discussions of diseases caused by fungi, protozoa, and helminths are dealt with in Part Four. If this chapter does not fit into the lecture portion of your course, it could be assigned as self-study in conjunction with Part Four or laboratory exercises. The Study Questions might be completed prior to the laboratory periods. The chapter test could be used as a posttest after completion of the laboratory exercises.

Answers

Review

1. Conidiospores are asexual spores formed by the aerial mycelia of one organism. Ascospores are sexual spores resulting from the fusion of the nuclei of two opposite mating strains of the same species of fungus.

2.

	Spore Type(s)	
Phylum	**Asexual**	**Sexual**
Zygomycota	Zygospore	Sporangiospore
Ascomycota	Blastoconidia, Arthrospore	Ascospore
Basidiomycota	Blastoconidia	Basidiospore

3.

Genus	**Mycosis**
Blastomyces	Systemic
Sporothrix	Subcutaneous
Microsporum	Cutaneous
Trichosporon	Superficial
Aspergillus	Systemic

4. a. *E. coli*

    b. *P. chrysogenum*

5. The alga produces carbohydrates for the lichen, and the fungus provides both the holdfast and protection from dessication.

6. As the first colonizers on newly exposed rock or soil, lichens are responsible for the chemical weathering of large inorganic particles and the consequent accumulation of soil.

7.

Phylum	**Cell wall composition**	**Special features**
Dinoflagellates*	Cellulose and silica	Some produce neurotoxins
Diatoms*	Pectin and silica	Produce much O_2; fossilized inclusions form petroleum
Red algae	Cellulose	Grow in deep water; source of agar
Brown algae	Cellulose and alginic acid	Source of algin
Green algae	Cellulose	Plantlike

*Unicellular

The green algae (Chlorophyta) could be placed in the plant kingdom. They have chlorophyll *b*, as do land plants, and have colonial forms. In the most advanced colonial form (*Volvox*), groups of *Chlamydomonas*-like cells live together; some are specialized for reproductive functions, which suggests a possible evolutionary route for the formation of plant tissue. The other algae are most often classified as protists.

8. Cellular slime molds exist as individual amoeboid cells. Plasmodial slime molds are multinucleate masses of protoplasm. Both survive adverse environmental conditions by forming spores.

9. Complete the following table.

Phylum	Method of Motility	One human parasite
Archaezoa	Flagella	*Giardia*
Microsporidia	None	*Nosema*
Rhizopoda	Pseudopods	*Entamoeba*
Apicomplexa	None	*Plasmodium*
Ciliophora	Cilia	*Balantidium*
Euglenozoa	Flagella	*Trypanosoma*

10. *Trichomonas* cannot survive for long outside of a host because it does not form a protective cyst. *Trichomonas* must be transferred from host to host quickly.

11. Asexual reproduction occurs in the human host and sexual reproduction takes place in the mosquito. The definitive host and the vector are the mosquito.

12. <u>Ingestion</u>.

13. This is a cestode. The encysted larva is called a cysticercus. Tapeworms are dorsoventrally flattened and have an incomplete digestive system.

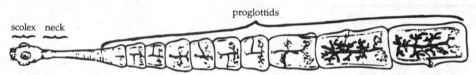

14. The male reproductive organs are in one individual, and the female reproductive organs in another. Nematodes belong to the Phylum Aschelminthes.

15.

Vector Type	Example	Disease
Mechanical	Housefly	Salmonellosis
Suitable for reproduction of parasite	*Ixodes*	Lyme disease
As a host	*Anopheles*	Malaria

Critical Thinking

1. Intermediate hosts: snail, fish
   Definitive host: human
   Phylum: Platyhelminthes
   Class: Trematode

2. Plasmodial slime molds have an internal transport system, called protoplasmic streaming, to ensure circulation.

3. Fish (larva) $\dfrac{\text{Ingestion of fish}>}{<\text{Ingestion of eggs}}$ Mammal (adult)

   Mammals (e.g., bears) are a more likely part of the freshwater ecosystem, so parasites would evolve to use mammalian hosts.

4. Information on related organisms can provide models for the laboratory culture and clinical treatment of infections. Knowledge about relationships can provide insight regarding evolution.

5. Phylum: Protozoa
   Class: Mastigophora
   Host: Human
   Vector: tsetse fly

Clinical Applications

1. *Taenia solium*; ingestion of tapeworm eggs excreted by a household member. Prevention: handwashing to break the fecal–oral cycle.

2. *Coccidioides immitis*; inhalation of arthrospores. Prevention: avoid working in contaminated soils.

3. Malaria; transmitted by bite of *Anopheles* mosquito.

4. a. From the Elastoplast bandage.

   b. Fungal spores can survive on the dry bandage.

5. *Blastomyces dermatitidis*, which may be treated with amphotericin B or an imidazole.

Case History: The Unfortunate Alaskan Fishing Trip

Background

The Alaska Department of Public Health was notified that foodborne illness had occurred in fishermen aboard a fishing boat off the Alaska peninsula. The fishermen had eaten baked fish, steamed clams and mussels, boiled rice, boiled potatoes, and green salad. No alcohol was consumed.

Data

Case	Symptoms	Onset (hr)	Foods Eaten					
			Clams	Mussels	Salmon	Halibut	Rice	Potatoes
1	None	—	1			x		x
2	Vomiting; numbness around mouth; lower back pain	2½	4–5			x	x	
3	None	—			x			x
4	Numbness of tongue and jaw; tingling of fingers	2	5			x		x
5	Numbness and tingling of face and hands; dizziness	1½						
6	None	—		1 raw		x	x	
7	None	—	1			x	x	
8	None	—		1 raw	x			x
9	Numbness and tingling around mouth, face, and fingers; cardiopulmonary arrest. Died.	½	25–30			x		

Case	Symptoms	Hr	Clams	Mussels	Salmon	Halibut	Rice	Potatoes
10	Vomiting; numbness of lips and fingers; lower back pain	2		4–5		x		x
11	Vomiting; numbness of throat, toes, and fingers; perioral numbness	1		12			x	x
12	Lower back pain	2	6–7		x		x	
13	Nausea	½	2			x	x	
14	Numbness of face; paralysis of legs	1	20–25					x
15	Vomiting; numbness and tingling of jaws and arms; loss of consciousness	1		18–24				x
16	None	—	1					x
17	None	—				x		x
18	Vomiting; numbness of mouth; tingling of fingers and toes	2	3–4		x			x
19	Vomiting; tingling of mouth, face, and fingers	2	6–7				x	x
20	Numbness of face and hands; dizziness	1½	10				x	
21	Paralysis of right arm; lower back pain	1½	15–20					x

Questions

1. Identify the etiologic agent of this outbreak of food poisoning.
2. Was it food infection or intoxication?
3. How did the food get contaminated, and what item was contaminated?
4. Briefly explain how you arrived at your conclusion.

Hints

1. Make a summary table of the persons not ill.
2. Make a table of the onset of symptoms following eating.

The Solution

1. Paralytic shellfish poisoning (PSP) caused by *Alexandrium* toxin.

2. Intoxication.

3. Mollusks can become toxic when toxin-producing dinoflagellates create massive algal blooms known as "red tides."

4. The diagnosis of PSP is based on patient exposure history and clinical manifestations.

Chapter 13 *Viruses, Viroids, and Prions*

Learning Objectives

1. Differentiate between a virus and a bacterium.
2. Describe the chemical composition and physical structure of an enveloped and a nonenveloped virus.
3. Define viral species.
4. Give an example of a family, genus, and common name for a virus.
5. Describe how bacteriophages are cultured.
6. Describe how animal viruses are cultured.
7. List three techniques that are used to identify viruses.
8. Describe the lytic cycle of T-even bacteriophages.
9. Describe the lysogenic cycle of bacteriophage lambda.
10. Compare and contrast the multiplication cycle of DNA- and RNA-containing animal viruses.
11. Define oncogene and transformed cell.
12. Discuss the relationship of DNA- and RNA-containing viruses to cancer.
13. Provide an example of a latent viral infection.
14. Differentiate between persistant viral infections and latent viral infections.
15. Discuss how a protein can be infectious.
16. Differentiate between virus, viroid, and prion.
17. Name a virus that causes a plant disease

NEW IN THIS EDITION

- A figure showing how prions can be infectious is included.
- The *MMWR* box on the AIDS risk to health care workers is updated.

CHAPTER SUMMARY

General Characteristics of Viruses

1. Depending on one's viewpoint, viruses may be regarded as exceptionally complex aggregations of nonliving chemicals or as exceptionally simple living microbes.
2. Viruses contain a single type of nucleic acid (DNA or RNA) and a protein coat, sometimes enclosed by an envelope composed of lipids, proteins, and carbohydrates.
3. Viruses are obligatory intracellular parasites. They multiply by using the host cell's synthesizing machinery to cause the synthesis of specialized elements that can transfer the viral nucleic acid to other cells.

Host Range

1. Host range refers to the spectrum of host cells in which a virus can multiply.
2. Most viruses infect only specific types of cells in one host species.
3. Host range is determined by the specific attachment site on the host cell's surface and the availability of host cellular factors.

Size

1. Viral size is ascertained by electron microscopy.
2. Viruses range from 20 to 14,000 nm in length.

Viral Structure

1. A virion is a complete, fully developed viral particle composed of nucleic acid surrounded by a coat.

Nucleic Acid

1. Viruses contain either DNA or RNA, never both, and the nucleic acid may be single- or double-stranded, linear or circular, or divided into several separate molecules.
2. The proportion of nucleic acid in relation to protein in viruses ranges from about 1% to about 50%.

Capsid and Envelope

1. The protein coat surrounding the nucleic acid of a virus is called the capsid.
2. The capsid is composed of subunits, capsomeres, which can be a single type of protein or several types.
3. The capsid of some viruses is enclosed by an envelope consisting of lipids, proteins, and carbohydrates.
4. Some envelopes are covered with carbohydrate-protein complexes called spikes.

General Morphology

1. Helical viruses (for example, Ebola virus) resemble long rods and their capsids are hollow cylinders surrounding the nucleic acid.
2. Polyhedral viruses (for example, adenovirus) are many-sided. Usually the capsid is an icosahedron.
3. Enveloped viruses are covered by an envelope and are roughly spherical but highly pleomorphic. There are also enveloped helical viruses (for example, *Influenzavirus*) and enveloped polyhedral viruses (for example, *Herpesvirus*).
4. Complex viruses have complex structures. For example, many bacteriophages have a polyhedral capsid with a helical tail attached.

Taxonomy of Viruses

1. Classification of viruses is based on type of nucleic acid, strategy for replication, and morphology.

2. Virus family names end in *-viridae;* genus names end in *-virus;* specific epithets have not been assigned.

3. A viral species is a group of viruses sharing the same genetic information and ecological niche.

Isolation, Cultivation, and Identification of Viruses

1. Viruses must be grown in living cells.
2. The easiest viruses to grow are bacteriophages.

Growth of Bacteriophages in the Laboratory

1. The plaque method mixes bacteriophages with host bacteria and nutrient agar.
2. After several viral multiplication cycles, the bacteria in the area surrounding the original virus are destroyed; the area of lysis is called a plaque.
3. Each plaque originates with a single viral particle; the concentration of viruses is given as plaque-forming units.

Growth of Animal Viruses in the Laboratory

1. Cultivation of some animal viruses requires whole animals.
2. Simian AIDS and feline AIDS provide models for study of human AIDS.
3. Some animal viruses can be cultivated in embryonated eggs.
4. Cell cultures are cells growing in culture media in the laboratory.
5. Primary cell lines and embryonic diploid cell lines grow for a short time in vitro.
6. Continuous cell lines can be maintained in vitro indefinitely.
7. Viral growth can cause cytopathic effects in the cell culture.

Viral Identification

1. Serological tests are used most often to identify viruses.
2. Viruses may be identified by RFLPs and PCR.

Viral Multiplication

1. Viruses do not contain enzymes for energy production or protein synthesis.
2. For a virus to multiply, it must invade a host cell and direct the host's metabolic machinery to produce viral enzymes and components.

Multiplication of Bacteriophages

1. During a lytic cycle, a phage causes the lysis and death of a host cell.
2. Some viruses can either cause lysis or have their DNA incorporated as a prophage into the DNA of the host cell. The latter situation is called lysogeny.
3. The T-even bacteriophages that infect *E. coli* have been studied extensively.
4. During the attachment phase of the lytic cycle, sites on the phage's tail fibers attach to complementary receptor sites on the bacterial cell.

5. In penetration, phage lysozyme opens a portion of the bacterial cell wall, the tail sheath contracts to force the tail core through the cell wall, and phage DNA enters the bacterial cell. The capsid remains outside.

6. In biosynthesis, transcription of phage DNA produces mRNA coding for proteins necessary for phage multiplication. Phage DNA is replicated, and capsid proteins are produced. During the eclipse period, separate phage DNA and protein can be found.

7. During maturation, phage DNA and capsids assemble into complete viruses.

8. During release, phage lysozyme breaks down the bacterial cell wall, and the multiplied phages are released.

9. The time from phage adsorption to release is called burst time (20 to 40 minutes). Burst size, the number of newly synthesized phages produced from a single infected cell, ranges from 50 to 200.

10. During the lysogenic cycle, prophage genes are regulated by a repressor coded for by the prophage. The prophage is replicated each time the cell divides.

11. Exposure to certain mutagens can lead to excision of the prophage and initiation of the lytic cycle.

12. Because of lysogeny, lysogenic cells become immune to reinfection with the same phage and may undergo phage conversion.

13. A lysogenic phage can transfer bacterial genes from one cell to another through transduction. Any genes can be transferred in generalized transduction, and specific genes can be transferred in specialized transduction.

Multiplication of Animal Viruses

1. Animal viruses attach to the plasma membrane of the host cell.

2. Penetration occurs by endocytosis or fusion.

3. Animal viruses are uncoated by viral or host cell enzymes.

4. The DNA of most DNA viruses is released into the nucleus of the host cell. Transcription of viral DNA and translation produce viral DNA and, later, capsid protein. Capsid protein is synthesized in the cytoplasm of the host cell.

5. DNA viruses include members of the families Adenoviridae, Poxviridae, Herpesviridae, Papovaviridae, and Hepadnaviridae.

6. Multiplication of RNA viruses occurs in the cytoplasm of the host cell. RNA-dependent RNA polymerase synthesizes a double-stranded RNA.

7. Picornaviridae + strand RNA acts as mRNA and directs the synthesis of RNA-dependent RNA polymerase.

8. Togaviridae + strand RNA acts as a template for RNA-dependent RNA polymerase, and mRNA is transcribed from a new – RNA strand.

9. Rhabdoviridae – strand RNA is a template for viral RNA-dependent RNA polymerase, which transcribes mRNA.

10. Reoviridae are digested in host-cell cytoplasm to release double-stranded RNA for viral biosynthesis.

11. Retroviridae carry reverse transcriptase (RNA-dependent DNA polymerase), which transcribes DNA from RNA.

12. After maturation, viruses are released. One method of release (and envelope formation) is budding. Nonenveloped viruses are released through ruptures in the host cell membrane.

Viruses and Cancer

1. The earliest relationship between cancer and viruses was demonstrated in the early 1900s, when chicken leukemia and chicken sarcoma were transferred to healthy animals by cell-free filtrates.

Transformation of Normal Cells into Tumor Cells

1. When activated, oncogenes transform normal cells into cancerous cells.
2. Viruses capable of producing tumors are called oncogenic viruses.
3. Several DNA viruses and retroviruses are oncogenic.
4. The genetic material of oncogenic viruses becomes integrated into the host cell's DNA.
5. Transformed cells lose contact inhibition, contain virus-specific antigens (TSTA and T antigen), exhibit chromosomal abnormalities, and can produce tumors when injected into susceptible animals.

DNA Oncogenic Viruses

1. Oncogenic viruses are found among the Adenoviridae, Herpesviridae, Poxviridae, and Papovaviridae.
2. The EB virus, a *Herpesvirus*, causes Burkitt's lymphoma and nasopharyngeal carcinoma. *Hepadnavirus* causes liver cancer.

RNA Oncogenic Viruses

1. Among the RNA viruses, only retroviruses seem to be oncogenic.
2. HTLV-1 and HTLV-2 have been associated with human leukemia and lymphoma.
3. The virus's ability to produce tumors is related to the production of reverse transcriptase. The DNA synthesized from the viral RNA becomes incorporated as a provirus into the host cell's DNA.
4. A provirus can remain latent, can produce viruses, or can transform the host cell.

Latent Viral Infections

1. A latent viral infection is one in which the virus remains in the host cell for long periods without producing an infection.
2. Examples are cold sores and shingles.

Persistant Viral Infections

1. Persistant viral infections are disease processes that occur over a long period and are generally fatal.
2. Persistant viral infections are caused by conventional viruses; viruses accumulate over a long period.

Prions

1. Prions are infectious proteins first discovered in the 1980s.
2. Prion diseases, such as CJD and mad cow disease, all involve degeneration of brain tissue.
3. Prion diseases are due to an altered protein; the cause can be a mutation in the normal gene for PrP or contact with an altered protein (PrP^{sc}).

Plant Viruses and Viroids

1. Plant viruses must enter plant hosts through wounds or with invasive parasites, such as insects.
2. Some plant viruses also multiply in insect (vector) cells.
3. Viroids are infectious pieces of RNA that cause some plant diseases, such as potato spindle tuber viroid disease.

THE LOOP

Specialized transduction is in this chapter; generalized transduction is in Chapter 8. Diseases caused by viruses are described in Part Four.

Answers

Review

1. The term filterable describes the property of passing through filters that retain bacteria. Viruses are too small to be seen with a light microscope, but their presence is known because material passed through a filter is still capable of causing a disease.

2. Viruses absolutely require living host cells to multiply.

3. A virus:

   (1) Contains DNA or RNA;

   (2) Has a protein coat surrounding the nucleic acid;

   (3) Multiplies inside a living cell using the synthetic machinery of the cell; and

   (4) Causes the synthesis of virions. A virion is a fully developed virus particle that transfers the viral nucleic acid to other cells and initiates multiplication.

4. The capsid of a helical virus is a hollow cylinder with a helical shape, which surrounds the nucleic acid (See Figure 13.4). An example of a helical virus is tobacco mosaic virus. Polyhedral viruses are many-sided (Figure 13.2). A polyhedral virus in the shape of an icosahedron is adenovirus. Polyhedral or helical viruses surrounded by an envelope are called enveloped viruses. An example of an enveloped helical virus is *Influenzavirus* (Figure 13.3), and herpes simplex is an enveloped polyhedral virus.

5. A sample of bacteriophage is mixed with host bacteria and melted nutrient agar. The mixture is then poured over a layer of nutrient agar in a Petri plate. Each phage infects a bacterium, multiplies, and releases new phages. These newly produced phages infect other bacteria, and more new viruses are produced. Following multiplication, the bacteria are destroyed. This produces a number of clearings or plaques in the layer of bacteria. The number of phages in the original sample can be estimated by counting the number of plaques.

6. Primary cell lines tend to die after a few generations. Continuous cell lines can be maintained through an indefinite number of generations. Continuous cell lines then allow long-term observations of viruses. Continuous cell lines are transformed cells.

7. A prophage gene codes for the cholera toxin. When phage DNA is incorporated into the cell's DNA, toxin can be produced.

8. **Adsorption**: The virus attaches to the cell membrane by means of spikes located on its envelope.

   Penetration: The virus gains entrance by pinocytosis, or its envelope may fuse with the plasma membrane of the host cell, allowing the virus to enter the cell.

   Uncoating: Uncoating refers to the separation of the capsid from the viral DNA.

   Biosynthesis: Viral DNA is released into the cell's nucleus, and transcription and translation from viral DNA occur. Viral DNA is synthesized.

   Maturation: Capsids form around strands of viral DNA.

   Release: The assembled capsid-containing nucleic acid pushes through the plasma membrane; a portion of the plasma membrane adheres to the capsid, thus forming the envelope.

9. a. Viruses cannot easily be observed in host tissues. Viruses cannot easily be cultured in order to be inoculated into a new host. Additionally, viruses are specific for their hosts and cells, making it difficult to substitute a laboratory animal for the third step of Koch's postulates.

   b. Some viruses can infect cells without inducing cancer. Cancer may not develop until long after infection. Cancers do not seem to be contagious.

10. Subacute sclerosing panencephalitis…common viruses…Students will have to suggest a mechanism to fill in the last blank; some examples are latent, in an abnormal tissue.

11. Provirus

    TSTA appear on the host cell surface, or T antigens appear in the nucleus. Transformed cells do not exhibit contact inhibition.

    RNA-containing oncogenic viruses produce a double-stranded DNA molecule using reverse transcriptase. The DNA is integrated into the host cell's DNA as a provirus. The provirus may transform the host cell into a tumor cell.

12. Prions are infectious proteins that appear to lack any nucleic acid. Viroids are infectious RNAs that do not have a protein coat. A prion causes CJD. A viroid causes potato spindle tuber viroid disease.

13. Of the rigid cell walls…vectors such as sap-sucking insects…plant protoplasts and insect cell cultures.

Critical Thinking

1. Outside of living cells, viruses are inert. They cannot ingest and metabolize nutrients, and they cannot reproduce. These are descriptions one might use for chemicals, not living organisms. However, inside a living cell, viruses can multiply. Clinically, since they cause infection and disease, they might be considered alive.

2. A virus is small and cannot hold as much DNA as a cell. Genes that code for proteins that serve two functions conserve space on a viral nucleic acid.

3. These two diseases provide animal models for the study of acquired immunodeficiencies and treatments. Study of the viruses (SIV and FIV) can provide more information regarding the evolution of retroviruses.

4. A prophage, provirus, or plasmid begins as a strand of DNA outside the cell's chromosome that can be integrated into the chromosome. Like a plasmid, a prophage carries genes that can be used by the cell but are not essential. Prophages and proviruses are replicated with the cell's chromosome and remain in progeny cells. Prophage DNA will form a circle and replicate itself in the cell's cytoplasm. Unlike a plasmid, prophages and proviruses are not transferred in conjugation, and when they replicate themselves, viruses are produced that can destroy the host cell.

Clinical Applications

1. Cytomegalovirus. The negative staining results indicated that the cause of her disease was not bacterial. The viral culture revealed the cause of her symptoms.

2. Herpes simplex virus. Presence of antibodies against this virus would confirm the etiology.

3. Hepatitis; these people acquired hepatitis A virus from contaminated ice-slushes.

Picornaviridae Hepatitis A Virus	Ingestion	+RNA, ss	Nonenveloped	
Hepadnaviridae Hepatitis B Virus	Injection	DNA, ds	Enveloped	Uses reverse transcriptase
Flaviviridae Hepatitis C Virus	Injection	+RNA, ss	Enveloped	

Case History: Encephalitis, Texas

Background

On May 30, a 22-year-old man complained of right hand weakness.

On June 1, he complained of right arm numbness.

On June 2, he exhibited several episodes of staring and unresponsiveness lasting 10 to 15 seconds. He consulted a physician in Mexico, who prescribed an unknown medication. That evening, he presented himself to a hospital emergency room in Texas complaining of right hand pain. He had been punctured by a catfish fin earlier in the week, so, based on this information, he was treated with ceftriaxone and tetanus toxoid.

On June 3, when he returned to the emergency room complaining of spasms, he was hyperventilating and had a white blood cell (WBC) count of 11,100 per mm^3. Although he was discharged after reporting some improvement, he began to have intermittent episodes of rigidity, breath holding, hallucinations, and difficulty swallowing. Eventually he refused liquids. That evening, he was admitted to the intensive-care unit of another hospital in Texas with a preliminary diagnosis of either encephalitis or tetanus. Manifestations included frequent spasms of the face, mouth, and neck; stuttering speech; hyperventilation; and a temperature of 37.8°C. His WBC count was 17,100 mm^3 with granulocytosis. He was sedated and observed.

On the morning of June 4, the patient was confused, disoriented, and areflexic (without reflexes). Although his neck was supple, muscle tonus was increased in his upper extremities. Analysis of cerebrospinal fluid indicated slightly elevated protein, slightly elevated glucose, and 1 WBC per 0.1 ml. An electroencephalogram showed abnormal activity. Because he had uncontrolled oral secretions, he was intubated. His temperature rose to 41.7°C, and he was sweating profusely.

On June 5, the man died.

The patient had worked as a phlebotomist for a blood bank and had donated blood on May 22. His platelets had been transfused before he became ill, but the remainder of his blood products were destroyed.

Questions

1. What was the purpose of the ceftriaxone? The tetanus toxoid?
2. What is granulocytosis?
3. What is the most likely cause of the man's illness and death?
4. What other information do you need to be sure?
5. How could he have been treated?
6. How should the platelet-recipient be treated?

The Solution

1. To prevent tetanus.
2. An increase in granulocytes (neutrophils, eosinophils, and basophils).
3. Rabies. As of this 1991 report, four cases of human rabies were known to be acquired in the United States since 1980, and this was the first case in Texas since 1985. On June 4, the patient's supervisor from work reported to hospital authorities that the man had suffered a bat bite on the right index finger. CSF, serum, and skin biopsy were tested for rabies; all of these samples were negative. Postmortem samples of brain tissue were positive for rabies by direct immunofluorescent antibody test.
4. A fluorescent-antibody test would confirm the diagnosis of rabies.
5. Treatment with antibodies against rabies (rabies immune globulin) before the symptoms began could have saved him.
6. With rabies immune globulin.

Chapter 14   *Principles of Disease and Epidemiology*

Learning Objectives

1. Define pathogen, etiology, infection, and disease.
2. Define normal and transient microbiota.
3. Compare commensalism, mutualism, and parasitism, and give an example of each.
4. Contrast normal and transient with opportunistic microbes.
5. List Koch's postulates.
6. Differentiate between a communicable and a noncommunicable disease.
7. Categorize diseases according to frequency of occurrence.
8. Categorize diseases according to severity.
9. Define herd immunity.
10. Identify four predisposing factors for disease.
11. Put the following terms in proper sequence in terms of the pattern of disease: period of decline, period of convalescence, period of illness, prodromal period, incubation period.
12. Define reservoir of infection.
13. Contrast human, animal, and nonliving reservoirs, and give one example of each.
14. Explain four methods of disease transmission.
15. Define nosocomial infections and explain their importance.
16. Define compromised host.
17. List several methods of disease transmission in hospitals.
18. Explain how nosocomial infections can be prevented.
19. List five probable reasons for emerging infectious diseases, and name one example for each reason.
20. Define epidemiology and describe three types of epidemiologic investigation.
21. Identify the function of the CDC.
22. Define the following terms: morbidity, mortality, and notifiable disease.

NEW IN THIS EDITION

- The pioneering epidemiologic works of Snow, Semmelweis, and Nightingale are included.
- A new Clinical Problem-Solving box describes an actual nosocomial outbreak of streptococcal toxic-shock syndrome.
- Morbidity data on nosocomial infections, AIDS, Lyme disease, and tuberculosis are updated.

CHAPTER SUMMARY

Introduction

1. Disease-causing microorganisms are called pathogens.
2. Pathogenic microorganisms have special properties that allow them to invade the human body or produce toxins.
3. When a microorganism overcomes the body's defenses, a state of disease results.

Pathology, Infection, and Disease

1. Pathology is the scientific study of disease.
2. Pathology is concerned with the etiology (cause), pathogenesis (development), and effects of disease.
3. Infection is the invasion and growth of pathogens in the body.
4. A host is an organism that shelters and supports the growth of pathogens.
5. Disease is an abnormal state in which part or all of the body is not properly adjusted or is incapable of performing normal functions.

Normal Microbiota

1. Animals, including humans, are usually germ-free in utero.
2. Microorganisms begin colonization in and on the surface of the body soon after birth.
3. Microorganisms that establish permanent colonies inside or on the body without producing disease make up the normal microbiota.
4. The transient microbiota is composed of microbes that are present for various periods and then disappear.

Relationships Between the Normal Microbiota and the Host

1. The normal microbiota can prevent pathogens from causing an infection; this phenomenon is known as microbial antagonism.
2. The normal microbiota and the host exist in symbiosis (living together).
3. The three types of symbiosis are commensalism (one organism benefits and the other is unaffected), mutualism (both organisms benefit), and parasitism (one organism benefits and one is harmed).

Opportunistic Organisms

1. Opportunists (opportunistic pathogens) do not cause disease under normal conditions but cause disease under special conditions.

Cooperation Among Microorganisms

1. In some situations, one microorganism makes it possible for another to cause a disease or to produce more severe symptoms; this is called cooperation.

Etiology of Infectious Diseases

Koch's Postulates

1. Koch's postulates are a method for establishing that specific microbes cause specific diseases.
2. Koch's postulates have the following requirements: (a) the same pathogen must be present in every case of the disease; (b) the pathogen must be isolated in pure culture; (c) the pathogen isolated from pure culture must cause the same disease in a healthy, susceptible laboratory animal; and (d) the pathogen must be reisolated from the inoculated laboratory animal.

Exceptions to Koch's Postulates

1. Koch's postulates are modified to establish etiologies of diseases caused by viruses and some bacteria, which cannot be grown on artificial media.
2. Some diseases, such as tetanus, have unequivocal signs and symptoms.
3. Some diseases, such as pneumonia and nephritis, may be caused by a variety of microbes.
4. Some pathogens, such as *S. pyogenes*, cause several different diseases.
5. Certain pathogens, such as HIV, cause disease in humans only.

Classifying Infectious Diseases

1. A patient may exhibit symptoms (subjective changes in body functions) and signs (measurable changes), which are used by a physician to make a diagnosis (identification of the disease).
2. A specific group of symptoms or signs that always accompanies a specific disease is called a syndrome.
3. Communicable diseases are transmitted directly or indirectly from one host to another.
4. A contagious disease is one that is easily spread from one person to another.
5. Noncommunicable diseases are caused by microorganisms that normally grow outside the human body and are not transmitted from one host to another.

The Occurrence of Disease

1. Disease occurrence is reported by incidence (number of people contracting the disease) and prevalence (number of cases at a particular time).
2. Diseases are classified by frequency of occurrence: sporadic, endemic, epidemic, and pandemic.

The Severity or Duration of a Disease

1. The scope of a disease can be defined as acute, chronic, subacute, or latent.
2. Herd immunity is the presence of immunity to a disease in most of the population.

The Extent of Host Involvement

1. A local infection affects a small area of the body; a systemic infection is spread throughout the body via the circulatory system.
2. A secondary infection can occur after the host is weakened from a primary infection.
3. An inapparent, or subclinical, infection does not cause any signs of disease in the host.

Patterns of Disease

Predisposing Factors

1. A predisposing factor is one that makes the body more susceptible to disease or alters the course of a disease.

2. Examples include gender, climate, age, and level of fatigue and nutrition.

The Development of Disease

1. The incubation period is the time interval between the initial infection and the first appearance of signs and symptoms.

2. The prodromal period is characterized by the appearance of the first mild signs and symptoms.

3. During the period of illness, the disease is at its height and all disease signs and symptoms are present.

4. During the period of decline, the signs and symptoms decrease.

5. During the period of convalescence, the body returns to its prediseased state, and health is restored.

The Spread of Infection

Reservoirs of Infection

1. A continual source of infection is called a reservoir of infection.

2. People who have a disease or are carriers of pathogenic microorganisms are human reservoirs of infection.

3. Zoonoses are diseases in wild and domestic animals that can be transmitted to humans.

4. Some pathogenic microorganisms grow in nonliving reservoirs, such as soil and water.

The Transmission of Disease

1. Transmission by direct contact involves close physical contact between the source of the disease and a susceptible host.

2. Transmission by fomites (inanimate objects) constitutes indirect contact.

3. Transmission via saliva or mucus in coughing or sneezing is called droplet transmission.

4. Transmission by a medium such as water, food, or air is called vehicle transmission.

5. Airborne transmission refers to pathogens carried on water droplets or dust for a distance greater than 1 meter.

6. Arthropod vectors carry pathogens from one host to another by both mechanical and biological transmission.

Portals of Exit

1. Just as pathogens have preferred portals of entry, they also have definite portals of exit.

2. Three common portals of exit are the respiratory tract via coughing or sneezing, the gastrointestinal tract via saliva or feces, and the urogenital tract via secretions from the vagina or penis.

3. Arthropods and syringes provide a portal of exit for microbes in blood.

Nosocomial (Hospital-Acquired) Infections

1. A nosocomial infection is one that is acquired during the course of stay in a hospital, nursing home, or other health-care facility.
2. About 5–15% of all hospitalized patients acquire nosocomial infections.

Microorganisms in the Hospital

1. Certain normal microbiota are often responsible for nosocomial infections when they are introduced into the body through such medical procedures as surgery and catheterization.
2. Opportunistic, drug-resistant gram-negative bacteria are the most frequent causes of nosocomial infections.

The Compromised Host

1. Patients with burns, surgical wounds, and suppressed immune systems are the most susceptible to nosocomial infections.

The Chain of Transmission

1. Nosocomial infections are transmitted by direct contact between staff members and patient and between patients.
2. Fomites such as catheters, syringes, and respiratory devices can transmit nosocomial infections.

The Control of Nosocomial Infections

1. Aseptic techniques can prevent nosocomial infections.
2. Hospital infection control staff are responsible for the overseeing of proper cleaning, storage, and handling of equipment and supplies.

Emerging Infectious Diseases

1. New diseases and diseases with increasing incidences are called emerging infectious diseases (EIDs).
2. EIDs can result from the use of antibiotics and pesticides, climatic changes, travel, the lack of vaccination, and insufficient case reporting.
3. The CDC, NIH, and WHO are responsible for surveillance and responses to emerging infectious diseases.

Epidemiology

1. The science of epidemiology is the study of the transmission, incidence, and frequency of disease.
2. Modern epidemiology began in the mid-1800s with the works of Snow, Semmelweis, and Nightingale.
3. Data about infected people are collected and analyzed in descriptive epidemiology.
4. In analytical epidemiology, a group of infected people is compared with an uninfected group.
5. Controlled experiments designed to test hypotheses are performed in experimental epidemiology.

6. Case reporting provides data on incidence and prevalence to local, state, and national health officials.

7. The Centers for Disease Control and Prevention (CDC) is the main source of epidemiologic information in the United States.

8. The CDC publishes the *Morbidity and Mortality Weekly Report* to provide information on morbidity (incidence) and deaths (mortality).

THE LOOP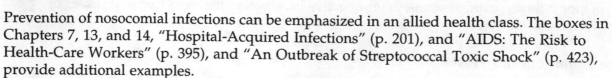

Prevention of nosocomial infections can be emphasized in an allied health class. The boxes in Chapters 7, 13, and 14, "Hospital-Acquired Infections" (p. 201), and "AIDS: The Risk to Health-Care Workers" (p. 395), and "An Outbreak of Streptococcal Toxic Shock" (p. 423), provide additional examples.

To expand on epidemiology, assign problems from the case histories of this guide. The following boxes illustrate the work of epidemiologists:

Nocosomial Infections	pp. 201, 423
Emerging Diseases	
Pfiesteria	p. 348
Streptococcal Toxic Shock Syndrome	p. 423
Mycobacterium ulcerans	p. 584
Rabies	p. 617
Shigellosis	p. 694
Leptospirosis	p. 724

Answers

Review

1. a. Etiology is the study of the cause of a disease, whereas pathogenesis is the manner in which the disease develops.
   b. Infection refers to the colonization of the body by a microorganism. Disease is any change from a state of health. A disease may, but does not always, result from infection.
   c. A communicable disease is a disease that is spread from one host to another, whereas a noncommunicable disease is not transmitted from one host to another.
2. Microorganisms that reside more or less permanently on the body are called normal microbiota. Microorganisms that are present for a few days or weeks are transient microbiota.
3. Symbiosis refers to different organisms living together. Commensalism is a symbiotic relationship in which one of the organisms is benefited and the other is unaffected. Corynebacteria living on the surface of the eye are commensals. Mutualism is a symbiosis in which both organisms are benefited. *E. coli* receives nutrients and a constant temperature in the large intestine and produces vitamin K and certain B vitamins that are useful for the human host. In parasitism, one organism benefits while the other is harmed. *Salmonella typhi* receives nutrients and warmth in the large intestine, and the human host experiences typhoid fever.
4. A reservoir of infection is a source of continual infection.

   Matching

b	Influenza
c	Rabies
a	Botulism

5. Koch's postulates establish the etiology of an infectious disease because the microorganism is removed from a sick organism, grown in a laboratory culture, and introduced into a healthy, susceptible organism. This shows that the microorganism caused the disease—not contact with a sick individual or environmental conditions.

   Some organisms are not easily seen in a host. Some microorganisms cannot be cultured on laboratory media. And some microorganisms are specific for one host. In a human, these pathogens will give rise to a group of signs and symptoms, but they will not cause the same disease in laboratory animals.
6. a. In transmission by direct contact, some kind of body contact between an infected individual and a susceptible host is required.
   b. Pathogens are transmitted from one host to another by fomites via indirect contact.
   c. Arthropod vectors can transmit pathogens mechanically where the pathogen is carried on external body parts. When an arthropod ingests a pathogen and the pathogen reproduces in the vector, it is called biological transmission.
   d. Droplet transmission also is a method of contact transmission. Pathogens are transmitted by droplets of saliva or mucus.
   e. Pathogens can be transmitted to a large number of individuals by food or water. This is called common-vehicle transmission.
   f. Airborne transmission refers to the spread of pathogens by droplet nuclei or dust.
7. Nutrition, fatigue, age, habits, lifestyle, occupation, preexisting illness, chemotherapy.
8. a. Acute
   b. Chronic
   c. Subacute

9. Hospital patients may be in a weakened condition and therefore predisposed to infection. Pathogenic microorganisms are generally transmitted to patients by contact and airborne transmission. The reservoirs of infection are the hospital staff, visitors, and other patients.

10. A disease constantly present in a population is an endemic disease. When many people acquire the disease in a relatively short time, it is an epidemic disease.

11. Epidemiology is the science dealing with when and where diseases occur and how they are transmitted in the human population. The Centers for Disease Control and Prevention (CDC) is a central source of epidemiological information.

12. Changes in body function felt by the patient are called symptoms. Symptoms such as weakness or pain are not measurable by a physician. Objective changes that the physician can observe and measure are called signs.

13. When microorganisms causing a local infection enter a blood or lymph vessel and are spread throughout the body, a systemic infection can result.

14. Mutualistic microorganisms are providing a chemical or environment that is essential for the host. In a commensal relationship, the microorganisms are obtaining nutrients from sloughed-off cells and secretions, which benefits the host by removing materials that might be invaded by pathogens. These organisms, however, are not essential; another microorganism might serve the function as well.

15. Incubation period, prodromal period, period of illness, period of decline (may be crisis), period of convalescence.

Critical Thinking

1. Koch provided reproducible steps using scientific methodology. De Bary was correct but did not provide proof or experimental procedures that would provide proof. Recall that Koch also developed culturing and staining procedures.

2. Nightingale collected data on infected persons (descriptive epidemiology) and learned that the place of most infections was in the combat area. She compared two groups of soldiers, those at home (control) and those in battle (experimental) (analytical epidemiology) to determine what factors contributed to disease. She then instituted sanitation measures in the Crimea, and the incidence of disease decreased (experimental epidemiology).

Disease	Transmission	Prevention
Cholera	Water	Proper disposal of sewage; disinfection of water.
Typhus	Body lice	Washing; disinfection of clothes and bedding.

   Also see: I. B. Cohen, "Florence Nightingale." *Scientific American* 250:128–137, March 1984.

3. a. Malaria—Vector
   b. Tuberculosis—Airborne
   c. Nosocomial infections—Any method, although vectorborne is unlikely
   d. Salmonellosis—Common vehicle
   e. Streptococcal pharyngitis—Direct contact
   f. Mononucleosis—Droplet
   g. Measles—Direct contact; airborne
   h. Hepatitis A—Common vehicle; direct contact
   i. Tetanus—Indirect contact
   j. Hepatitis B—Indirect contact
   k. Chlamydial urethritis—Direct contact

4.

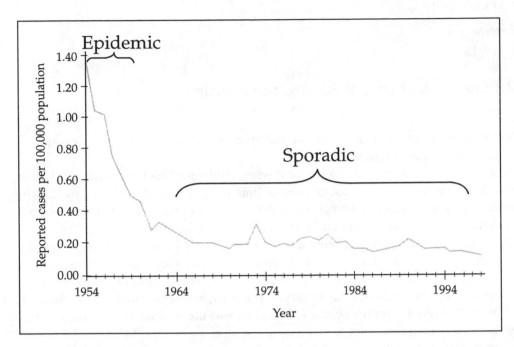

Endemic level ≈ 0.10/100,000. Information for the world population would have to be provided to indicate whether a pandemic state exists. Typhoid fever is transmitted by the fecal–oral route, usually through water.

5. Cholera is transmitted by the fecal–oral route, usually through contaminated water. *V. cholerae* was brought in contaminated fish carried by tourists. People in the United States contracted cholera eating these fish. *V. cholerae* entered coastal waters when ships from South American waters released bilge water.

Clinical Applications

1. Mistakes: Exposure to nasopharyngeal secretions and failure to get antibiotic therapy.

    Meningitis is transmitted by the respiratory route.

2. Probable source: contaminated cleaning water.

    Pseudomonads can tolerate a wide range of environmental conditions (e.g., low temperatures) and can grow on unusual carbon sources, including many detergents and disinfectants.

3. February 7 to March 9: incubation period.
    March 9: prodromal period.
    March 10 to March 17: period of illness.
    March 17: period of decline by crisis.
    Next 2 weeks: period of convalescence.

    Psittacosis is caused by *Chlamydia psittaci.*

4. The infection is probably transmitted from patient to patient on the unwashed hands of hospital staff. Hospital staff must wash their hands between patients, wear gloves when appropriate, and wash their hands after removing gloves.

5. *Mycobacterium* is usually transmitted by the respiratory route. The hospital source seems to be the water system. The boiler and pipes need to be cleaned and disinfected. Biofilms must be prevented from growing in the boiler and pipes.

Learning With Technology

1. *Serratia marcescens*
2. *Enterococcus faecalis*

Case History: An Outbreak of Food Poisoning, San Francisco

Background

An outbreak of food poisoning occurred, affecting half the members of five families who ate at a restaurant in San Francisco at 3 p.m. on June 20th.

On June 20th, a woman was admitted to a hospital with "chilliness," nausea, abdominal cramps, and watery diarrhea. The next morning, she complained of limb numbness and difficulty swallowing and breathing. Examination was unremarkable except for slight weakness of the upper extremities and diminished deep tendon reflexes. Laboratory analysis yielded no viral or bacterial infection.

An investigation was started to trace her contacts. During the investigation, it became apparent that the symptoms were due to a meal shared by 32 persons on June 20th. Rice was purchased from a produce market in a 50 kg bag. The rice was boiled the morning of the dinner and kept warm in foil-covered pots. Chicken was purchased from a supermarket. The chicken was cut up, browned in oil, boiled for two hours, and the flesh pulled from the bones. The chicken was mixed with rice noodles. A whole, gutted pig was roasted in a conventional hot air oven until the meat was white. Rice noodles were purchased from a supermarket, boiled, and pan-fried. Whole, ungutted jackfish were purchased from a seaman who caught the fish at Midway Island. The fish were frozen in the ship's freezer until return to the Port of Oakland on June 19; they were gutted and cut into steaks, which were deep-fried and served in vinegar/herb sauce. The fish heads and viscera were boiled with vegetables for 15 minutes to make chowder (escabeche).

Questions

1. On one page, identify the etiologic agent of this outbreak of food poisoning.
2. Was it food infection or intoxication?
3. How did the food get contaminated, and what item was contaminated?
4. Briefly explain how you arrived at your conclusion. How did you eliminate the other major causes of food poisoning?

Hints

1. Make a summary table of the persons not ill.
2. Make a table of the onset of symptoms following eating.

Data

			Symptoms								Foods Eaten							Onset of Symptoms		Duration
Case	Age	Sex	C	M	N V	A	B	R	N	V	1	2	3	4	5	6	7	Date	Hr	Days
1	66	M	x	x	x	x	x			x	x	x	x	x	x		20	7p	1
2	32	M		x	x	x					x		x	x	x	x	20	6p	1.5
3	29	F								x	x	x	x	x	x	x			
4	55	F											x	x	x	x	x			
5	39	F								x	x			x	x	x			
6	25	M								x	x			x	x	x			
7	32	M									x	x			x	x	x			

Data (*continued*)

Case	Age	Sex	C	M	NV	A	B	R	N	V	1	2	3	4	5	6	7	Date	Hr	Days
			Symptoms								**Foods Eaten**							**Onset of Symptoms**		**Duration**
8		F		x		x	x			x	x	x			x	x		20	8p	1
9	25											x	x				x				
10	9	M		x	x	x		x			x		x	x		x	x				
11	10	M	x					x	x		x	x		x	x			21	8p	1
12	20	F										x	x	x		x	x	20	5p	14
13	34	F										x	x	x	x		x				
14	42	F									x		x	x							
15	13	F										x	x				x				
16	12	M											x	x	x	x					
17	66	F	x	x		x		x			x	x	x	x	x	x	x	20	6p	1
18	49	M		x	x	x	x					x		x	x	x		20	7p	1
19	32	F	x					x	x		x	x		x		x	x	20	5p	14
20	33	M	x	x		x		x			x	x	x		x		x	20	6p	1.5
21	25	M		x	x		x	x	x			x	x	x	x			21	6a	1.5
22	25	F	x		x	x				x	x	x	x	x	x	x	x	20	5p	0.5
23	40	F								x		x	x		x			22	11a	1
24	43	M	x	x	x	x	x	x	x	x	x	x		x		x	x	20	5p	14
25	30	F				x					x		x	x	x			20	5p	1
26	44	M		x	x	x		x			x	x		x	x	x	x	20	5p	3
27	49	F										x	x	x		x				
28	50	F										x	x	x	x	x				
29	32	M										x		x	x	x				
30	33	M	x		x			x	x		x	x			x		x	20	5p	14
31	52	F			x	x					x	x			x	x	x	20	7p	2
32	36	F	x	x	x	x					x	x			x		x	20	9p	1

Legend:
Symptoms: C = Chills, M = Malaise, NV = Nausea, vomiting, A = Abdominal pain,
B = Blurred vision, R = Respiratory difficulty, N = Numbness, V = Vertigo.
Food: 1-Fried fish, 2-Chowder, 3-Pork, 4-Chicken/noodles, 5-Rice, 6-Fruit, 7-Chocolate cake.

The Solution
1. Ciguatera poisoning.
2. Intoxication.
3. Ciguateratoxin is derived from the dinoflagellate *Gambierdiscus toxicus,* which herbivorous coral reef fish consume. Jackfish eat the herbivorous fish.
4. Diagnosis is clinical, based on the combination of gastrointestinal and neurological symptoms. The disease is endemic in many areas of the Caribbean and South Pacific.

Chapter 15 *Microbial Mechanisms of Pathogenicity*

Learning Objectives

1. Identify the principal portals of entry.
2. Define LD_{50} and ID_{50}.
3. Using examples, explain how microbes adhere to host cells.
4. Explain how capsules and cell wall components contribute to pathogenicity.
5. Compare the effects of leukocidins, hemolysins, coagulases, kinases, hyaluronidase, and collagenase.
6. Describe how bacteria use the host cell's cytoplasm to enter the cell.
7. Contrast the nature and effects of exotoxins and endotoxins.
8. Outline the mechanisms of action of diphtheria toxin, botulinum toxin, tetanus toxin, cholera toxin, and lipid A.
9. Identify the importance of the LAL assay.
10. Using examples, describe the role of plasmids and lysogeny in pathogenicity.
11. List nine cytopathic effects of viral infections.
12. Discuss the causes of symptoms in fungal, protozoan, helminthic, and algal diseases.

NEW IN THIS EDITION

- Trichothecene mycotoxins are included.

CHAPTER SUMMARY

Introduction

1. Pathogenicity is the ability of a pathogen to produce a disease by overcoming the defenses of the host.
2. Virulence is the degree of pathogenicity.

How Microorganisms Enter a Host

1. The specific route by which a particular pathogen gains access to the body is called its portal of entry.

Portals of Entry

1. Many microorganisms can penetrate mucous membranes of the conjunctiva and the respiratory, gastrointestinal, and genitourinary tracts.
2. Microorganisms that are inhaled with droplets of moisture and dust particles gain access to the respiratory tract.

3. The respiratory tract is the most common portal of entry.
4. Microorganisms that gain access via the genitourinary tract can enter the body through mucous membranes.
5. Microorganisms enter the gastrointestinal tract via food, water, and contaminated fingers.
6. Most microorganisms cannot penetrate intact skin; they enter hair follicles and sweat ducts.
7. Some fungi infect the skin itself.
8. Some microorganisms can gain access to tissues by inoculation through the skin and mucous membranes in bites, injections, and other wounds. This route of penetration is called the parenteral route.

The Preferred Portal of Entry

1. Many organisms can cause infections only when they gain access through their specific portal of entry.

Numbers of Invading Microbes

1. Virulence can be expressed as LD_{50} (lethal dose for 50% of the inoculated hosts) or ID_{50} (infectious dose for 50% of the inoculated hosts).

Adherence

1. Surface projections on a pathogen called adhesins (ligands) adhere to complementary receptors on the host cells.
2. Ligands can be glycoproteins or lipoproteins and are frequently associated with fimbriae.
3. Mannose is the most common receptor.

How Bacterial Pathogens Penetrate Host Defenses

Capsules

1. Some pathogens have capsules that prevent them from being phagocytized.

Components of the Cell Wall

1. Proteins in the cell wall can facilitate adherence or prevent a pathogen from being phagocytized.
2. Some microbes can reproduce inside phagocytes.

Enzymes

1. Leukocidins destroy neutrophils and macrophages.
2. Hemolysins lyse red blood cells.
3. Local infections can be protected in a fibrin clot caused by the bacterial enzyme coagulase.
4. Bacteria can spread from a focal infection by means of kinases (which destroy blood clots), hyaluronidase (which destroys a mucopolysaccharide that holds cells together), and collagenase (which hydrolyzes connective tissue collagen).

Penetration Via the Host Cell Cytoskeleton

1. Salmonellae produce invasins, proteins that cause the actin of the host cell's cytoskeleton to form a basket to carry the bacteria into the cell.

How Bacterial Pathogens Damage Host Cells

Direct Damage

1. Host cells can be destroyed when pathogens metabolize and multiply inside the host cells.

The Production of Toxins

1. Poisonous substances produced by microorganisms are called toxins; toxemia refers to the presence of toxins in the blood. The ability to produce toxins is called toxigenicity.
2. Exotoxins are produced by bacteria and released into the surrounding medium. Exotoxins, not the bacteria, produce the disease symptoms.
3. Antibodies produced against exotoxins are called antitoxins.
4. Cytotoxins include diphtheria toxin (which inhibits protein synthesis) and erythrogenic toxins (which damage capillaries).
5. Neurotoxins include botulinum toxin (which prevents nerve transmission) and tetanus toxin (which prevents inhibitory nerve transmission).
6. *Vibrio cholerae* toxin and staphylococcal enterotoxin are enterotoxins, which induce fluid and electrolyte loss from host cells.
7. Endotoxins are lipopolysaccharides (LPS)—the lipid A component of the cell wall of gram-negative bacteria.
8. Bacterial cell death, antibiotics, and antibodies may cause the release of endotoxins.
9. Endotoxins cause fever (by inducing release of interleukin-1) and shock (because of a TNF-induced decrease in blood pressure).
10. Endotoxins allow bacteria to cross the blood-brain barrier.
11. The *Limulus* amoebocyte lysate (LAL) test is used to detect endotoxins in drugs and on medical devices.

Plasmids, Lysogeny, and Pathogenicity

1. Plasmids may carry genes for antibiotic resistance, toxins, capsules, and fimbriae.
2. Lysogenic conversion can result in bacteria with virulence factors, such as toxins or capsules.

Pathogenic Properties of Nonbacterial Microorganisms

Viruses

1. Viruses avoid the host's immune response by growing inside cells.
2. Viruses gain access to host cells because they have attachment sites for receptors on the host cell.
3. Signs of viral infections are called cytopathic effects (CPE).
4. Some viruses cause cytocidal effects (cell death), and others cause noncytocidal effects.
5. Cytopathic effects include the stopping of mitosis, lysis, the formation of inclusion bodies, cell fusion, antigenic changes, chromosomal changes, and transformation.

Fungi, Protozoa, Helminths, and Algae

1. Symptoms of fungal infections can be caused by capsules, toxins, and allergic responses.

2. Symptoms of protozoal and helminthic diseases can be caused by damage to host tissue or by the metabolic waste products of the parasite.

3. Some protozoa change their surface antigens while growing in a host so the host's antibodies don't kill the protozoa.

4. Some algae produce neurotoxins that cause paralysis when ingested by humans.

THE LOOP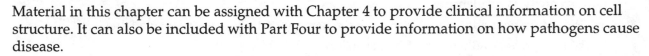

Material in this chapter can be assigned with Chapter 4 to provide clinical information on cell structure. It can also be included with Part Four to provide information on how pathogens cause disease.

Answers

Review

1. Mucous membranes: Microorganisms can adhere to and then penetrate mucous membranes.

   Skin: Microorganisms can penetrate unbroken skin through hair follicles and sweat ducts.

   Parenteral route: Pathogens can be introduced into tissues beneath the skin and mucous membranes by punctures, injections, bites, and cuts.

2. The ability of a microorganism to produce a disease is called pathogenicity. The degree of pathogenicity is virulence.

3. a. Would prevent adherence by making the mannose attachment site unavailable.

   b. Would prevent adherence of *N. gonorrhoeae*.

   c. *S. pyogenes* would not be able to attach to host cells and would be more susceptible to phagocytosis.

4. Cytopathic effects are observable changes produced in cells infected by viruses. Five examples are:

   a. Cessation of mitosis.

   b. Autolysis.

   c. The presence of inclusion bodies.

   d. Cell fusion producing syncytia.

   e. Transformation.

5.

	Exotoxin	Endotoxin
Bacterial source	Gram +	Gram −
Chemistry	Proteins	Lipid A
Toxigenicity	High	Low
Pharmacology	Destroy certain cell parts or physiological functions	Systemic, fever, weakness, aches, and shock
Example	Botulinum toxin	Salmonellosis

6. Encapsulated bacteria can resist phagocytosis and continue growing. *Streptococcus pneumoniae* and *Klebsiella pneumoniae* produce capsules that are related to their virulence. M protein found in the cell walls of *Streptococcus pyogenes* and A protein in the cell walls of *Staphylococcus aureus* help these bacteria resist phagocytosis.

7. Hemolysins are enzymes that cause the lysis of red blood cells. Hemolysis might supply nutrients for bacterial growth. Leukocidins destroy neutrophils and macrophages that are active in phagocytosis. This decreases host resistance to infection. Coagulase is an enzyme that causes the fibrinogen in blood to clot. The clot may protect the bacterium from phagocytosis and other host defenses. Bacterial kinases break down fibrin. Kinases can destroy a clot that was made to isolate the bacteria, thus allowing the bacteria to spread. Hyaluronidase dissolves the hyaluronic acid that binds cells together. This could allow the bacteria to spread through tissues.

8. Pathogenic fungi do not have specific virulence factors; capsules, metabolic products, toxins, and allergic responses contribute to the virulence of pathogenic fungi. Some fungi produce toxins that, when ingested, produce disease. Protozoa and helminths elicit symptoms by destroying host tissues and producing toxic metabolic wastes.

9. *Legionella.*

10. Botulinum toxin is more potent than *Salmonella* toxin. A much smaller amount of botulinum toxin will kill 50 percent of the inoculated hosts.

11. Food infection refers to a disease that results from pathogens entering through the gastrointestinal route. The pathogens infect the gastrointestinal tract and produce endotoxins while they are growing. Food intoxication results from ingestion of a toxin formed in food. Pathogens grow in the food and excrete an exotoxin. The pathogens do not infect the host; symptoms are due to the toxin.

12. Viruses avoid the host's immune response by growing inside host cells; some can remain latent in a host cell for prolonged periods. Some protozoa avoid the immune response by mutations that change their antigens.

Critical Thinking

1. Toxins can be carried on plasmids. (Enterotoxins are carried on plasmids.) Bacteriophages could introduce chromosomally carried toxins via lysogeny. (Perhaps the *Shigella*-like toxins originated in this way.)

   Increased incidence in summer months suggests fecal–oral transmission. The disease is associated with the seasonal use of recreational waters and the lack of rainfall to provide clean water.

2. During the summer (highest light intensity), a smaller dose produces symptoms.

3.

Yersinia	Avoids destruction by complement	Plague
Helicobacter	Neutralizes stomach acid	Peptic Disease Syndrome
Rhinovirus	Gains access to host cells to avoid the immune reponse	Common cold
Salmonella	Gains access to the host cells to avoid the immune response (mimics substrate for receptor on host cells	Gastroenteritis

4. If normal microbiota are killed by the sulfonamide and the *Salmonella* are sulfonamide-resistant, the *Salmonella* have an easier time growing.

Clinical Applications

1. *Clostridium tetani* growing at the site of the wound produced an exotoxin. Her pain and spasms were due to an infection. This disease, tetanus, cannot be transmitted to another person.

2. a. Infection. *Vibrio parahaemolyticus.*
   b. Intoxication. Ciguatera.

3. *Pseudomonas* is a gram-negative bacterium. Endotoxin from killed bacteria was present in the water.

4. *Salmonella* used the host's cytoskeleton to enter the cell. If the drug that inhibits cell division affects arrangement of the cytoskeleton, *Salmonella* will not be able to enter the cell.

Case History: Food Poisoning, New Mexico

Background

The New Mexico Health Department was consulted by an Albuquerque physician regarding two patients, a husband and wife, who had become ill within 45 minutes of eating dinner. Their symptoms included nausea, vomiting, diarrhea, headache, fever, flushing, and rapid pulse rate. An investigation found that the couple had shared a meal of grilled mahi mahi, pasta, salad, water, and wine. In a hospital emergency room, both patients were treated with antihistamines and ipecac. Their symptoms resolved within 36 hours of the onset of illness.

The fish had been imported from Taiwan through California and shipped frozen to the Albuquerque distributor, where it was thawed and sold from iced refrigerator cases. The patients had frozen the fish after they bought it. Later, they thawed it for three hours at room temperature and then grilled the still icy fish.

Questions

1. Identify the etiologic agent of this outbreak of food poisoning.

2. Was it food infection or intoxication?

3. How did the food get contaminated, and what item was contaminated?

4. Briefly explain how you arrived at your conclusion.

Data

Food Eaten	Husband	Wife	Daughter	Dog
Fish	x	x		x
Pasta	x	x	x	
Salad	x	x	x	
Wine	x	x	x	
Water	x	x	x	x
Ill	Yes	Yes	No	Yes

The Solution

1. Scombroid fish poisoning.

2. Intoxication.

3. Of all varieties of fish, the scombroid species and certain other dark-meat fish are the most likely to develop high levels of histamine. When fresh scombroid fish are not continuously iced or refrigerated, bacteria may convert the amino acid histidine, which occurs naturally in the muscle of the fish, to histamine. Since histamine is resistant to heat, cooking the fish generally will not prevent illness. Histamine levels may not be correlated with any obvious signs of decomposition of the fish. Thus, prompt and proper refrigeration or icing from the time the fish is caught until it is preserved, processed, or cooked is essential to prevent scombroid fish poisoning. Antihistamines may be useful for symptomatic treatment.

   Because histamine is metabolized by intestinal microbiota, even large doses of ingested pure histamine usually do not cause symptoms. Thus, although histamine is a marker for fish that could cause scombroid fish poisoning, the actual mechanism for the poisoning must depend on an additional cofactor. Experimental evidence indicates that other substances produced in fish by putrefactive bacteria inhibit the metabolism of histamine and permit its absorption and circulation.

4. Recovery after treatment with antihistamines. Fish samples obtained from the store yielded histamine levels of 3 mg/100 g of sample.

Chapter 16   *Nonspecific Defenses of the Host*

NEW IN THIS EDITION

*   A new box describes nitric oxide activation in macrophages.
*   The role of normal microbiota in nonspecific resistance is discussed.

CHAPTER SUMMARY

Introduction

1. The ability to ward off disease through body defenses is called resistance.
2. Lack of resistance is called susceptibility.
3. Nonspecific resistance refers to all body defenses that protect the body against any kind of pathogen.
4. Specific resistance (immunity) refers to defenses (antibodies) against specific microorganisms.

Skin and Mucous Membranes

Mechanical Factors

1. The structure of intact skin and the waterproof protein keratin provide resistance to microbial invasion.
2. Some pathogens, if present in large numbers, can penetrate mucous membranes.
3. The lacrimal apparatus protects the eyes from irritating substances and microorganisms.
4. Saliva washes organisms from teeth and gums.
5. Mucus traps many microorganisms that enter the respiratory and gastrointestinal tracts; in the lower respiratory tract, the ciliary escalator moves mucus up and out.
6. The flow of urine moves microorganisms out of the urinary tract, and vaginal secretions move microorganisms out of the vagina.

Chemical Factors

1. Sebum contains unsaturated fatty acids, which inhibit the growth of pathogenic bacteria. Some bacteria commonly found on the skin can metabolize sebum and cause the inflammatory response associated with acne.
2. Perspiration washes microorganisms off the skin.
3. Lysozyme is found in tears, saliva, nasal secretions, and perspiration.
4. The high acidity (pH 1.2 to 3.0) of gastric juice prevents microbial growth in the stomach.
5. Normal microbiota prevent the growth of many pathogens.

Normal Microbiota and Nonspecific Resistance

1. Normal microbiota change the environment which can prevent the growth of pathogens.

Phagocytosis

1. Phagocytosis is the ingestion of microorganisms or particulate matter by a cell.
2. Phagocytosis is performed by phagocytes—certain types of white blood cells or derivatives of them.

Formed Elements in Blood

1. Blood consists of plasma (fluid) and formed elements (cells and cell fragments).
2. Leukocytes (white blood cells) are divided into three categories: granulocytes (neutrophils, basophils, and eosinophils), lymphocytes, and monocytes.
3. During many infections, the number of leukocytes increases (leukocytosis); some infections are characterized by leukopenia (a decrease in leukocytes).
4. Phagocytes are activated by bacterial components (for example, lipid A) and cytokines.

Actions of Phagocytic Cells

1. Among the granulocytes, neutrophils are the most important phagocytes.
2. Enlarged monocytes become wandering macrophages and fixed macrophages.

3. Fixed macrophages are located in selected tissues and are part of the mononuclear phagocytic system.
4. Granulocytes predominate during the early stages of infection, whereas monocytes predominate as the infection subsides.

Mechanism of Phagocytosis

1. Chemotaxis is the process by which phagocytes are attracted to microorganisms.
2. The phagocyte then adheres to the microbial cells; adherence may be facilitated by opsonization—coating the microbe with plasma proteins.
3. Pseudopods of phagocytes engulf the microorganism and enclose it in a phagocytic vesicle to complete ingestion.
4. Many phagocytized microorganisms are killed by lysosomal enzymes and oxidizing agents.
5. Some microbes are not killed by phagocytes and can even reproduce in phagocytes.

Inflammation

1. Inflammation is a bodily response to cell damage; it is characterized by redness, pain, heat, swelling, and sometimes loss of function.

Vasodilation and Increased Permeability of Blood Vessels

1. The release of histamine, kinins, and prostaglandins causes vasodilation and increased permeability of blood vessels.
2. Blood clots can form around an abscess to prevent dissemination of the infection.

Phagocyte Migration and Phagocytosis

1. Phagocytes have the ability to stick to the lining of the blood vessels (margination).
2. They also have the ability to squeeze through blood vessels (emigration).
3. Pus is the accumulation of damaged tissue and dead microbes, granulocytes, and macrophages.

Tissue Repair

1. A tissue is repaired when the stroma (supporting tissue) or parenchyma (functioning tissue) produces new cells.
2. Stromal repair by fibroblasts produces scar tissue.

Fever

1. Fever is an abnormally high body temperature produced in response to a bacterial or viral infection.
2. Bacterial endotoxins and interleukin-1 can induce fever.
3. A chill indicates a rising body temperature; crisis (sweating) indicates that the body's temperature is falling.

Antimicrobial Substances

The Complement System

1. The complement system consists of a group of serum proteins that activate one another to destroy invading microorganisms.

2. C1 binds to antigen–antibody complexes to eventually activate C3 protein. Factor B, factor D, factor P, and C3 bind to certain cell wall polysaccharides to activate C3b.

3. C3 activation can result in cell lysis, inflammation, and opsonization.

4. Complement is deactivated by host-regulatory proteins.

5. Complement deficiencies can result in increased susceptibility to disease.

Interferons

1. Interferons (IFNs) are antiviral proteins produced in response to viral infection.

2. There are three types of human interferon: α–IFN, β–IFN, and γ–IFN. Recombinant interferons have also been produced.

3. The mode of action of α–IFN and β–IFN is to induce uninfected cells to produce antiviral proteins (AVPs) that prevent viral replication.

4. Interferons are host-cell-specific but not virus-specific.

5. Gamma-IFN activates neutrophils to kill bacteria.

THE LOOP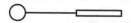

If you prefer to include the complement system with antigen–antibody reactions, pages 467–469 can be assigned with Chapter 17. Review questions 12, 13, 14, and Critical Thinking question 3, concerning complement, can be assigned when complement is covered.

Answers

Review

1. a. The ability of the human body to ward off diseases.
   b. The lack of resistance to an infectious disease.
   c. Host defenses that tend to protect the body from any kind of pathogen.

2.

	Mechanical	**Chemical**
Skin	Dry, packed cells	Sebum
Eyes	Tears	Lysozyme
Digestive tract	Movement out	HCl
Respiratory tract	Ciliary escalator	
Urinary tract	Movement out	
Genital tract	Movement out	Acidic in female

3. See Table 16.1.

4. Phagocytosis is the ingestion of a microorganism or any foreign particulate matter by a cell.

5. Granulocytes have granules in the cytoplasm. Among the granulocytes, neutrophils have the most prominent phagocytic activity. Monocytes are agranulocytes (without granules) that develop into macrophages.

   When an infection occurs, granulocytes migrate to the infected area. Monocytes follow the granulocytes to the infected tissue. During migration, monocytes enlarge and develop into actively phagocytic cells called macrophages. Macrophages phagocytize dead or dying bacteria.

6. Phagocytic cells that migrate to the infected area are called wandering macrophages. Fixed macrophages remain in certain tissues and organs.

7. Refer to Figures 16.8 and 16.9.

8. Inflammation is the body's response to tissue damage. The characteristic symptoms of inflammation are redness, pain, heat, and swelling.

9. The functions of inflammation are:
   (1) To destroy the injurious agent, if possible, and to remove it and its by-products from the body;
   (2) If destruction is not possible, to confine or wall off the injurious agent and its by-products by forming an abscess;
   (3) To repair or replace tissues damaged by the injurious agent or its by-products.

10. Leukocytic pyrogen, released from phagocytic granulocytes, has the ability to raise body temperature. The higher temperature is believed to inhibit the growth of some microorganisms. The higher temperature speeds up body reactions and may help body tissues to repair themselves more quickly.

11. The chill is an indication that body temperature is rising. Shivering and cold skin are mechanisms for increasing internal temperature. Crisis indicates body temperature is falling. The skin becomes warm as circulation is returned to it when the body attempts to dissipate extra heat.

12. Complement is a group of proteins found in normal blood serum. See Figures 16.10 and 16.11.

13. Activation of complement can result in immune adherence and phagocytosis, local inflammation, and cell lysis.

14. Endotoxin binds C3b, which activates C5–C9 to cause cell lysis. This can result in free cell wall fragments, which bind more C3b, resulting in C5–C9 damage to host cell membranes.

15. Interferons are antiviral proteins produced by infected cells in response to viral infections. Alpha-IFN and β-IFN induce uninfected cells to produce antiviral proteins. Gamma-IFN is produced by lymphocytes and activates neutrophils to kill bacteria.

Critical Thinking

1. Transferrin binds available iron so bacteria can't have it to grow. A bacterium might respond with increased siderophores to take up iron.

2. The inflammatory response is usually a beneficial response. Exceptions to this are hypersensitivities and autoimmune diseases, which are discussed in Chapter 19. Each of the drugs has side effects; while reducing inflammation, another undesirable condition might result.

3.

Organism	How does this strategy avoid destruction by the complement?	Disease
Group A streptococci	No C5–C9	Streptococcal sore throat
Haemophilus influenzae type b	Hides LPS, which can activate C	Meningitis
Pseudomonas aeruginosa	Binds C in solution instead of on cell surface	Septicemia; pyelonephritis
Trypanosoma cruzi	C5–C9 doesn't get activated	Chagas' disease

4.

Microorganisms	Effect	Disease
Influenzavirus	Kills host cell	Influenza
M. tuberculosis	Prevents digestion in phagocytes	Tuberculosis
T. gondii	Prevents digestion in phagocytes	Toxoplasmosis
Trichophyton	Digests keratin	Athlete's foot
T. cruzi	Prevents digestion in phagocytes	Chagas' disease

Clinical Applications

1. Kinins cause vasodilation and increased permeability of blood vessels. Symptoms should include increased secretions from the nose and eyes. Rhinoviruses cause the common cold.

2. The proportions of white blood cells may change during diseases. The results of a differential count can be used to diagnose diseases. A patient with mononucleosis will have an increased number of monocytes. Neutropenia: decreased neutrophils. Eosinophilia: increased eosinophils.

3. Phagocytosis is inhibited.

4. Neutrophils will not phagocytize and they will die prematurely.

Case History: Complement Evasion

The Problem

If a microbe is to be a successful parasite, it must avoid destruction by the host's complement system. The following list provides examples of known complement-evading techniques. Complete the table to identify the disease and method of action.

Organism	Strategy	Disease	How does this strategy avoid destruction by the complement?
Legionella pneumophila	Uses C3b receptors to enter monocytes		
Pseudomonas aeruginosa	Produces proteases		
Salmonella enterica	Activates C3 on long-chain LPS molecules		
Schistosoma mansoni	Sheds glycocalyx molecules		

The Solution

Organism	How does this strategy avoid destruction by the complement?
Legionella pneumophila	Binds cells instead of C3
Pseudomonas aeruginosa	Degrades complement
Salmonella enterica	C5–C9 complex is too far away from cell surface to cause damage
Schistosoma mansoni	Binds C in solution instead of on cell surface

Chapter 17 *Specific Defenses of the Host: The Immune Response*

Learning Objectives

1. Differentiate between immunity and nonspecific resistance.
2. Contrast the four types of acquired immunity.
3. Differentiate between humoral (antibody-mediated) and cell-mediated immunity.
4. Define antigen and hapten.
5. Explain the function of antibodies and describe their structural and chemical characteristics.
6. Name one function for each of the five classes of antibodies.
7. Name the function of B cells.
8. Define apoptosis, and give a potential medical application.
9. Describe the clonal selection theory.
10. Explain how an antibody reacts with an antigen; identify the consequences of the reaction.
11. Distinguish a primary from a secondary immune response.
12. Define monoclonal antibodies and identify their advantage over conventional antibody production.
13. Identify at least one function of each of the following in cell-mediated immunity: cytokines, interleukins, interferons.
14. Describe at least one function for each of the following: T_H cell, T_C cell, T_D cell, T_S cell, APC, MHC, activated macrophage, NK cell.
15. Compare and contrast cell-mediated and humoral immunity.
16. Compare and contrast T-dependent antigens and T-independent antigens.
17. Describe the role of antibodies and NK cells in antibody-dependent cell-mediated cytotoxicity.

NEW IN THIS EDITION

- The discussion of apoptosis has been revised.
- Figure 17.18 has been redrawn.

CHAPTER SUMMARY

Introduction

1. An individual's genetically predetermined resistance to certain diseases is called innate resistance.
2. Individual resistance is affected by gender, age, nutritional status, and general health.

Immunity

1. Immunity is the ability of the body to specifically counteract foreign organisms or substances called antigens.
2. Immunity results from the production of antibodies and specialized lymphocytes.

Types of Acquired Immunity

1. Acquired immunity is specific resistance to infection developed during the life of the individual.
2. A person may develop or acquire immunity after birth.

Naturally Acquired Immunity

1. Immunity resulting from infection is called naturally acquired active immunity; this type of immunity may be long-lasting.
2. Antibodies transferred from a mother to a fetus (transplacental transfer) or to a newborn in colostrum result in naturally acquired passive immunity in the newborn; this type of immunity can last up to a few months.

Artificially Acquired Immunity

1. Immunity resulting from vaccination is called artificially acquired active immunity and can be long-lasting.
2. Vaccines can be prepared from attenuated, inactivated, or killed microorganisms and toxoids.
3. Artificially acquired passive immunity refers to humoral antibodies acquired by injection; this type of immunity can last for a few weeks.
4. Antibodies made by a human or other mammal may be injected into a susceptible individual.
5. Serum containing antibodies is often called antiserum.
6. When serum is separated by gel electrophoresis, antibodies are found in the gamma fraction of the serum and are termed gamma globulin or immune serum globulin.

The Duality of the Immune System

1. Humoral immunity is found in body fluids.
2. Cell-mediated immunity is due to certain types of lymphocytes.

Humoral (Antibody-Mediated) Immunity

1. The humoral immune system involves antibodies produced by B cells in response to a specific antigen.
2. Antibodies primarily defend against bacteria, viruses, and toxins in blood plasma and lymph.

Cell-Mediated Immunity

1. The cell-mediated immune system depends on T cells and does not involve antibody production.
2. Cellular immunity is primarily a response to intracellular viruses, multicellular parasites, transplanted tissue, and cancer cells.

Antigens and Antibodies

The Nature of Antigens

1. An antigen (or immunogen) is a chemical substance that causes the body to produce specific antibodies or sensitized T cells.
2. As a rule, antigens are foreign substances; they are not part of the body's chemistry.
3. Most antigens are proteins, nucleoproteins, lipoproteins, glycoproteins, or large polysaccharides with a molecular weight greater than 10,000.
4. Antibodies are formed against specific regions on the surface of an antigen called antigenic determinant groups.
5. Most antigens have many different determinants.
6. A hapten is a low-molecular-weight substance that combines with an antibody but cannot cause the formation of antibodies unless combined with a carrier molecule.

The Nature of Antibodies

1. An antibody or immunoglobulin is a protein produced by B cells in response to the presence of an antigen and is capable of combining specifically with the antigen.
2. An antibody has at least two identical antigen-binding (valence) sites.

Antibody Structure

1. A single bivalent antibody unit is a monomer.
2. Most antibody monomers consist of four polypeptide chains. Two are heavy chains and two are light chains.
3. Within each chain is a variable (V) region, where antigen binding occurs, and a constant (C) region, which serves as a basis for distinguishing the classes of antibodies.
4. An antibody monomer is Y- or T-shaped; the variable regions form the tips and the constant regions form the base and Fc fragment.
5. The Fc region can attach to a host cell or complement.

Immunoglobulin Classes

1. IgG antibodies are the most prevalent in serum; they provide naturally acquired passive immunity, neutralize bacterial toxins, participate in complement fixation, and enhance phagocytosis.
2. IgM antibodies consist of five monomers held by a joining chain; they are involved in agglutination and complement fixation.
3. Serum IgA antibodies are monomers; secretory IgA antibodies are dimers that protect mucosal surfaces from invasion by pathogens.
4. IgD antibodies are antigen receptors on B cells.
5. IgE antibodies bind to mast cells and basophils and are involved in allergic reactions.

B Cells and Humoral Immunity

1. Humoral immunity involves antibodies that are produced by B cells.
2. Bone marrow stem cells give rise to B cells.
3. Mature B cells migrate to lymphoid organs.
4. A mature B cell recognizes an antigen with antigen receptors.

Apoptosis

1. Lymphocytes that are not needed undergo apoptosis, or programmed cell death, and are destroyed by phagocytes.

Activation of Antibody-Producing Cells by Clonal Selection

1. According to the clonal selection theory, a B cell becomes activated when an antigen reacts with antigen receptors on its surface.
2. Recombination events in the gene coding for the variable region result in the ability to produce more than 100 million different antibody molecules.
3. The activated B cell produces a clone of plasma cells and memory cells.
4. Plasma cells secrete antibody. Memory cells recognize pathogens from previous encounters.
5. T cells and B cells that react with self antigens are destroyed during fetal development; this is called clonal deletion.

Antigen–Antibody Binding and Its Results

1. An antigen binds to the antigen-binding site (variable regions) of an antibody to form an antigen–antibody complex.
2. IgG antibodies inactivate viruses and neutralize bacterial toxins.
3. Agglutination of cellular antigens occurs when an IgG or IgM antibody combines with two cells.
4. Antigen–antibody complexes involving IgG and IgM antibodies can fix complement, resulting in the lysis of a bacterial (antigenic) cell.

Immunological Memory

1. The amount of antibody in serum is called the antibody titer.
2. The response of the body to the first contact with an antigen is called the primary response. It is characterized by the appearance of IgM followed by IgG.
3. Subsequent contact with the same antigen results in a very high antibody titer and is called the secondary, anamnestic, or memory response. The antibodies are primarily IgG.

Monoclonal Antibodies and Their Uses

1. Hybridomas are produced in the laboratory by fusing a cancerous cell with an antibody-secreting plasma cell.
2. A hybridoma cell culture produces large quantities of the plasma cell's antibody, called monoclonal antibodies.
3. Monoclonal antibodies are used in serologic identification tests, to prevent tissue rejections, and to treat septic shock.
4. Immunotoxins can be made by combining a monoclonal antibody and a toxin; the toxin will then kill a specific antigen.

T Cells and Cell-Mediated Immunity

1. Cell-mediated immunity involves specialized lymphocytes called T cells, which respond to intracellular antigens.

Chemical Messengers of Immune Cells: Cytokines

1. Cells of the immune system communicate with each other by means of chemicals called cytokines.
2. Interleukins (IL) are cytokines that serve as communicators between leukocytes.
3. Interferons are cytokines that protect against viruses.
4. Chemokines cause leukocytes to move to the site of infection.
5. Cytokines may be useful in treating tumors.

Cellular Components of Immunity

1. T cells are responsible for cell-mediated immunity.
2. After differentiation in the thymus gland, T cells migrate to lymphoid tissue.
3. T cells differentiate into effector T cells when they are stimulated by an antigen.
4. Some effector T cells become memory cells.

Types of T Cells

1. T cells are classified according to their functions and cell-surface receptors, called CDs.
2. The antigen must be processed by an antigen-presenting cell (APC) and positioned on the surface of the APC.
3. The major histocompatibility complex (MHC) consists of cell-surface proteins that are unique to each individual and provide self molecules.
4. A T cell recognizes antigens in association with MHC on an APC causing the APC to release IL-1.
5. After binding to an APC, helper T (T_H) or CD4 cells secrete IL-2 to activate other T_H cells specific for that antigen.
6. Cytotoxic T (T_C) or CD8 cells release perforin to lyse cells carrying the target antigen and MHC.
7. Delayed hypersensitivity T cells are associated with certain types of allergic reactions and transplant rejection.
8. Suppressor T cells appear to regulate the immune response.

Nonspecific Cellular Components

1. Macrophages that are stimulated by ingesting an antigen or by cytokines become activated to have enhanced phagocytic ability.
2. Natural killer (NK) cells lyse viral-infected and tumor cells. They are not T cells and are not antigenically specific.

The Interrelation of Cell-Mediated and Humoral Immunity

1. T_H cells activate B cells to produce antibodies against T-dependent antigens.
2. Antigens that directly activate B cells are called T-independent antigens.
3. In antibody-dependent cell-mediated cytotoxicity (ADCC), NK cells, macrophages, and other leukocytes lyse antibody-coated cells.
4. ADCC is useful against helminthic parasites.

THE LOOP

Complement is included in Chapter 16 (pp. 467–469), but can be assigned with this chapter.

Answers

Review

1.   The ability to produce antibodies against microorganisms and their toxins provides a type of resistance called immunity.

2.   a. Nonspecific defenses are designed to protect you against any kind of microorganism. Immunity or specific resistance involves the production of antibodies. Antibodies are directed against specific microorganisms.

   b. Humoral immunity is due to antibodies (and B cells). Cell-mediated immunity is due to T cells.

   c. Active immunity refers to antibodies produced by the individual who carries them. Passive immunity refers to antibodies produced by another source and then transferred to the individual who needs the antibodies.

   d. Acquired immunity is the resistance to infection obtained during the life of the individual. Acquired immunity results from the production of antibodies. Innate resistance refers to the resistance of species or individuals to certain diseases that is not dependent on antigen-specific immunity such as antibodies.

   e. Natural immunity is acquired naturally, i.e., from mother to newborn, or following an infection. Artificial immunity is acquired from medical treatment, i.e., by injection of antibodies or by vaccination.

   f. T-dependent antigens: Certain antigens must combine with self-antigens to be recognized by T_H cells and then by B cells. T-independent antigens can elicit an antibody response without T cells.

   g. T cells can be classified by their surface antigens: T_H cells possess the CD4 antigen; T_C and T_S cells have the CD8 antigen.

3.   a. Artificially acquired active immunity.

   b. Naturally acquired active immunity.

   c. Naturally acquired passive immunity.

   d. Artificially acquired passive immunity.

4.   An antigen is a chemical substance that causes the body to produce specific antibodies and can combine with these antibodies. A hapten is a low-molecular-weight substance that is not antigenic unless it is attached to a carrier molecule. Once an antibody has been formed against the hapten, the hapten alone will react with the antibodies independently of its carrier.

5.   An antibody is a protein produced by the body in response to the presence of an antigen; it is capable of combining specifically with that antigen. Antibodies are proteins and usually consist of four polypeptide chains. Two of the chains are identical and are called heavy (H) chains. The other two chains are identical to each other but are of lower molecular weight and are called light (L) chains. The variable portions of the H and L chains are where antigen binding occurs. The variable portion is different for each kind of antibody. The remaining constant portions of each chain are identical for all of the antibodies in one class of antibody. Refer to Figure 17.5 for the structure of IgG antibodies.

6.   Each person has a population of B cells with receptors for different antigens. When the appropriate antigen contacts the antigen receptor on a B cell, the cell proliferates to produce a clone of cells. Plasma cells in this clone produce antibodies specific to the antigen that caused their formation.

7.   See Figures 17.7, 17.12, 17.13, 17.15, 17.17, and 17.18.

8. Cytotoxic T cells (T$_C$) destroy target cells upon contact. Delayed hypersensitivity T cells (T$_D$) produce lymphokines. Helper T cells (T$_H$) interact with an antigen to "present" it to a B cell for antibody formation. Suppressor T cells (T$_S$) inhibit the conversion of B cells into plasma cells. Lymphokines cause an inflammatory response. An example of a cytokine is macrophage chemotactic factor, which attracts macrophages to the infection site. See Table 17.2 for functions of other cytokines.

9. a. Area *a* shows the primary response to the antigen. Area *b* shows the anamnestic response, in which the antibody titer is greater and remains high longer than in the primary response. The booster dose stimulated the memory cells to respond to the antigen.

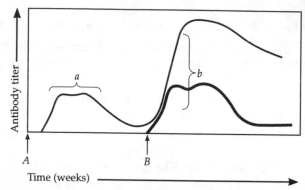

10. Neutralize toxins, inactivate viruses, fix complement to initiate cytolysis.

11. Surface recognition sites for antigen peptides and MHC proteins.

12. NK cells lyse target cells (usually tumor cells and virus-infected cells) on contact.

13. Both would prevent attachment of the pathogen; (a) interferes with the attachment site on the pathogen and (b) interferes with the pathogen's receptor site.

14. See Figure 17.10.

15. The person recovered because s/he produced antibodies against the pathogen. The memory response will continue to protect the person against that pathogen.

16. Human gamma globulin is the fraction of human serum in which antibodies are found. If antibodies against hepatitis are in the gamma globulin, this would be artificially acquired passive immunity.

Critical Thinking

1. T$_C$ cells secreted TNF and IFN, which diffuse through liver cells and stimulate these cells to produce antiviral proteins.

2. a. IL-2 stimulates proliferation and differentiation of T cells and natural killer (NK) cells. NK cells and some T cells are effective against cancer cells.

   b. IL-2 stimulates the immune response, which increases the unwanted immune response as well.

3. Having had an *M. tuberculosis* infection and recovered (naturally acquired active immunity); vaccination with BCG (artificially acquired passive immunity).

Clinical Applications

1. Antibiotics and immunity can cause gram-negative cells to lyse, releasing cell wall fragments. This exposes the body to more endotoxin. The woman's life-threatening condition was due to endotoxin shock. Monoclonal antibodies removed the cell walls.

2. Increased susceptibility to infection due to decreased antibody formation.

3. He could not produce the secretory component of IgA.

4. The mechanism is called antibody enhancement. Immune complexes of antibodies and viruses attach to cells, facilitating viral penetration.

Case History: Jules Bordet's Discovery

Background

In 1885, Jules Bordet tested the effects of immune serum on a bacterial culture. Goat serum was taken from an animal vaccinated against *Vibrio cholerae* cells. The goat serum had been heated to 58°C for 1 hour. Guinea pig serum was from a non-immune animal. Each tube contained 12 drops of the specified serum and about 9000 *V. cholerae* cells. Plate counts were made at intervals to determine the number of bacteria in each tube.

Data

	Number of *V. cholerae*		
Hour of counting	Heated goat serum	Guinea pig serum	8 drops guinea pig + 4 drops goat serum
6:00 PM	8640	9600	10,200
7:30 PM	4320	2160	0
10:00 PM	6480	3600	0
10:00 AM	very high	very high	0

Questions
1. How is serum obtained from blood?
2. What accounts for the results Bordet observed?

The Solution
1. Serum is the fluid left after blood clots.
2. Bordet discovered complement. (Goat serum) antibodies alone or (guinea pig) serum alone did not kill the bacteria overnight. (Goat serum) antibodies acted with (guinea pig) complement to kill the bacteria within 1.5 hours.

Chapter 18 *Practical Applications of Immunology*

Learning Objectives

1. Define vaccine.
2. Explain why vaccination works.
3. Differentiate between the following, and provide an example of each: attenuated, inactivated, toxoid, subunit, and conjugated vaccines.
4. Contrast subunit vaccines and nucleic acid vaccines.
5. Compare and contrast the production of whole-agent vaccines, recombinant vaccines, and DNA vaccines.
6. Define adjuvant.
7. Explain how antibodies are used to diagnose diseases.
8. Explain how precipitation and immunodiffusion tests work.
9. Differentiate between direct and indirect tests.
10. Differentiate between agglutination and precipitation tests.
11. Define hemagglutination.
12. Explain how a neutralization test works.
13. Differentiate between precipitation and neutralization tests.
14. Explain the basis for the complement-fixation test.
15. Compare and contrast direct and indirect fluorescent-antibody tests.
16. Explain how direct and indirect ELISA tests work.
17. Explain the importance of monoclonal antibodies.

NEW IN THIS EDITION

- A discussion on vaccine safety is included.
- Adjuvants are defined.

CHAPTER SUMMARY

Vaccines

1. Edward Jenner developed the modern practice of vaccination when he inoculated people with cowpox virus to protect them from smallpox.

Principles and Effects of Vaccination

1. Herd immunity results when most of a population is immune to a disease.

Types of Vaccines and Their Characteristics

1. Attenuated whole-agent vaccines consist of inactivated or attenuated microorganisms; attenuated virus vaccines generally provide lifelong immunity.
2. Inactivated whole-agent vaccines consist of killed bacteria or viruses.
3. Toxoids are inactivated toxins.
4. Subunit vaccines consist of antigenic fragments of a microorganism; these include recombinant vaccines and acellular vaccines.
5. Conjugated vaccines combine the desired antigen with a protein that boosts the immune response.
6. Nucleic acid vaccines or DNA vaccines are being developed. These cause the recipient to make the antigenic protein associated with MHC I.

The Development of New Vaccines

1. Viruses for vaccines may be grown in animals, cell cultures, or chick embryos.
2. Recombinant vaccines and nucleic acid vaccines do not need to be grown in cells or animals.
3. Adjuvants improve the effectiveness of some antigens.

Safety of Vaccines

1. Vaccines are the safest and most effective means of controlling infectious diseases.

Diagnostic Immunology

1. Many tests based on the interactions of antibodies and antigens have been developed to determine the presence of antibodies or antigens in a patient.

Precipitation Reactions

1. The interaction of soluble antigens with IgG or IgM antibodies (precipitins) leads to precipitation reactions.
2. Precipitation reactions depend on the formation of lattices and occur best when antigen and antibody are present in optimal proportions. Excesses of either component decrease lattice formation and subsequent precipitation.
3. The precipitin ring test is performed in a small tube.
4. Immunodiffusion procedures involve precipitation reactions carried out in an agar gel medium.
5. Immunoelectrophoresis combines electrophoresis with immunodiffusion for the analysis of serum proteins.

Agglutination Reactions

1. The interaction of particulate antigens with antibodies leads to agglutination reactions.
2. Diseases may be diagnosed by combining the patient's serum with a known antigen.
3. Diseases can be diagnosed by a rising titer or seroconversion (from no antibodies to the presence of antibodies).
4. Direct agglutination reactions can be used to determine antibody titer.
5. Antibodies cause visible agglutination of soluble antigens affixed to latex spheres in indirect or passive agglutination tests.

6. Hemagglutination reactions involve agglutination reactions using red blood cells. Hemagglutination reactions are used in blood typing, diagnosis of certain diseases, and identification of viruses.

Neutralization Reactions

1. In these reactions, the harmful effects of a bacterial exotoxin or virus are eliminated by a specific antibody.

2. An antitoxin is an antibody produced in response to a bacterial exotoxin or a toxoid that neutralized the exotoxin.

3. In a virus neutralization test, the presence of antibodies against a virus can be detected by the antibodies' ability to prevent cytopathic effects of viruses in cell cultures.

4. Antibodies against certain viruses can be detected by their ability to interfere with viral hemagglutination in viral hemagglutination inhibition tests.

Complement Fixation Reactions

1. Complement fixation reactions are serological tests based on the depletion of a fixed amount of complement in the presence of an antigen–antibody reaction.

Fluorescent-Antibody Techniques

1. Immunofluorescence techniques use antibodies labeled with fluorescent dyes.

2. Direct fluorescent-antibody tests are used to identify specific microorganisms.

3. Indirect fluorescent-antibody tests are used to demonstrate the presence of antibody in serum.

4. A fluorescence-activated cell sorter can be used to detect and count cells labeled with fluorescent antibodies.

Enzyme-Linked Immunosorbent Assay (ELISA)

1. ELISA techniques use antibodies linked to an enzyme such as horseradish peroxidase or alkaline phosphatase.

2. Antigen–antibody reactions are detected by enzyme activity. If the indicator enzyme is present in the test well, an antigen–antibody reaction has occurred.

3. The direct ELISA is used to detect antigens against a specific antibody bound in a test well.

4. The indirect ELISA is used to detect antibodies against an antigen bound in a test well.

Radioimmunoassay

1. In RIA, antibodies against the compound are combined with a radioactively labeled antigen and a sample containing an unknown amount of antigen.

2. Analysis of radioactivity in the resulting antigen–antibody complexes indicates the amount of antigen in the sample.

The Future of Diagnostic Immunology

1. The use of monoclonal antibodies will continue to make new diagnostic tests possible.

THE LOOP

This chapter can be used as a reference for Part Four or with Chapter 9 in a discussion of biotechnology.

Answers

Review

1. a. Whole-agent. Live, avirulent virus that can cause the disease if it mutates back to its virulent state.

 b. Whole-agent; (heat-) killed bacteria.

 c. Subunit; (heat- or formalin-) inactivated toxin.

 d. Subunit

 e. Subunit

 f. Conjugated

 g. Nucleic acid

2. If excess antibody is present, an antigen will combine with several antibody molecules. Excess antigen will result in an antibody combining with several antigens. Refer to Figure 18.2.

3. Particulate antigens react in agglutination reactions. The antigens can be cells or soluble antigens bound to synthetic particles. Soluble antigens take part in precipitation reactions.

4. a. Some viruses are able to agglutinate red blood cells. This is used to detect the presence of large numbers of virions capable of causing hemagglutination (e.g., *Influenzavirus*).

 b. Antibodies produced against viruses that are capable of agglutinating red blood cells will inhibit the agglutination. Hemagglutination inhibition can be used to detect the presence of antibodies against these viruses.

 c. This is a procedure to detect antibodies that react with soluble antigens by first attaching the antigens to insoluble latex spheres. This procedure may be used to detect the presence of antibodies that develop during certain mycotic or helminthic infections.

5. See Figure 18.10.

6. a. Direct test (see Figure 18.10a)

 b. Indirect test (see Figure 18.10b)

7. An indirect ELISA test is used to detect the presence of antibodies. A known antigen is fixed to a small well, and the patient's serum is added. Patient's antibodies will react with the antigen in the well. Antihuman immunoglobulins bound to an enzyme are added to the well. The antihuman immunoglobulins will bind to the patient's antibodies. Substrate for the enzyme is then added and a positive reaction indicating presence of the antibody in the patient's serum is shown by the enzyme–substrate reaction.

 A direct ELISA test is used to detect the presence of an antigen. Antibodies are fixed to a small well, and the unknown antigen is added. If the antigen reacts with the antibodies, the antigen will be bound to the well. Antibodies specific for the antigen are then added to the well. This second layer of antibodies is bound to an enzyme. Substrate for the enzyme is then added, and a positive reaction indicating the identity of the antigen is shown by the enzyme–substrate reaction (see Figure 18.12).

8. a. Direct test

    b. Indirect test

    The direct test provides definitive proof.

9. **Matching**

_e_	Precipitation
_d, f_	Immunoelectrophoresis
_a_	Agglutination
_g_	Radioimmunoassay
_c_	Complement fixation
_f_	Neutralization
_b, d_	ELISA

10.

_e_	Agglutination
_c_	Complement fixation
_a_	ELISA
_f_	FA
_b_	Neutralization
_d_	Precipitation

Critical Thinking

1. The live vaccine may revert to a more virulent form; exogenous protein contaminants in viral vaccines; the inherent instability of certain live viral preparations.

2. Viruses multiply only in cells of a particular species. Protection of that species through immunization could prevent further growth of the virus. Many pathogenic bacteria are capable of growth in different species or in nonliving reservoirs.

3. Traditionally, by vaccinating a large animal such as a horse or a goat, and purifying the antibodies from its blood. Now these antibodies can be obtained in vitro by monoclonal antibody techniques.

4. The antibodies (called reagin) are not specific. The disease is syphilis.

Clinical Applications

1. (a) is proof of a disease state. (b) could indicate disease, prior disease and recovery, or vaccination. The disease is tuberculosis.

2. No reaction; the antibodies will neutralize the toxin. This is a neutralization reaction. The disease is scarlet fever.

3. Patient A probably has the disease. Patient B does not have and never had the disease. Patient C recovered from the disease. Patient D acquired the disease between days 7 and 14; an example of seroconversion. The disease is legionellosis.

4. A high level of radioactive Ag in Ag–Ab complexes means the Ab combined with the known estrone because there was no estrone in the soil; the horses are not pregnant. A low level of radioactive Ag in Ag–Ab complexes means the Ab combined with estrone in the soil; the horses are pregnant.

Case History: *Pseudomonas*

Background

Pseudomonas aeruginosa is able to colonize the respiratory mucosa and cause recurring pneumonia in cystic fibrosis patients. The following experiments were performed to determine how *P. aeruginosa* can colonize mucous membranes.

0.1 ml broth culture was added to 0.4 ml serum and incubated for 1 hour; 0.1 ml of the same broth culture was added to 0.4 ml isotonic saline and incubated for 1 hour. Plate counts were performed on 1-ml samples to determine the number of bacteria.

Dilution	*Serratia* in serum Number of colonies	*Serratia* in saline Number of colonies
$1:10^2$	92	Too many to count
$1:10^4$	10	Too many to count
$1:10^6$	0	293

Dilution	*Pseudomonas* in serum Number of colonies	*Pseudomonas* in saline Number of colonies
$1:10^2$	Too many to count	Too many to count
$1:10^4$	600	150
$1:10^6$	65	12

Immunoelectrophoresis was then used to compare normal serum to the serum incubated with *Serratia* and *Pseudomonas*. Serum was placed in a well in agarose. Serum components were separated by electrophoresis. After electrophoresis, antihuman serum was added to the center trough. As antibodies diffused out of the trough, precipitation bands formed at zones of equivalence. The dark band at the cathode end of the gel is albumin; other bands indicate other serum proteins, such as the various immunoglobulins.

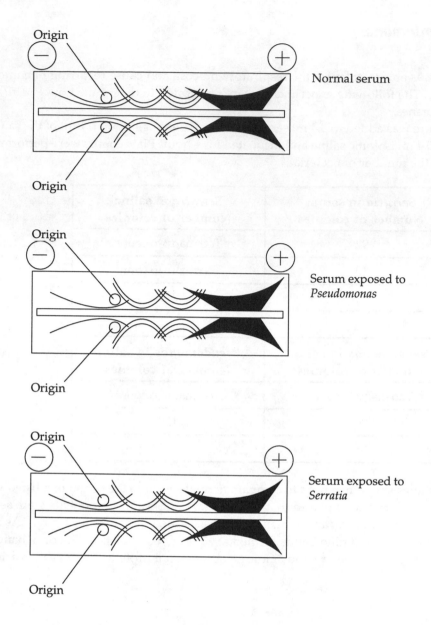

Questions

1. Calculate the number of bacteria in each serum and saline sample.

2. What was the purpose of the saline?

3. Provide an explanation for *Pseudomonas'* ability to colonize mucous membranes.

The Solution

1. $\text{Bacteria/ml} = \dfrac{\text{Number of Colonies}}{\text{Dilution} \times \text{Amount plated}}$

	Serum	Saline
Serratia sample	9.2×10^3	2.93×10^8
Pseudomonas sample	6.5×10^7	1.50×10^6

2. Saline was the control.

3. The IgA band is missing from the serum exposed to *Pseudomonas*. *P. aeruginosa* can degrade IgA antibodies found on mucous membranes.

Chapter 19 — *Disorders Associated with the Immune System*

Learning Objectives

1. Define hypersensitivity.
2. Describe the mechanism of anaphylaxis.
3. Compare and contrast systemic and localized anaphylaxis.
4. Explain how allergy skin tests work.
5. Define desensitization and blocking antibody.
6. Describe the mechanism of cytotoxic reactions and how drugs can induce them.
7. Describe the bases of the ABO and Rh blood group systems.
8. Explain the relationship between blood groups and blood transfusions and hemolytic disease of the newborn.
9. Describe the mechanism of immune complex reactions.
10. Describe the mechanism of cell-mediated reactions, and name two examples.
11. Describe a mechanism for self-tolerance.
12. Give an example of a type I, type II, type III, and type IV autoimmune disease.
13. Define HLA complex, and explain its importance in disease susceptibility and tissue transplants.
14. Explain how rejection of a transplant occurs.
15. Define privileged site.
16. Define autograft, isograft, allograft, and xenograft.
17. Explain how graft-versus-host disease occurs.
18. Explain how rejection of a transplant is prevented.
19. Compare and contrast congenital and acquired immune deficiencies.
20. Describe the immune responses to cancer and how cancer cells evade immune responses.
21. Define immunotherapy, and give two examples.
22. Give two examples of how emerging infectious diseases arise.
23. Explain the attachment of HIV to a host cell.
24. List two ways in which HIV avoids the host's antibodies.
25. Describe the stages of HIV infection.
26. Describe the effects of HIV infection on the immune system.
27. Describe how HIV infection is diagnosed.
28. List the routes of HIV transmission.
29. Identify geographic patterns of HIV transmission.
30. List the current methods of preventing and treating HIV infection.

NEW IN THIS EDITION

- The *MMWR* box on AIDS drugs is updated.
- The discussion of autoimmunity is revised.
- The general discussion of AIDS and incidence and prevalence data on AIDS are updated.

CHAPTER SUMMARY

Introduction

1. Hay fever, transplant rejection, and autoimmunity are examples of harmful immune reactions.
2. Infection, immunosuppression, and the carrier state are examples of failure of the immune system.
3. Superantigens activate many T-cell receptors, resulting in the release of excessive cytokines that can cause adverse host responses.

Hypersensitivity

1. Hypersensitivity reactions represent immunologic responses to an antigen (allergen) that lead to tissue damage rather than immunity.
2. Hypersensitivity reactions occur when a person has been sensitized to an antigen.
3. Hypersensitivity reactions can be divided into four classes: Types I, II, and III are immediate reactions based on humoral immunity, and type IV is a delayed reaction based on cell-mediated immunity.

Type I (Anaphylactic) Reactions

1. Anaphylactic reactions involve the production of IgE antibodies that bind to mast cells and basophils to sensitize the host.
2. The binding of two adjacent IgE antibodies to an antigen causes the target cell to release chemical mediators, such as histamine, leukotrienes, and prostaglandins, which cause the observed allergic reactions.
3. Systemic anaphylaxis may develop in minutes after injection or ingestion of the antigen; this may result in circulatory collapse and death.
4. Localized anaphylaxis is exemplified by hives, hay fever, and asthma.
5. Skin testing is useful in determining sensitivity to an antigen.
6. Desensitization to an antigen can be achieved by repeated injections of the antigen, which leads to the formation of blocking (IgG) antibodies and T_S cells.

Type II (Cytotoxic) Reactions

1. Type II reactions are mediated by IgG or IgM antibodies and complement.
2. The antibodies are directed toward foreign cells or host cells. Complement fixation may result in cell lysis. Macrophages and other cells may also damage the antibody-coated cells.

The ABO Blood Group System

1. Human blood may be grouped into four principal types, designated by A, B, AB, and O.
2. The presence or absence of two carbohydrate antigens designated A and B on the surface of the red blood cell determines a person's blood type.
3. Naturally occurring antibodies are present or absent in serum against the opposite AB antigen.
4. Incompatible blood transfusions lead to the complement-mediated lysis of red blood cells.

The Rh Blood Group System

1. Approximately 85% of the human population possesses another blood group antigen, designated the Rh antigen; these individuals are designated Rh+.
2. The absence of this antigen in certain individuals (Rh-) can lead to sensitization upon exposure to it.
3. An Rh+ person can receive Rh+ or Rh- blood transfusions.
4. When an Rh- person receives Rh+ blood, that person will produce anti-Rh antibodies.
5. Subsequent exposure to Rh+ cells will result in a rapid hemolytic reaction.
6. An Rh- mother carrying an Rh+ fetus will produce anti-Rh antibodies.
7. Subsequent pregnancies involving Rh incompatibility may result in hemolytic disease of the newborn.
8. The disease may be prevented by passive immunization of the mother with anti-Rh antibodies.

Drug-Induced Cytotoxic Reactions

1. In the disease thrombocytopenic purpura, platelets are destroyed by antibodies and complement.
2. Agranulocytosis and hemolytic anemia result from antibodies against one's own blood cells coated with drug molecules.

Type III (Immune Complex) Reactions

1. Immune complex diseases occur when IgG antibodies and soluble antigen form small complexes that lodge in the basement membranes of cells.
2. Subsequent complement fixation results in inflammation.
3. Glomerulonephritis are immune complex diseases.

Type IV (Cell-Mediated) Reactions

1. Delayed hypersensitivity responses are due primarily to T_D-cell proliferation.
2. Sensitized T cells secrete cytokines in response to the appropriate antigen.
3. Cytokines attract and activate macrophages and initiate tissue damage.
4. The tuberculin skin test and allergic contact dermatitis are examples of delayed hypersensitivities.

Autoimmune Diseases

1. Autoimmunity results from loss of self-tolerance.
2. Self-tolerance occurs during fetal development; T cells that will target host cells are eliminated (clonal deletion) or inactivated (clonal anergy).
3. Type I autoimmunity may be due to antibodies against infectious agents that attack self.
4. Graves' disease and myasthenia gravis are type II autoimmune reactions in which antibodies react to cell-surface antigens.
5. Systemic lupus erythematosus and rheumatoid arthritis are type III autoimmune reactions in which deposition of immune complexes results in tissue damage.
6. Hashimoto's disease and insulin-dependent diabetes are type IV autoimmune reactions mediated by T cells.

Reactions Related to the Human Leukocyte Antigen (HLA) Complex

1. Histocompatibility antigens located on cell surfaces express genetic differences among individuals; these antigens are coded for by MHC or HLA gene complexes.
2. To prevent rejection of transplants, HLA and ABO blood group antigens of the donor and recipient are matched as closely as possible.

Reactions to Transplantation

1. Transplants recognized as foreign antigens may be lysed by T cells and attacked by macrophages and complement-fixing antibodies.
2. Transplantation to a privileged site (such as the cornea) or of a privileged tissue (such as pig heart valves) does not cause an immune response.
3. Four types of transplants have been defined on the basis of genetic relationships between the donor and the recipient: autografts, isografts, allografts, and xenografts.
4. Xenografts are subject to hyperacute rejection.
5. Bone marrow (with immunocompetent cells) can cause graft-versus-host disease.
6. Successful transplant surgery often requires immunosuppressive treatment, which increases the host's susceptibility to infections and cancer.
7. Cyclosporine, an immunosuppressant drug, inhibits the secretion of IL-2 so cellular immunity is suppressed.

Immune Deficiencies

1. Immune deficiencies can be congenital or acquired.
2. Congenital immune deficiencies are due to defective or absent genes.
3. A variety of drugs, cancers, and infectious diseases can cause acquired immune deficiencies.

The Immune System and Cancer

1. Cancer cells are normal cells that have undergone transformation, divide uncontrollably, and possess tumor-associated antigens.
2. The response of the immune system to cancer is called immunological surveillance.
3. T_C cells recognize and lyse cancerous cells.

4.   Cancer cells can escape detection and destruction by the immune system.

5.   Cancer cells may suppress T cells or grow faster than the immune system can respond.

Immunotherapy

1.   Tumor necrosis factor (TNF) and other cytokines are being tested as cancer treatments.

2.   Immunotoxins are chemical poisons linked to a monoclonal antibody; the antibody selectively locates the cancer cell for release of the poison.

3.   A vaccine consisting of tumor antigens has been effective in controlling one type of cancer in poultry.

Acquired Immunodeficiency Syndrome (AIDS)

The Origin of AIDS

1.   HIV originated in central Africa and was brought to other countries by modern transportation and unsafe sexual practices.

HIV Infection

1.   AIDS is the final stage of HIV infection.

2.   HIV is a retrovirus with single-stranded RNA, reverse transcriptase, and a phospholipid envelope with gp120 spikes.

3.   HIV spikes attach to CD4 and coreceptors on host cells; the CD4 receptor is found on helper T cells, macrophages, and dendritic cells.

4.   Viral RNA is transcribed in DNA by reverse transcriptase. The viral DNA becomes integrated into the host chromosome to direct synthesis of new viruses or to remain latent as a provirus.

5.   HIV evades the immune system in latency, in vacuoles, by using cell-cell fusion, and by antigenic change.

6.   HIV infection is categorized by symptoms: Categories A (asymptomatic) and B (selected symptoms) are reported as AIDS if CD4 T cells fall below 200 cells/mm^3; Category C (AIDS indicator conditions) is reported as AIDS.

7.   HIV infection is also categorized by CD4 T-cell numbers: <200 cells/mm^3 is reported as AIDS.

8.   The progression from HIV infection to AIDS takes about 10 years.

9.   The life of an AIDS patient can be prolonged by proper treatment of opportunistic infections.

Diagnostic Methods

1.   HIV antibodies are detected by ELISA. HIV antigens are dectected by Western blotting.

2.   Human tissues and organs are tested for HIV before transplantation.

HIV Transmission

1.   HIV is transmitted transplacentally and by sexual contact, breast milk, contaminated needles, artificial insemination, and blood transfusion.

2.   In developed countries, blood transfusions are not a likely source of infection because blood is tested for HIV antibodies.

AIDS Worldwide

1. In the United States, Canada, western Europe, Australia, northern Africa, and parts of South America, transmission has been by IDU and male-to-male sexual contact. Heterosexual transmission is increasing.

2. In sub-Saharan Africa and the rest of South America, transmission is primarily heterosexual contact.

3. In eastern Europe and Asia, transmission is by IDU and heterosexual contact.

Prevention and Treatment of AIDS

1. The use of condoms and sterile needles prevents the transmission of HIV.

2. Vaccine development is hampered by lack of a nonhuman host for HIV.

3. The nucleotide analogs AZT, ddI, and ddC inhibit reverse transcriptase.

4. Protease inhibitors block this viral enzyme.

THE LOOP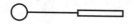

This chapter can be assigned with pages 21–22 and 395 for coverage of AIDS.

Answers

Review

1. The immune state that results in altered immunologic reactions leading to pathogenic changes in tissue.

2. **Mediator** **Function**

   Histamine Increases blood capillary permeability, mucus secretion, and smooth muscle contraction.

   Leukotrienes Increase blood capillary permeability and smooth muscle contraction.

   Prostaglandins Increase smooth muscle contraction and mucus secretion.

3. Recipient's antibodies will react with donor's tissues.

4. The recipient will experience symptoms due to lysis of the donor RBCs. Hemolysis occurs because the antigen (donor RBCs)–antibody reaction fixes complement.

5. This condition develops when an Rh⁻ mother becomes sensitized to the Rh+ antigens of her fetus. The mother's anti-Rh antibodies (IgG) can cross the placenta and react with fetal RBCs, causing their destruction. This condition can be prevented by passive immunization of the Rh⁻ mother with anti-Rh antibodies shortly after birth. These anti-Rh antibodies combine with fetal Rh+ RBCs, which may have entered maternal circulation, and enhance their clearance, thereby reducing the sensitization of the mother's immune system to this antigen.

6. Refer to Figure 19.7.

   a. The observed symptoms are due to lymphokines.

   b. When a person contacts poison oak initially, the antigen (catechols on the leaves) binds to tissue cells, is phagocytized by macrophages, and is presented to receptors on the surface of T cells. Contact between the antigen and the appropriate T cell stimulates the T cell to proliferate and become sensitized. Subsequent exposure to the antigen results in sensitized T cells releasing lymphokines, and a delayed hypersensitivity occurs.

   c. Small repeated doses of the antigen are believed to cause the production of IgG (blocking) antibodies.

7. Autografts and isografts are the most compatible. Xenografts are the least compatible.

8. a. Compatible. There are no Rh antigens on the donor's RBCs.

   b. Incompatible. The recipient will produce anti-Rh antibodies. If the recipient receives Rh+ RBCs in a subsequent transfusion, a hemolytic reaction will develop.

   c. Incompatible. The recipient has anti-A antibodies that will result in lysis of the donor's RBCs.

9. Autoimmunity is a humoral (types I, II, and III) or cell-mediated (type IV) immune response against a person's own tissue antigen. During development, T cells that recognize self may not be eliminated. During adulthood, inactive T cells may become active or antibodies could cross-react with host cell antigens. New or altered antigens may be formed on the surface of host cells. These antigens may result from the use of certain drugs, or from infections by certain viruses.

   Antibodies to cell-membrane antigens of certain group A streptococci cross-react with human heart tissue. Severe, recurrent infections caused by β-hemolytic group A streptococci sometimes lead to the development of rheumatoid arthritis long after the streptococcal infection has subsided.

10. Type I   Antibodies against microbes react with self

    Type II   Antibodies react with self.

    Type III Antibody–complement complexes deposit in tissues.

    Type IV T cells destroy self cells.

    See Table 19.3.

11. Natural

    Inherited

    Viral infections, most notably HIV

   Artificial

    Induced by immunosuppression drugs

   Result: Increased susceptibility to various infections depending on the type of immune deficiency.

12. Tumor cells have tumor-specific antigens such as TSTA and T antigen. Sensitized T_C cells may react with tumor-specific antigens, initiating lysis of the tumor cells.

13. Some malignant cells can escape the immune system by antigen modulation or immunological enhancement. Immunotherapy might trigger immunological enhancement. The body's defense against cancer is cell-mediated and not humoral. Transfer of lymphocytes could cause graft-versus-host disease.

14. AIDS is the last stage of an HIV infection. HIV is transmitted by sexual contact, by intravenous drug use, across the placenta, and in mother's milk. HIV is prevented by using condoms for hetero- and homosexual intercourse and oral and anal copulation, and by not re-using needles.

Critical Thinking

1. During embryonic development, clones of lymphocytes called forbidden clones react with "self" antigens. These forbidden clones are suppressed by specific classes of T lymphocytes. Lymphocytes that can react with foreign ("nonself") antigens are left to function in the mature immune system.

2. When a human is passively immunized with horse serum, antibodies may be produced against the horse immunoglobulins (antibodies). Immune complexes form between horse immunoglobulins and the antibodies formed against them. Symptoms are due to complement fixation.

3. Yes, they make antibodies. They are more likely to have T-independent antibodies. Their anti-HIV antibodies are ineffective because the virus can remain in the host cell, can be transmitted by cell-to-cell fusion, and can undergo antigenic changes to its surface proteins.

4. Anti-AIDS drugs are nucleoside analogs or enzyme inhibitors.

Clinical Applications

1. The infections are long-lasting, allowing sufficient time for sensitizing and shocking exposures to the fungal antigens.

2. a. An immediate hypersensitivity

   b. The mediators of anaphylaxis; refer to Review question 2

   c. Skin tests

   d. Some workers will not produce IgE antibodies against the conidiospores.

3. Epinephrine is used to treat symptoms of type I hypersensitivity, systemic anaphylaxis. People with hypersensitivity to eggs may experience anaphylaxis from this vaccine.

4. The woman made IgG antibodies in response to the B antigen in the transfusion. IgG antibodies can cross the placenta. A normal type A+ person has anti-B antibodies of the IgM type that cannot cross the placenta.

Case History: HIV Transmission

Background

A dentist showed symptoms of HIV infection in late 1986, and was diagnosed with AIDS in 1987. At the time of the AIDS diagnosis, AZT therapy was begun, discontinued for a short period in late 1987, then restarted and continued until his practice closed in 1989. All of the dentist's employees, including the dental hygienists, tested negative for HIV antibodies.

The dentist wrote an open letter to his former patients, which prompted 591 persons to be tested for HIV antibodies. The following list summarizes the seropositive individuals.

Patient	Sex	Clinical Status	HIV Risk Factor	Dental Visits No.	Dental Visits Dates
A	F	AIDS	No	6	1987–89
B	F	$CD4 < 500/mm^3$	No	21	1987–89
C	M	$CD4 < 200/mm^3$	Unknown	14	1984–89
D	M	AIDS	Yes	19	1985–89
E	F	$CD4 < 500/mm^3$	No*	10	1988
F	M	AIDS	Yes[†]	5	1988
G	M	$CD4 < 400/mm^3$	No	2	1988

* Patient F was an infrequent sex partner of patient E. Their last contact was in the fall of 1988.
[†]Tested seronegative in October and December 1988, positive in December 1990.

Questions
1. Analyses of DNA and amino acid sequences from HIV isolated from the patients strongly suggest that five of the patients were infected by the dentist. Which two were not? How did you arrive at your conclusion?

2. How can transmission of HIV and hepatitis B by health-care workers be prevented?

The Solution
1. Patients D and F were probably not infected by the dentist.

2. See the box in Chapter 13 of *Microbiology: An Introduction.*

Chapter 20 *Antimicrobial Drugs*

Learning Objectives

1. Identify the contributions of Paul Ehrlich and Alexander Fleming to chemotherapy.
2. Name the microbes that produce most of the antibiotics.
3. Describe the problems of chemotherapy for viral, fungal, protozoan, and helminthic infections.
4. Define the following terms: spectrum of activity, broad-spectrum drugs, superinfection.
5. Identify five modes of action of antimicrobial drugs.
6. Explain why the drugs described in this section are specific for bacteria.
7. List the advantages of each of the following over penicillin: semisynthetic penicillins, cephalosporins, and vancomycin.
8. Explain why INH and ethambutol are antimycobacterial agents.
9. Describe how each of the following inhibits protein synthesis: aminoglycosides, tetracyclines, chloramphenicol, macrolides.
10. Compare the method of action of polymyxin B, bacitracin, and neomycin.
11. Describe how rifamycins and quinolones kill bacteria.
12. Describe how sulfa drugs inhibit microbial growth.
13. Explain the modes of action of currently used antifungal drugs.
14. Explain the modes of actions of currently used antiviral drugs.
15. Explain the modes of actions of currently used antiprotozoan and antihelminthic drugs.
16. Describe two tests for microbial susceptibility to chemotherapeutic agents.
17. Describe the mechanisms of drug resistance.
18. Compare and contrast synergism and antagonism.
19. Identify three areas of research on new chemotherapeutic agents.

NEW IN THIS EDITION

- The new group of antibiotics, streptogramins, is included.
- The discussion of global misuse of antibiotics has been expanded.

CHAPTER SUMMARY

Introduction

1. An antimicrobial drug is a chemical substance that destroys disease-causing microorganisms with minimal damage to host tissues.
2. Chemotherapeutic agents include chemicals that combat disease in the body.

The History of Chemotherapy

1. Paul Ehrlich developed the concept of chemotherapy to treat microbial diseases; he predicted development of chemotherapeutic agents that would kill pathogens without harming the host.
2. Sulfa drugs came into prominence in the late 1930s.
3. Alexander Fleming discovered the first antibiotic, penicillin, in 1929; its first clinical trials were done in 1940.

The Spectrum of Antimicrobial Activity

1. Antibacterial drugs affect many targets in a prokaryotic cell.
2. Fungal, protozoan, and helminthic infections are more difficult to treat because these organisms have eukaryotic cells.
3. Narrow-spectrum drugs affect only a select group of microbes—gram-positive cells, for example; broad-spectrum drugs affect a large number of microbes.
4. Small, hydrophilic drugs can affect gram-negative cells.
5. Antimicrobial agents should not cause excessive harm to normal microbiota.
6. Superinfections occur when a pathogen develops resistance to the drug being used or when normally resistant microbiota multiply excessively.

The Action of Antimicrobial Drugs

1. General action is either by directly killing microorganisms (bactericidal) or by inhibiting their growth (bacteriostatic).
2. Some agents, such as penicillin, inhibit cell wall synthesis in bacteria.
3. Other agents, such as chloramphenicol, erythromycin, tetracyclines, and streptomycin, inhibit protein synthesis by acting on 70S ribosomes.
4. Agents such as polymyxin B cause injury to plasma membranes.
5. Rifampin and the quinolones inhibit nucleic acid synthesis.
6. Agents such as sulfanilamide act as antimetabolites by competitively inhibiting enzyme activity.

A Survey of Commonly Used Antimicrobial Drugs

Antibacterial Antibiotics: Inhibitors of Cell Wall Synthesis

1. All penicillins contain a β-lactam ring. Penicillins inhibit peptidoglycan synthesis.
2. Natural penicillins produced by *Penicillium* are effective against gram-positive cocci and spirochetes.
3. Penicillinases (β-lactamases) are bacterial enzymes that destroy natural penicillins.
4. Semisynthetic penicillins are made in the laboratory by adding different side chains onto the β-lactam ring made by the fungus.
5. Semisynthetic penicillins are resistant to penicillinases and have a broader spectrum of activity than natural penicillins.
6. The monobactam aztreonam affects only gram-negative bacteria.
7. Cephalosporins inhibit cell wall synthesis and are used against penicillin-resistant strains.

8. Carbapenems are broad-spectrum antibiotics that inhibit cell wall synthesis.
9. Polypeptides such as bacitracin and polymyxin B are applied topically to treat superficial infections.
10. Bacitracin inhibits cell wall synthesis primarily in gram-positive bacteria.
11. Vancomycin inhibits cell wall synthesis and may be used to kill penicillinase-producing staphylococci.
12. Isoniazid (INH) inhibits mycolic acid synthesis in mycobacteria. INH is administered with rifampin or ethambutol to treat tuberculosis.
13. The antimetabolite ethambutol is used with other drugs to treat tuberculosis.

Inhibitors of Protein Synthesis

1. Aminoglycosides, tetracyclines, chloramphenicol, and macrolides inhibit protein synthesis at 70S ribosomes.

Injury to Plasma Membrane

1. Polymyxin B and bacitracin cause damage to plasma membranes.

Inhibitors of Nucleic Acid (DNA/RNA) Synthesis

1. Rifamycin inhibits mRNA synthesis; it is used to treat tuberculosis.
2. Quinolones and fluoroquinolones inhibit DNA gyrase for treatment of urinary tract infections.

Competitive Inhibitors of the Synthesis of Essential Metabolites

1. Sulfonamides competitively inhibit folic acid synthesis.
2. TMP-SMX competitively inhibits dihydrofolic acid synthesis for treatment of urinary tract and intestinal infections.

Antifungal Drugs

1. Polyenes, such as nystatin and amphotericin B, combine with plasma membrane sterols and are fungicidal.
2. Azoles interfere with sterol synthesis and are used to treat cutaneous and systemic mycoses.
3. Griseofulvin interferes with eukaryotic cell division and is used primarily to treat skin infections caused by fungi.

Antiviral Drugs

1. Nucleoside and nucleotide analogs such as acyclovir, AZT, ddI, and ddC inhibit DNA or RNA synthesis.
2. Protease inhibitors, such as indinavir and saquinavir, block activity of an HIV enzyme essential for assembly of a new viral coat.
3. Alpha-interferons inhibit the spread of viruses to new cells.

Antiprotozoan and Antihelminthic Drugs

1. Chloroquine, quinacrine, diiodohydroxyquin, pentamidine, and metronidazole are used to treat protozoan infections.
2. Chloroquine and quinacrine stop DNA synthesis by intercalation.
3. Antihelminthic drugs include niclosamide, mebendazole, praziquantel, and piperazine.
4. Mebendazole disrupts microtubules; pyantel pamoate paralyzes intestinal roundworms.

Tests to Guide Chemotherapy

1. These tests are used to determine which chemotherapeutic agent is most likely to combat a specific pathogen.
2. These tests are used when susceptibility cannot be predicted or when drug resistance arises.

The Disk-Diffusion Method

1. In this test, also known as the Kirby–Bauer test, a bacterial culture is inoculated on an agar medium, and filter paper disks impregnated with chemotherapeutic agents are overlaid on the culture.
2. After incubation, the absence of microbial growth around a disk is called a zone of inhibition.
3. The diameter of the zone of inhibition, when compared with a standardized reference table, is used to determine whether the organism is sensitive, intermediate, or resistant to the drug.
4. The minimum inhibitory concentration (MIC) is the lowest concentration of chemotherapeutic agent capable of preventing microbial growth; MIC can be estimated using the E test.

Broth Dilution Tests

1. In the broth dilution test, the microorganism is grown in liquid media containing different concentrations of a chemotherapeutic agent.
2. The lowest concentration of chemotherapeutic agent that kills bacteria is called the minimum bactericidal concentration (MBC).

The Effectiveness of Chemotherapeutic Agents

Drug Resistance

1. Resistance may be due to enzymatic destruction of a drug, prevention of penetration of the drug to its target site, or cellular or metabolic changes at target areas.
2. Hereditary drug resistance, called resistance (R) factor, is carried by plasmids and transposons.
3. Resistance can be minimized by the discriminate use of drugs in appropriate concentrations and dosages.

Effects of Combinations of Drugs

1. Some combinations of drugs are synergistic; they are more effective when taken together.
2. Some combinations of drugs are antagonistic; when taken together, both drugs become less effective than when taken alone.

The Future of Chemotherapeutic Agents

1. Many bacterial diseases, previously treatable with antibiotics, have become resistant to antibiotics.

2. Chemicals produced by plants and animals are providing new antimicrobial agents, including antimicrobial peptides.

3. New antimicrobials include DNA that is complementary to specific genes in a pathogen; the DNA will bind to the pathogen's DNA or mRNA and inhibit protein synthesis.

THE LOOP

This chapter can be covered with disinfectants, antiseptics, and physical methods of controlling microorganisms (Chapter 7).

Answers

Review

1.

Antimicrobial Agents	Synthetic or Antibiotic	Method of Action	Principal Use
Isoniazid	Synthetic	Vitamin B$_6$ analog	Tuberculosis
Sulfonamides	Synthetic	Inhibit folic acid synthesis	Gram-negative bacteria
Ethambutol	Synthetic	Competitive inhibitor	Tuberculosis
Trimethoprim	Synthetic	Inhibits folic acid synthesis	*Pneumocystis*
Fluoroquinolones	Synthetic	Inhibit DNA synthesis	Urinary tract infections
Penicillin, natural	Antibiotic	Inhibits cell wall synthesis	Gram-positive bacteria
Penicillin, semisynthetic	Antibiotic	Inhibits cell wall synthesis	Broad spectrum; penicillin-resistant bacteria
Cephalosporins	Antibiotic	Inhibit cell wall synthesis	Penicillin-resistant bacteria
Carbapenems	Antibiotic	Inhibit cell wall synthesis	Broad spectrum
Aminoglycosides	Antibiotic	Inhibit protein synthesis	Gram-negative bacteria
Tetracyclines	Antibiotic	Inhibit protein synthesis	Broad spectrum
Chloramphenicol	Antibiotic	Inhibits protein synthesis	*Salmonella*
Macrolides	Antibiotic	Inhibit protein synthesis	Gram-negative bacteria
Polypeptides	Antibiotic	Inhibit cell wall synthesis; injure plasma membrane	Gram-positive bacteria; gram-negative bacteria
Vancomycin	Antibiotic	Inhibits cell wall synthesis	Penicillin-resistant *Staphylococcus*
Rifamycins	Antibiotic	Inhibit mRNA synthesis	Tuberculosis
Polyenes	Antibiotic	Injure plasma membrane	Fungicide
Griseofulvin	Antibiotic	Inhibits mitosis	Antifungal
Amantadine	Synthetic	Blocks viral entry or uncoating	Influenza A
Zidovudine	Synthetic	Inhibits DNA synthesis	AIDS
Niclosamide	Synthetic	Inhibits oxidative phosphorylation	Tapeworms

2. A chemotherapeutic agent is a substance taken into the body to combat disease. A synthetic chemotherapeutic agent is prepared in a laboratory, whereas antibiotics are produced naturally by bacteria and some fungi.

3. a. Ehrlich discovered the first chemotherapeutic agent (salvarsan, which was used to treat syphilis).

    b. Fleming discovered the antibiotic penicillin.

4. The drug should be toxic to the undesired microorganisms and not harmful to the host (selective toxicity).
    The drug should be active against many microorganisms (broad spectrum).
    The drug should not produce hypersensitivity in the host.
    The drug should not produce drug resistance in the host.
    The drug should not harm normal microbiota.

5. Because a virus uses the host cell's metabolic machinery, it is difficult to damage the virus without damaging the host. Fungi, protozoa, and helminths possess eukaryotic cells. Therefore, antiviral, antifungal, antiprotozoan, and antihelminthic drugs must also affect eukaryotic cells.

6. Pyrimidine (idoxuridine) and purine (acyclovir) analogs.
    Prevent release of nucleic acid from viruses into the host cell (amantadine).
    Inhibition of infection of cells (interferon).
    Enzyme inhibitors (indinavir).

7. In the broth dilution test, a series of cultures is prepared in a microtiter plate. To each well of liquid medium, the test organism and a different concentration of chemotherapeutic agent are added. The plate is incubated for 16–20 hours and observed for the presence of microbial growth.

    The minimal inhibitory concentration (MIC) is the lowest concentration of chemotherapeutic agent capable of preventing growth of the test organism. The lowest concentration of the agent that results in no growth in a subculture is the minimal bactericidal concentration (MBC).

    In the agar dilution method, bacterial colonies are replica-plated onto nutrient media plus varying concentrations of antimicrobial agents. The MIC is determined by measuring the colony growth. In the tube dilution test, both the MIC and the MBC can be determined. The agar dilution method has the advantage of ease of inoculation and media preparation.

8. In the disk-diffusion test, filter paper disks impregnated with chemotherapeutic agents are overlaid on an inoculated agar medium. During incubation, the agents diffuse from the disk and a zone of inhibition is observed in the area immediately around the disks. The zone of inhibition indicates susceptibility of the test organism to the agent tested.

9. Drug resistance is the lack of susceptibility of a microorganism to a chemotherapeutic agent. Drug resistance may develop when microorganisms are constantly exposed to an antimicrobial agent. The development of drug-resistant microorganisms can be minimized by judicious use of antimicrobial agents; following directions on the prescription; or by administering two or more drugs simultaneously.

10. a. Prevention of resistant strains of microorganisms;

    b. Take advantage of the synergistic effect;

    c. Provide therapy until a diagnosis is made; and

    d. Lessen the toxicity of individual drugs by reducing the dosage of each in combination.

11. a. Like polymyxin B, causes leaks in the plasma membrane.

    b. Interferes with translation.

12. a. Inhibits formation of peptide bond.

    b. Prevents translocation of ribosome along mRNA.

    c. Interferes with attachment of tRNA to mRNA-ribosome complex.

    d. Changes shape of 30S portion of ribosome, resulting in misreading mRNA.

13. DNA polymerase adds bases to the 3' –OH.

Critical Thinking

1. a. No. Human cells lack cell walls.

    b. No. It inhibits a viral enzyme.

    c. Yes. It affects mitochondrial ribosomes.

    d. Probably not. Sterols protect human membranes.

2. Cells infected by viruses may be rapidly metabolizing (anabolizing) in order to synthesize viruses. These cells are more likely to incorporate base analogs than normal, uninfected cells.

3. The carbon source can't enter the cells.

4. a. A and D were equally effective.

    b. A or D would be recommended depending upon their respective side effects.

    c. You can't tell. A subculture from the zone of inhibition would have to be done to determine whether A was bactericidal or bacteriostatic.

5. *S. griseus* makes streptomycin during idiophase. The inactivating enzyme is necessary to protect the bacterium from the streptomycin.

6. MIC, 50 μg; MBC, 100 μg.

Clinical Applications

1. Tetracycline induces lag phase, and penicillin requires log phase. In some infections, streptomycin enters more easily through a penicillin-damaged cell wall.

2. The bacteria were resistant to nalidixic acid but susceptible to sulfonamide.

3. Many bacteria were killed initially, so the patient started to feel better. Recall that bacteria die logarithmically, so it takes quite a while to kill an entire population. The concentration of penicillin dropped in his body when he stopped taking penicillin, so the surviving bacteria were able to grow.

Case History: Determining the Method of Action

Background

Assume you have discovered a new antibiotic. The effectiveness of this drug is shown below. To determine the method of action of this chemical, you culture a bacterium in two bioreactors containing nutrient broth, add the new drug to the test broth after 20 hours, and analyze both media. Protease activity is easily measured because protease is an extracellular enzyme.

Data

Antibiotic Disk Test

Bacterium	Zone of Inhibition (mm)
Bacillus subtilis	0
Escherichia coli	10
Mycobacterium phlei	0
Pseudomonas aeruginosa	8
Salmonella typhimurium	9
Staphylococcus aureus	5
Streptococcus pyogenes	5

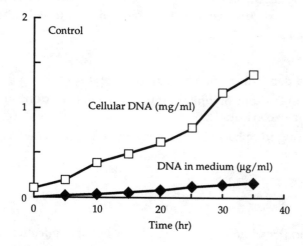

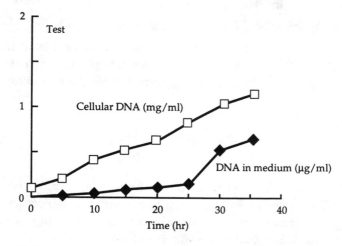

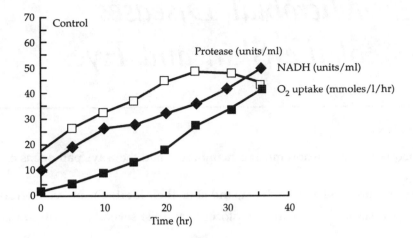

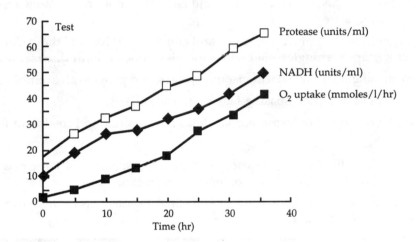

Questions
1. What bacterium did you use as your test organism?

2. Why were these particular parameters measured? What do they tell you about cell activity?

3. What is the method of action of this antibiotic? How can you tell?

The Solution
1. The test organism should be gram-negative as the drug is much less effective against gram-positives.

2. Cellular DNA is proportional to growth.
   Extracellular DNA indicates cells are lysing.
   The concentration of this protein is proportional to the rate of transcription.
   The concentration of NADH indicates the rate of catabolism.
   The rate of oxygen uptake is a measure of electron transport activity.

3. The decrease in protease suggests that protein synthesis is stopped.

Chapter 21 *Microbial Diseases of the Skin and Eyes*

Learning Objectives

1. Describe the structure of the skin and mucous membranes and the ways pathogens can invade the skin.
2. Provide examples of normal skin microbiota, and state their locations and ecological roles.
3. Differentiate between staphylococci and streptococci, and list several skin infections caused by each.
4. List the causative agent, method of transmission, and clinical symptoms of *Pseudomonas*, dermatitis, otitis externa, acne.
5. List the causative agent, method of transmission, and clinical symptoms of these skin infections: warts, smallpox, chickenpox, shingles, cold sores, measles, rubella, Fifth disease, roseola.
6. Differentiate between cutaneous and subcutaneous mycoses, and provide an example of each.
7. List the causative agent of and predisposing factors for candidiasis.
8. List the causative agent, method of transmission, clinical symptoms, and treatment for scabies.
9. Define conjunctivitis.
10. List the causative agent, method of transmission, and clinical symptoms of these eye infections: neonatal gonorrheal ophthalmia, inclusion conjunctivitis, trachoma.
11. List the causative agent, method of transmission, and clinical symptoms of these eye infections: herpetic keratitis, *Acanthamoeba* keratitis.

NEW IN THIS EDITION

- A new Clinical Problem-Solving box describes an emerging necrotizing mycobacterial disease.
- Roseola and Fifth disease are included.
- All morbidity and mortality data are updated to January, 2000.

CHAPTER SUMMARY

Introduction

1. The skin is a physical and chemical barrier against microorganisms.
2. Moist areas of the skin (such as the axilla) support larger populations of bacteria than dry areas (such as the scalp).

Structure and Function of the Skin

1. The outer portion of the skin, called the epidermis, contains keratin, a waterproof coating.
2. The inner portion of the skin, the dermis, contains hair follicles, sweat ducts, and oil glands that provide passageways for microorganisms.

3. Sebum and perspiration are secretions of the skin that can inhibit growth of organisms.
4. Sebum and perspiration provide nutrients for some microorganisms.
5. Body cavities are lined with epithelial cells. When these cells secrete mucus, they constitute the mucous membrane.

Normal Microbiota of the Skin

1. Microorganisms that live on skin are resistant to desiccation and high concentrations of salt.
2. Gram-positive cocci predominate on the skin.
3. The normal skin microbiota are not completely removed by washing.
4. Members of the genus *Propionibacterium* metabolize oil from the oil glands and colonize hair follicles.
5. *Pityrosporum ovale* yeast grows on oily secretions and may be the cause of dandruff.

Microbial Diseases of the Skin

1. Vesicles are small fluid-filled lesions; bullae are larger than 1 cm; macules are flat, reddened lesions; papules are raised lesions; and pustules are raised lesions containing pus.

Bacterial Diseases of the Skin

Staphylococcal Skin Infections

1. Staphylococci are gram-positive bacteria that often grow in clusters.
2. The majority of skin microbiota consist of coagulase-negative *S. epidermidis*.
3. Almost all pathogenic strains of *S. aureus* produce coagulase.
4. Pathogenic *S. aureus* can produce enterotoxins, leukocidins, and exfoliative toxin.
5. Many strains of *S. aureus* produce penicillinase; these are treated with vancomycin.
6. Localized infections (sties, pimples, and carbuncles) result from *S. aureus* entering openings in the skin.
7. Impetigo of the newborn is a highly contagious superficial skin infection caused by *S. aureus*.
8. Toxemia occurs when toxins enter the bloodstream; staphylococcal toxemias include scalded skin syndrome and toxic shock syndrome.

Streptococcal Skin Infections

1. Streptococci are gram-positive cocci that often grow in chains.
2. Streptococci are classified according to their hemolytic enzymes and cell wall (M) antigens.
3. Group A β-hemolytic streptococci (including *S. pyogenes*) are the pathogens most important to humans.
4. Group A β-hemolytic streptococci produce a number of virulence factors: erythrogenic toxin, deoxyribonuclease, streptokinases, and hyaluronidase.
5. Erysipelas (reddish patches) and impetigo (isolated pustules) are skin infections caused by *S. pyogenes*.
6. Invasive group A β-hemolytic streptococci cause severe and rapid tissue destruction.

Infections by Pseudomonads

1. Pseudomonads are gram-negative rods that are aerobes found primarily in soil and water. They are resistant to many disinfectants and antibiotics.

2. *Pseudomonas aeruginosa* is the most prominent species; it produces an endotoxin and several exotoxins.

3. Diseases caused by *P. aeruginosa* include otitis externa, respiratory infections, burn infections, and dermatitis.

4. Infections have a characteristic blue-green pus caused by the pigment pyocyanin.

5. Fluoroquinolones are useful in treating *P. aeruginosa* infections.

Acne

1. *Propionibacterium acnes* can metabolize sebum trapped in hair follicles.

2. Metabolic end-products (fatty acids) cause an inflammatory response known as acne.

3. Tretinoin, benzoyl peroxide, tetracycline, and isotretinoin are used to treat acne.

Viral Diseases of the Skin

Warts

1. Papillomaviruses cause skin cells to proliferate and produce a benign growth called a wart or papilloma.

2. Warts are spread by direct contact.

3. Warts may regress spontaneously or be removed chemically or physically.

Smallpox (Variola)

1. Variola virus causes two types of skin infections: variola major and variola minor.

2. Smallpox is transmitted by the respiratory route, and the virus is moved to the skin via the bloodstream.

3. The only host for smallpox is humans.

4. Smallpox has been eradicated as a result of vaccination efforts by the WHO.

Chickenpox (Varicella) and Shingles (Herpes Zoster)

1. Varicella-zoster virus is transmitted by the respiratory route and is localized in skin cells, causing a vesicular rash.

2. Complications of chickenpox include encephalitis and Reye's syndrome.

3. After chickenpox, the virus can remain latent in nerve cells and subsequently activate as shingles.

4. Shingles (herpes zoster) is characterized by a vesicular rash along the affected cutaneous sensory nerves.

5. The virus can be treated with acyclovir. An attenuated live vaccine is used.

Herpes Simplex

1. Herpes simplex infection of mucosal cells results in cold sores and occasionally encephalitis.

2. The virus remains latent in nerve cells, and cold sores can recur when the virus is activated.

3. HSV-I is transmitted primarily by oral and respiratory routes.

4. Herpes encephalitis occurs when herpes simplex viruses infect the brain.

5. Acyclovir has proven successful in treating herpes simplex.

Measles (Rubeola)

1. Measles is caused by measles virus and transmitted by the respiratory route.
2. Vaccination provides effective long-term immunity.
3. After the virus has incubated in the upper respiratory tract, macular lesions appear on the skin, and Koplik's spots appear on the oral mucosa.
4. Complications of measles include middle ear infections, pneumonia, encephalitis, and secondary bacterial infections.

Rubella

1. The rubella virus is transmitted by the respiratory route.
2. A red rash and light fever might occur in an infected individual; the disease can be asymptomatic.
3. Congenital rubella syndrome can affect a fetus when a woman contracts rubella during the first trimester of her pregnancy.
4. Damage from congenital rubella syndrome includes stillbirth, deafness, eye cataracts, heart defects, and mental retardation.
5. Vaccination with live rubella virus provides immunity of unknown duration.

Fifth Disease and Roseola

1. Human parvovirus B19 causes Fifth Disease and HHV-6 causes roseola.

Fungal Diseases of the Skin

Cutaneous Mycoses

1. Fungi that colonize the outer layer of the epidermis cause dermatomycoses.
2. *Microsporum, Trichophyton,* and *Epidermophyton* cause dermatomycoses called ringworm, or tinea.
3. These fungi grow on keratin-containing epidermis, such as hair, skin, and nails.
4. Ringworm and athlete's foot are usually treated with topical antifungal chemicals.
5. Diagnosis is based on the microscopic examination of skin scrapings or fungal culture.

Subcutaneous Mycoses

1. Sporotrichosis results from a soil fungus that penetrates the skin through a wound.
2. The fungi grow and produce subcutaneous nodules along the lymphatic vessels.

Candidiasis

1. *Candida albicans* causes infections of mucous membranes and is a common cause of thrush (in oral mucosa) and vaginitis.
2. *C. albicans* is an opportunistic pathogen that may proliferate when normal bacterial microbiota are suppressed.
3. Topical antifungal chemicals may be used to treat candidiasis.

Parasitic Infestation of the Skin

1. Scabies is caused by a mite burrowing and laying eggs in the skin.
2. Topical application of gamma benzene hexachloride is used to treat scabies.

Microbial Diseases of the Eye

1. Conjunctivitis is caused by several bacteria and can be transmitted by improperly disinfected contact lenses.

Bacterial Diseases of the Eye

1. Bacterial microbiota of the eye usually originate from the skin and upper respiratory tract.
2. Neonatal gonorrheal ophthalmia is caused by the transmission of *Neisseria gonorrhoeae* from an infected mother to an infant during its passage through the birth canal.
3. All newborn infants are treated with an antibiotic to prevent the growth of *Neisseria* and *Chlamydia*.
4. Inclusion conjunctivitis is an infection of the conjunctiva caused by *Chlamydia trachomatis*. It is transmitted to infants during birth and is transmitted in unchlorinated swimming water.
5. In trachoma, which is caused by *C. trachomatis,* scar tissue forms on the cornea.
6. Trachoma is transmitted by hands, fomites, and perhaps flies.

Other Infectious Diseases of the Eye

1. Inflammation of the cornea is called keratitis.
2. Herpetic keratitis causes corneal ulcers. The etiology is HSV-I that invades the central nervous system and can recur.
3. Trifluridine is an effective treatment for herpes keratitis.
4. *Acanthamoeba* protozoa, transmitted via water, can cause keratitis.

THE LOOP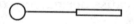

For a taxonomic approach, pages can be assigned as follows:

Answers

Review

1. Bacteria usually enter through inapparent openings in the skin. Fungal pathogens (except subcutaneous) often grow on the skin itself. Viral infections of the skin (except warts and herpes simplex) most often gain access to the body through the respiratory tract.

2. *Staphylococcus aureus; Streptococcus pyogenes.*

3.

Disease	Etiology	Symptoms	Treatment	Notes
Impetigo	*Staphylococcus aureus*	Vesicles that rupture and crust over	Hexachlorophene	May be epidemic
Erysipelas	*Streptococcus pyogenes*	Thickened red patches, swollen at margins	Penicillin	May be endogenous

4.

Disease	Etiological Agent	Clinical Symptoms	Method of Transmission
Acne	*P. acnes*	Infected oil glands	Direct contact
Pimples	*S. aureus*	Infected hair follicles	Direct contact
Warts	Papovavirus	Benign tumor	Direct contact
Chickenpox	Herpesvirus	Vesicular rash	Respiratory route
Fever blisters	Herpesvirus	Recurrent "blisters"	Direct contact
Measles	Paramyxovirus	Papular rash, Koplik's spots	Respiratory route
Rubella	Togavirus	Macular rash	Respiratory route

5. Both are fungal infections. Sporotrichosis is a subcutaneous mycosis; athlete's foot is a cutaneous mycosis.

6. a. Conjunctivitis is an inflammation of the conjunctiva, and keratitis is an inflammation of the cornea.

    b. See pp. 595–597.

7. Candidiasis is caused by *Candida albicans*. The yeast is able to grow when the normal microbiota are suppressed or when the immune system is suppressed. The yeast can be transferred from another person or be transient microbiota. White patches in the mouth or bright red areas of the skin and mucous membranes are signs of infection. Antifungal agents such as miconazole are used to treat candidiasis. Systemic infections are treated with oral ketoconazole.

8. The test determines the woman's susceptibility to rubella. If the test is negative, she is susceptible to the disease. If she acquires the disease during pregnancy the fetus could become infected. A susceptible woman should be vaccinated.

9.
Symptoms	Disease
Koplik's spots	Measles
Macular rash	Measles
Vesicular rash	Chickenpox
Small, spotted rash	German measles
"Blisters"	Cold sore
Corneal ulcer	Keratoconjunctivitis

10. The central nervous system can be invaded following keratoconjunctivitis; this results in encephalitis.

11. Attenuated measles, mumps, and rubella viruses.

12. Varicella-zoster virus appears to remain latent in nerve cells following recovery from a childhood infection of chickenpox. Later, the virus may be activated and cause a vesicular rash (shingles) in the area of the nerve.

13. To prevent neonatal gonorrheal ophthalmia. This is caused by *N. gonorrhoeae* contracted by the newborn during passage through the birth canal.

14. Trachoma.

15. Scabies is an infestation of mites in the skin. It is treated with permethrin insecticide or gamma benzene hexachloride.

Critical Thinking

1. *S. aureus* is adapted for surviving on the human skin, which has a high concentration of NaCl. Microorganisms that are not adapted to this hypertonic environment will not be able to tolerate the 7.5% NaCl in mannitol salt agar.

2. Most warts regress spontaneously. Removal of warts is usually for cosmetic reasons. Occasionally warts are painful when they are located where pressure is placed on them (e.g., plantar warts on the sole of the foot).

3. The infections were transmitted by the contact lenses or cosmetics. Cosmetics are inoculated with microbes each time they are used. Some of the microbes grow, resulting in large inoculations of the eyes. Contact lenses can be improperly cleaned (i.e., not using an antiseptic) or contaminated by fingers.

4. The virus had one host—humans. It was not found in soil, water, or nonhuman organisms. Polio and measles meet this criterion.

Clinical Applications

1. *Pseudomonas aeruginosa.* This bacterium is common in soil and is resistant to many antibiotics.

2. Toxic shock syndrome due to growth of *Staphylococcus* at the injection site.

3. The symptoms of toxic shock syndrome were caused by toxins produced from the secondary infection (*S. aureus*).

Learning with Technology

1. *Staphylococcus aureus*

2. *Micrococcus luteus*

Case History: Wrestling with Skin Infections

Background

A wrestling camp held July 2 through July 28 was attended by 175 male high school wrestlers from throughout the United States. On July 19, seven wrestlers were referred to a local urgent care facility because of complaints of painful vesicles on various parts of their bodies [head or neck (3), extremities (2), trunk (1)] and conjunctiva (1). Bacterial and fungal cultures from the skin lesions were negative.

A questionnaire was administered to wrestlers by telephone following the conclusion of camp. Sixty-one wrestlers met the case definition of the presence of cutaneous vesicles. The athletes had onset during the camp session or within one week after leaving camp (see the figure).

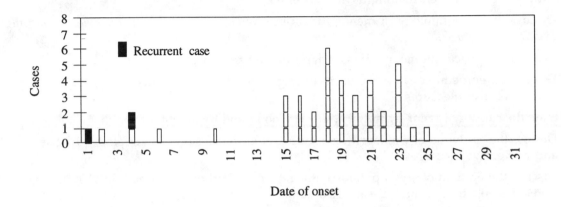

Date of onset

Athletes who reported wrestling with a participant with a rash were more likely to have the infection. Thirty-eight wrestlers interviewed reported a past history of oral cold sores. The attack rate was 24% for wrestlers who reported a past history of oral cold sores and 38% for wrestlers without a history of oral cold sores.

Questions

1. What diseases do you suspect?
2. How was this disease transmitted?
3. How is this disease treated?
4. Provide a possible explanation of the lower attack rate in wrestlers with a history of oral cold sores.
5. How can such outbreaks be prevented?

The Solution

1. Herpes gladiatorum
2. Direct contact
3. Acyclovir
4. Individuals with a history of oral herpes may have circulating antibodies that will prevent a new infection or recurrence.
5. Control methods shoud include education of athletes and trainers regarding herpes gladiatorum, routine skin examinations before wrestling contact, and exclusion of wrestlers with suspicious skin lesions. This outbreak might have been prevented if athletes with such lesions had been promptly excluded from contact competition.

Chapter 22 — *Microbial Diseases of the Nervous System*

Learning Objectives

1. Define the following terms: central nervous system; blood-brain barrier.
2. Differentiate between meningitis and encephalitis.
3. Discuss the epidemiology of meningitis caused by *H. influenzae, S. pneumoniae, N. meningitidis,* and *L. monocytogenes.*
4. Explain how bacterial meningitis is diagnosed and treated.
5. Discuss the epidemiology of tetanus, including mode of transmission, etiology, disease symptoms, and preventive measures.
6. State the causative agent, symptoms, suspect foods, and treatment for botulism.
7. Discuss the epidemiology of leprosy, including mode of transmission, etiology, disease symptoms, and preventive measures.
8. Discuss the epidemiology of poliomyelitis, rabies, and arboviral encephalitis, including mode of transmission, etiology, and disease symptoms.
9. Compare the Salk and Sabin vaccines.
10. Compare preexposure and postexposure treatments for rabies.
11. Explain how arboviral encephalitis can be prevented.
12. Identify the causative agent, vector, symptoms, and treatment for cryptococcosis.
13. Identify the causative agent, vector, symptoms, and treatment for African trypanosomiasis and *Naegleria* meningoencephalitis.
14. List the characteristics of diseases caused by prions.

NEW IN THIS EDITION

- All morbidity and mortality data are updated to January, 2000.
- A new Clinical Problem-Solving box describes an actual case of rabies.
- Recent data comparing the incidence of meningitis attributed to different organisms is included.
- A new chart illustrates the effect of the Hib vaccine on the incidence of *Haemophilus influenzae* meningitis.

CHAPTER SUMMARY

Structure and Function of the Nervous System

1. The central nervous system (CNS) consists of the brain, which is protected by the skull bones, and the spinal cord, which is protected by the backbone.
2. The peripheral nervous system (PNS) consists of the nerves that branch from the CNS.

3. The CNS is covered by three layers of membranes called meninges: the dura mater, arachnoid, and pia mater. Cerebrospinal fluid (CSF) circulates between the arachnoid and pia mater in the subarachnoid space and in the ventricles of the brain.

4. The blood-brain barrier normally prevents many substances, such as antibiotics, from entering the brain.

5. Microorganisms can enter the CNS through trauma, along peripheral nerves, and through the bloodstream and lymphatic system.

6. An infection of the meninges is called meningitis. An infection of the brain is called encephalitis.

Bacterial Diseases of the Nervous System

Bacterial Meningitis

1. Meningitis can be caused by viruses, bacteria, fungi, and protozoa.

2. The three major causes of bacterial meningitis are *Haemophilus influenzae*, *Streptococcus pneumoniae*, and *Neisseria meningitidis*.

3. Nearly 50 species of opportunistic bacteria can cause meningitis.

Haemophilus influenzae Meningitis

1. *H. influenzae* is part of the normal throat microbiota.

2. *H. influenzae* requires blood factors for growth; there are six types of *H. influenzae* based on capsule differences.

3. *H. influenzae* type b is the most common cause of meningitis in children under 4 years old.

4. A conjugated vaccine directed against the capsular polysaccharide antigen is available.

Neisseria Meningitis (Meningococcal Meningitis)

1. *N. meningitidis* causes meningococcal meningitis. This bacterium is found in the throats of healthy carriers.

2. The bacteria probably gain access to the meninges through the bloodstream. The bacteria may be found in leukocytes in the CSF.

3. Symptoms are due to endotoxin. The disease occurs most often in young children.

4. Military recruits are vaccinated with purified capsular polysaccharide to prevent epidemics in training camps.

Streptococcus pneumoniae Meningitis (Pneumococcal Meningitis)

1. *S. pneumoniae* is commonly found in the nasopharynx.

2. Hospitalized patients and young children are most susceptible to *S. pneumoniae* meningitis. It is rare but has a high mortality rate.

3. The vaccine for pneumococcal pneumonia may provide some protection against pneumococcal meningitis.

Diagnosis and Treatment of the Most Common Types of Bacterial Meningitis

1. Cephalosporins may be administered before identification of the pathogen.

2. Diagnosis is based on gram stain and serological tests of the bacteria in CSF.

3. Cultures are usually made on blood agar and incubated in an atmosphere containing reduced oxygen levels.

Listeriosis

1. *Listeria monocytogenes* causes meningitis in newborns, the immunosuppressed, pregnant women, and cancer patients.
2. Acquired by ingestion of contaminated food, it may be asymptomatic in healthy adults.
3. *L. monocytogenes* can cross the placenta and cause spontaneous abortion and stillbirth.

Tetanus

1. Tetanus is caused by a localized infection of a wound by *Clostridium tetani*.
2. *C. tetani* produces the neurotoxin tetanospasmin, which causes the symptoms of tetanus: spasms, contraction of muscles controlling the jaw, and death resulting from spasms of respiratory muscles.
3. *C. tetani* is an anaerobe that will grow in unclean wounds and wounds with little bleeding.
4. Acquired immunity results from DTP immunization that includes tetanus toxoid.
5. Following an injury, an immunized person may receive a booster of tetanus toxoid. An unimmunized person may receive (human) tetanus immune globulin.
6. Debridement (removal of tissue) and antibiotics may be used to control the infection.

Botulism

1. Botulism is caused by an exotoxin produced by *C. botulinum* growing in foods.
2. Serological types of botulinum toxin vary in virulence, with type A being the most virulent.
3. The toxin is a neurotoxin that inhibits the transmission of nerve impulses.
4. Blurred vision occurs in 1–2 days; progressive flaccid paralysis follows for 1–10 days, resulting in respiratory and cardiac failure.
5. *C. botulinum* will not grow in acidic foods or in an anaerobic environment.
6. Endospores are killed by proper canning. The addition of nitrites to foods inhibits outgrowth after endospore germination.
7. The toxin is heat labile and is destroyed by boiling (100°C) for 5 minutes.
8. Infant botulism results from the growth of *C. botulinum* in an infant's intestines.
9. Wound botulism occurs when *C. botulinum* grows in anaerobic wounds.
10. For diagnosis, mice protected with antitoxin are inoculated with toxin from the patient or foods.

Leprosy

1. *Mycobacterium leprae* causes leprosy, or Hansen's disease.
2. *M. leprae* has never been cultured on artificial media. It can be cultured in armadillos.
3. The tuberculoid form of the disease is characterized by loss of sensation in the skin surrounded by nodules. The lepromin test is positive.
4. Laboratory diagnosis is based on observation of acid-fast rods in lesions or fluids and the lepromin test.
5. In the lepromatous form, disseminated nodules and tissue necrosis occur. The lepromin test is negative.
6. Leprosy is not highly contagious and is spread by prolonged contact with exudates.
7. Untreated individuals often die of secondary bacterial complications, such as tuberculosis.

8. Patients with leprosy are made noncommunicable within 4–5 days with sulfone drugs and then treated as outpatients.

9. Leprosy occurs primarily in the tropics.

Viral Diseases of the Nervous System

Poliomyelitis

1. The symptoms of poliomyelitis are usually headache, sore throat, fever, stiffness of the back and neck, and occasionally paralysis (less than 1% of cases).

2. Poliovirus is found only in humans and is transmitted by the ingestion of water contaminated with feces.

3. Poliovirus first invades lymph nodes of the neck and small intestine. Viremia and spinal cord involvement may follow.

4. Diagnosis is based on isolation of the virus from feces and throat secretions.

5. The Salk vaccine (an inactiviated polio vaccine, IPV) involves injection of formalin-inactivated viruses and boosters every few years. The Sabin (oral polio vaccine, OPV) contains three attenuated live strains of poliovirus and is administered orally.

6. Polio will be eliminated through vaccination.

Rabies

1. Rabies virus (a rhabdovirus) causes an acute, usually fatal, encephalitis called rabies.

2. Rabies may be contracted through the bite of a rabid animal, by inhalation of aerosols, or invasion through minute skin abrasions. The virus multiplies in skeletal muscle and connective tissue.

3. Encephalitis occurs when the virus moves along peripheral nerves to the CNS.

4. Symptoms of rabies include spasms of mouth and throat muscles, followed by extensive brain and spinal cord damage and death.

5. Laboratory diagnosis may be made by direct immunofluorescent tests of saliva, serum, and CSF or brain smears.

6. Reservoirs for rabies in the United States include skunks, bats, foxes, and raccoons. Domestic cattle, dogs, and cats may get rabies. Rodents and rabbits seldom get rabies.

7. Current postexposure treatment includes administration of human rabies immune globulin (RIG) along with multiple intramuscular injections of vaccine.

9. Preexposure treatment consists of vaccination.

Arboviral Encephalitis

1. Symptoms of encephalitis are chills, headache, fever, and eventually coma.

2. Many types of arboviruses transmitted by mosquitoes cause encephalitis.

3. The incidence of arboviral encephalitis increases in the summer months when mosquitoes are most numerous.

4. Horses are frequently infected by EEE and WEE viruses.

5. Diagnosis is based on serological tests.

6. Control of the vector is the most effective way to control encephalitis.

Fungal Disease of the Nervous System

Cryptococcus neoformans Meningitis (Cryptococcosis)

1. *Cryptococcus neoformans* is an encapsulated yeastlike fungus that causes *Cryptococcus neoformans* meningitis.
2. The disease may be contracted by inhalation of dried infected pigeon droppings.
3. The disease begins as a lung infection and may spread to the brain and meninges.
4. Immunosuppressed individuals are most susceptible to *Cryptococcus neoformans* meningitis.
5. Diagnosis is based on latex agglutination for cryptococcal antigens in serum or CSF.

Protozoan Diseases of the Nervous System

African Trypanosomiasis

1. African trypanosomiasis is caused by the protozoan *Trypanosoma brucei gambiense* and *T. b. rhodesiense* and transmitted by the bite of the tsetse fly (*Glossina*).
2. The disease affects the nervous system of the human host, causing lethargy and eventually coma. It is commonly called sleeping sickness.
3. Vaccine development is hindered by the protozoan's ability to change its surface antigens.

Naegleria Meningoencephalitis

1. Encephalitis caused by the protozoan *N. fowleri* is almost always fatal.
2. The protozoan invades the brain from the nasal mucosa.

Nervous System Diseases Caused by Prions

1. Diseases of the CNS that progress slowly and cause spongiform degeneration are caused by prions.
2. Sheep scrapie and bovine spongiform encephalopathy (BSE) are examples of diseases caused by prions that are transferable from one animal to another.
3. Creutzfeldt–Jakob disease and kuru are human diseases similar to scrapie. Kuru occurs in isolated groups of cannibals who eat brains.
4. Prions are self-replicating proteins with no nucleic acids.

THE LOOP

For a taxonomic approach, pages can be assigned as follows:

Answers

Review

1. Meningitis is an infection of the meninges; encephalitis is an infection of the brain itself.

2.

Causative Agent	Susceptible Population	Mode of Transmission	Treatment
N. meningitidis	Children; military recruits	Respiratory	Penicillin
H. influenzae	Children	Respiratory	Rifampin
S. pneumoniae	Children; elderly	Respiratory	Penicillin
L. monocytogenes	Anyone	Foodborne	Penicillin
C. neoformans	Immunosuppressed individuals	Respiratory	Amphotericin B

3. *"Haemophilus"* refers to the requirement of this genus for growth factors found in blood (X and V factors) (Chapter 11).*"Influenzae"* because it was thought to be the causative agent of influenza.

4. The symptoms of tetanus are not due to bacterial growth (infection and inflammation) but to neurotoxin.

5.

	Salk	**Sabin**
Composition	Formalin-inactivated viruses	Live, attenuated viruses
Advantages	No reversion to virulence	Oral administration
Disadvantages	Booster dose needed; injected	Reversion to virulence

6. a. Vaccination with tetanus toxoid.

   b. Immunization with antitetanus toxin antibodies.

7. "Cleaned" because *C. tetani* is found in soil that might contaminate a wound. "Deep puncture" because it is likely to be anaerobic. "No bleeding" because a flow of blood ensures an aerobic environment and some cleansing.

8. *Clostridium botulinum.* Canned foods. Paralysis. Supportive respiratory care; antitoxin. Anaerobic, non-acidic environment. Diagnosis is made by detecting toxin in foods or patient by inoculating mice with suspect samples. Prevention: use of adequate heat in canning; boiling food before consumption to inactivate toxin.

9. Etiology—*Mycobacterium leprae.*
   Transmission—Direct contact.
   Symptoms—Nodules on the skin; loss of sensation.
   Treatment—Dapsone and rifampin.
   Prevention—BCG vaccine.
   Susceptible—People living in the tropics; genetic predisposition.

10. Etiology—Picornavirus (poliovirus).
    Transmission—Ingestion of contaminated water.
    Symptoms—Headache, sore throat, fever, nausea; rarely paralysis.
    Prevention—Sewage treatment.

    These vaccinations provide artificially acquired active immunity because they cause the production of antibodies, but they do not prevent or reverse damage to nerves.

11. Etiology—Rhabdovirus.
    Transmission—Bite of infected animal; inhalation.
    Reservoirs—Skunks, bats, foxes, raccoons.
    Symptoms—Muscle spasms, hydrophobia, CNS damage.

12. Postexposure treatment—Passive immunization with antibodies followed by active immunization with HDCV. Preexposure treatment—Active immunization with HDCV.

    Following exposure to rabies, antibodies are needed immediately to inactivate the virus. Passive immunization provides these antibodies. Active immunization will provide antibodies over a longer period of time, but they are not formed immediately.

13.

Disease	Etiology	Vector	Symptoms	Treatment
Arboviral encephalitis	Togaviruses, Arboviruses	Mosquitoes (*Culex*)	Headache, fever, coma	Immune serum
African trypanosomiasis	*Trypanosoma brucei gambiense*, *T. b. rhodesiense*	Tsetse fly	Decreased physical activity and mental acuity	Suramin; melarsoprol

14. Most antibiotics cannot cross the blood-brain barrier.

15. The causative agent of Creutzfeldt-Jakob disease (CJD) is transmissible. Although there is some evidence for an inherited form of the disease, it has been transmitted by transplants. Similarities with viruses are (1) the prion cannot be cultured by conventional bacteriological techniques and (2) the prion is not readily seen in patients with CJD.

Critical Thinking

1. The term rusty nail implies that the sharp object has been contaminated with soil and possibly *C. tetani*. *C. tetani* can grow in deep puncture wounds, and a nail is capable of producing such a wound.

2. Both diseases are caused by species of *Mycobacterium*.

3. The only cases of polio in the United States during the last ten years have been caused by OPV.

Clinical Applications

1. *Haemophilus influenzae* meningitis, treated with rifampin.

2. *Cryptococcus neoformans*. Need microscopic observation of the fungus from cerebrospinal fluid or culture.

3. *Naegleria fowleri* meningitis; treated with amphotericin B, miconazole, and rifampin.

Learning with Technology

Haemophilus influenzae

Case History: An Outbreak of Food Poisoning

Background

On October 15, a 40-year-old man was admitted to the hospital. He had a "splitting" headache, his legs were unsteady, and his vision was blurred. During examination, it was apparent that there was something wrong with his throat. It wasn't sore, but it felt stiff and tight, and it was almost impossible for him to speak.

Over the next seven days, 28 persons with similar symptoms were admitted to the hospital. Twelve of these patients required ventilatory support; no deaths were reported. During the investigation, it became apparent that the illnesses were due to meals consumed between October 14 and 16 at one restaurant. Detailed food histories were obtained from the patients. An additional case-control study was conducted on well people who had consumed food at the restaurant during the same three-day period. Repeated news media announcements aided health personnel in locating 18 other people who had eaten virtually the same foods at the restaurant.

The meals consisted of the following foods:

Patty Melt. Frozen hamburger patties purchased from a restaurant distributor. Patties were removed from the freezer and fried as ordered. Presliced pasteurized American cheese purchased from a distributor was kept refrigerated, and a slice was melted on each cooked hamburger patty. Meat and cheese were served on rye bread purchased from a local bakery.

Sauteed Onions. Onions were purchased fresh from a farmer. Fresh whole onions were sliced and then sauteed with margarine, paprika, garlic salt, and a chicken-based powder. After the initial cooking, the onions were held uncovered in a pan on a warm stove (<60°C) along with a large volume of melted margarine; they were not reheated before serving.

French-Fried Potatoes. Precut frozen potatoes were deep-fried in two- or three-serving batches as needed.

Potato Salad. Potatoes were purchased from a farmer. They had been stored in a root cellar and were transported loose in a pickup truck. The potatoes were washed, peeled, diced, and boiled. Cooled, drained potato cubes were mixed with oil, vinegar, dry mustard, and garlic salt, and kept refrigerated. Individual servings were removed from the container as needed.

Lettuce and Tomato Salad. Produce was delivered every other day. Lettuce and tomatoes were cut in the morning, refrigerated, and mixed with oil and vinegar for serving.

Questions
1. On one page, identify the etiologic agent of this outbreak of food poisoning.
2. Was it food infection or intoxication?
3. What item was contaminated, and how did it become contaminated?
4. Briefly explain how you arrived at your conclusion. How did you eliminate the other major causes of food poisoning?

Hints
1. Make a summary table of the persons not ill.
2. Make a table of the onset of symptoms following eating.

Data

Case	Sex	Age	R	H	V	Symptoms								Foods Eaten				
						W	D	N	C	Dd	Dy	M	Rd	1	2	3	4	5
1	M	25	26	20	N	x			x	x		x	x	x	x			x
2	F	20	14	15	Y	x			x	x	x	x	x	x	x	x		
3	M	40	15	17	Y	x				x	x	x	x	x	x		x	
4	M	55	14	17	Y	x			x		x	x	x	x	x	x		
5	F	72	15	16	N	x			x	x	x	x	x	x	x		x	
6	M	43	15	17	N	x				x	x			x	x	x	x	
7	F	37	16	17	Y	x			x	x	x	x	x	x	x			x
8	F	51	15	16	Y	x			x		x	x	x	x	x			x
9	F	25	14	16	N	x			x	x	x	x		x	x		x	x
10	M	40	16	—	—	x			x	x				x				x
11	F	35	14	—	—		x							x		x		

Case	Sex	Age	R	H	V	W	D	N	C	Dd	Dy	M	Rd	1	2	3	4	5
12	F	39	15	—	—		x							x	x		x	
13	M	54	15	18	N	x			x					x	x	x		x
14	F	34	15	18	Y	x		x	x		x	x	x	x	x	x		
15	M	45	14	16	Y	x			x		x	x	x	x			x	
16	M	27	16	—	—									x	x			x
17	F	37	14	15	Y	x		x	x	x	x	x	x	x			x	x
18	M	34	14	—	—									x			x	
19	M	30	15	—	—				x					x		x		
20	F	22	16	20	Y	x		x	x	x	x	x		x	x			
21	F	39	14	17	N	x		x	x	x	x			x	x			x
22	M	45	16	—	—	x								x	x			x
23	M	53	14	—	—				x					x	x		x	
24	F	41	15	20	N	x	x		x									x
25	F	42	14	16	N	x			x	x	x			x	x		x	
26	F	54	16	18	Y	x		x	x	x	x	x		x	x	x		
27	M	42	16	—	—	x			x					x	x		x	x
28	F	43	16	19	N	x		x	x	x	x						x	
29	M	42	15	—	—	x			x	x			x	x			x	
30	F	65	15	—	—	x			x					x	x	x		
31	F	33	16	21	N	x				x		x		x	x	x	x	
32	M	52	14	—	—			x	x					x			x	
33	F	26	14	—	—	x	x			x	x			x	x	x		
34	F	40	15	18	N	x		x	x		x						x	
35	F	22	15	19	Y	x		x		x	x	x	x	x	x	x		
36	M	63	16	20	N	x		x	x					x	x		x	x
37	M	35	15	—	—									x	x		x	x
38	M	56	15	17	Y	x		x	x		x	x		x	x			x
39	F	60	15	17	N	x		x	x					x	x	x		
40	M	37	15	—	—			x						x			x	
41	F	28	15	19	N	x			x					x	x			x
42	M	19	15	—	—									x	x	x		x
43	F	28	15	—	—			x						x			x	x
44	F	55	15	17	N	x				x							x	x
45	M	28	15	—	—									x			x	
46	F	45	15	18	N	x				x				x	x	x		

Legend:
R = Date at restaurant, H = Date hospitalized, V = Ventilated?
Symptoms: W = Weakness, lassitude, D = Diarrhea, N = Nausea, vomiting, C = Constipation, Dd = Diplopia, Dy = Dysphagia, dysphonia, M = Muscle weakness, Rd = Respiratory difficulty.
Food: 1–Patty Melt, 2–Sauteed Onions, 3–French-Fried Potatoes, 4–Potato Salad, 5–Lettuce and Tomato Salad.

The Solution

1. *Clostridium botulinum.*

2. Intoxication.

3. The onions were contaminated by *C. botulinum* from soil.

4. Of the 28 patients, 24 recalled eating the patty melt. All 24 patients, but only 10 of 18 controls, reported eating the sauteed onions. The original batch of sauteed onions was not available for culture or toxin testing, but type A botulinal toxin was detected in an extract made from washings of a discarded foil wrapper used by one of the patients to take a patty melt home. Type A botulinal spores were cultured from 5 of 75 skins of whole onions taken from the restaurant. No other ingredients of the sauteed onions contained toxin or spores.

Chapter 23 — Microbial Diseases of the Cardiovascular and Lymphatic Systems

Learning Objectives

1. Identify the role of the cardiovascular and lymphatic systems in spreading and eliminating infections.

2. List the signs and symptoms of septicemia, and explain the importance of infections that develop into septicemia.

3. Differentiate between septic shock and puerperal sepsis.

4. Describe the epidemiologies of bacterial endocarditis and rheumatic fever.

5. Discuss the epidemiology of tularemia.

6. Discuss the epidemiology of brucellosis.

7. Discuss the epidemiology of anthrax.

8. Discuss the epidemiology of gas gangrene.

9. List three pathogens acquired by animal bites.

10. Compare and contrast the causative agents, vectors, reservoirs, symptoms, treatments, and preventive measures for plague, relapsing fever, and Lyme disease.

11. Describe the epidemiologies of epidemic typhus, endemic murine typhus, and spotted fevers.

12. Describe the epidemiologies of Burkitt's lymphoma and infectious mononucleosis.

13. Compare and contrast the causative agents, vectors, reservoirs, and symptoms for yellow fever, dengue, and dengue hemorrhagic fever.

14. Compare and contrast the causative agents, modes of transmission, reservoirs, and symptoms for Ebola hemorrhagic fever and *Hantavirus* pulmonary syndrome.

15. Compare and contrast the causative agents, modes of transmission, reservoirs, symptoms, and treatments for American trypanosomiasis, toxoplasmosis, malaria, leishmaniasis, and babesiosis.

16. Discuss the worldwide effects of these diseases on health.

17. Diagram the life cycle of *Schistosoma*, and show where the cycle can be interrupted to prevent human disease.

NEW IN THIS EDITION

- All morbidity and mortality data are updated to January, 2000.
- A new Clinical Problem-Solving box describes the emergence of malaria in the United States.
- Leishmaniasis is discussed.

CHAPTER SUMMARY

Introduction

1. The heart, blood, and blood vessels make up the cardiovascular system.
2. Lymph, lymph vessels, lymph nodes, and lymphoid organs constitute the lymphatic system.

Structure and Function of the Cardiovascular and Lymphatic Systems

1. The heart circulates substances to and from tissue cells.
2. Blood is a mixture of plasma and cells.
3. Plasma transports dissolved substances. Red blood cells carry oxygen. White blood cells are involved in the body's defense against infection.
4. Fluid that filters out of capillaries into spaces between tissue cells is called interstitial fluid.
5. Interstitial fluid enters lymph capillaries and is called lymph; vessels called lymphatics return lymph to the blood.
6. Lymph nodes contain fixed macrophages, B cells, and T cells.

Bacterial Diseases of the Cardiovascular and Lymphatic Systems

Septicemia, Sepsis, and Septic Shock

1. The growth of microorganisms in blood is called septicemia. Signs include lymphangitis (inflamed lymph vessels).
2. Septicemia can lead to septic shock, characterized by decreased blood pressure.
3. Septicemia usually results from a focus of infection in the body.
4. Gram-negative rods are usually implicated. Endotoxin causes the symptoms.

Puerperal Sepsis

1. Puerperal sepsis begins as an infection of the uterus following childbirth or abortion; it can progress to peritonitis or septicemia.
2. *Streptococcus pyogenes* is the most frequent cause.
3. Oliver Wendell Holmes and Ignaz Semmelweis demonstrated that puerperal sepsis was transmitted by the hands and instruments of midwives and physicians.
4. Puerperal sepsis is now uncommon because of modern hygienic techniques and antibiotics.

Bacterial Endocarditis

1. The inner layer of the heart is the endocardium.
2. Subacute bacterial endocarditis is usually caused by α-hemolytic streptococci, staphylococci, or enterococci.
3. The infection arises from a focus of infection, such as a tooth extraction.
4. Preexisting heart abnormalities are predisposing factors.
5. Signs include fever, anemia, and heart murmur.
6. Acute bacterial endocarditis is usually caused by *Staphylococcus aureus*.
7. The bacteria cause rapid destruction of heart valves.

Rheumatic Fever

1. Rheumatic fever is an autoimmune complication of group A β-hemolytic streptococcal infections.
2. Rheumatic fever is expressed as arthritis or inflammation of the heart. It can result in permanent heart damage.
3. Antibodies against group A β-hemolytic streptococci react with streptococcal antigens deposited in joints or heart valves or cross-react with the heart muscle.
4. Rheumatic fever can be a sequel to a streptococcal infection, such as streptococcal sore throat. Streptococci might not be present at the time of rheumatic fever.
5. Prompt treatment of streptococcal infections can reduce the incidence of rheumatic fever in the United States.
6. Rheumatic fever is treated with anti-inflammatory drugs.

Tularemia

1. Tularemia is caused by *Francisella tularensis*. The reservoir is small wild mammals, especially rabbits.
2. Signs include ulceration of the site of entry, followed by septicemia and pneumonia.
3. Humans contract tularemia by handling diseased carcasses, eating undercooked meat of diseased animals, and being bitten by certain vectors (such as deer flies).
4. *F. tularensis* is resistant to phagocytosis.
5. Laboratory diagnosis is based on an agglutination test on isolated bacteria.

Brucellosis (Undulant Fever)

1. Brucellosis can be caused by *Brucella abortus, B. melitensis,* and *B. suis.*
2. Elk and bison constitute the reservoir in the United States; worldwide, domesticated animals (cattle, pigs, goats, and camels) are the reservoir.
3. The bacteria enter through minute breaks in the mucosa or skin, reproduce in macrophages, and spread via lymphatics to the liver, spleen, or bone marrow.
4. Signs include malaise and fever that spikes each evening (undulant fever).
5. A vaccine for cattle is available.
6. Diagnosis is based on serological tests.

Anthrax

1. *Bacillus anthracis* causes anthrax. In soil, endospores can survive for up to 60 years.
2. Grazing animals acquire an infection after ingesting the endospores.
3. Humans contract anthrax by handling hides from infected animals. The bacteria enter through cuts in the skin or through the respiratory tract.
4. Entry through the skin results in a pustule that can progress to septicemia. Entry through the respiratory tract can result in pneumonia.
5. Diagnosis is based on isolation and identification of the bacteria.

Gangrene

1. Soft tissue death from ischemia is called gangrene.
2. Microorganisms grow on nutrients released from gangrenous cells.

3. Gangrene is especially susceptible to the growth of anaerobic bacteria such as *Clostridium perfringens,* the causative agent of gas gangrene.

4. *C. perfringens* can invade the wall of the uterus during improperly performed abortions.

5. Debridement, hyperbaric chambers, and amputation are used to treat gas gangrene.

Systemic Diseases Caused by Bites and Scratches

1. *Pasteurella multocida,* introduced by the bite of a dog or cat, can cause septicemia.

2. Anaerobic bacteria such as *Clostridium, Bacteroides,* and *Fusobacterium* infect deep animal bites.

3. Cat-scratch disease is caused by *Bartonella henselae.*

Vector-Transmitted Diseases

Plague

1. Plague is caused by *Yersinia pestis.* The vector is usually the rat flea (*Xenopsylla cheopis*).

2. Reservoirs for plague include European rats and North American rodents.

3. Signs of bubonic plague include bruises on the skin and enlarged lymph nodes (buboes).

4. The bacteria can enter the lungs and cause pneumonic plague.

5. Laboratory diagnosis is based on isolation and identification of the bacteria.

6. Antibiotics are effective in treating plague, but they must be administered promptly after exposure to the disease.

Relapsing Fever

1. Relapsing fever is caused by *Borrelia* species and transmitted by soft ticks.

2. The reservoir for the disease is rodents.

3. Signs include fever, jaundice, and rose-colored spots. Signs recur three or four times after apparent recovery.

4. Laboratory diagnosis is based on the presence of spirochetes in the patient's blood.

Lyme Disease (Lyme Borreliosis)

1. Lyme disease is caused by *Borrelia burgdorferi* and is transmitted by a tick (*Ixodes*).

2. Lyme disease is prevalent on the Atlantic coast of the United States.

3. Deer and field mice are the animal reservoirs.

4. Diagnosis is based on serological tests and clinical symptoms.

Other Tickborne Diseases

1. Ehrlichiosis is caused by *Ehrlichia* species.

Typhus

1. Typhus is caused by rickettsias, obligate intracellular parasites of eukaryotic cells.

Epidemic Typhus
1. The human body louse *Pediculus humanus corporis* transmits *Rickettsia prowazekii* in feces, which are deposited while the louse is feeding.

2. Epidemic typhus is prevalent in crowded and unsanitary living conditions that allow the proliferation of lice.

3. The signs of typhus are rash, prolonged high fever, and stupor.

4. Tetracyclines and chloramphenicol are used to treat typhus.

Endemic Murine Typhus

1. Endemic murine typhus is a less severe disease caused by *Rickettsia typhi* and transmitted from rodents to humans by the rat flea.

Spotted Fevers

1. *Rickettsia rickettsii* is a parasite of ticks (*Dermacentor* species) in the southeastern United States, Appalachia, and the Rocky Mountain states.

2. The rickettsia may be transmitted to humans, in whom it causes tickborne typhus fever.

3. Chloramphenicol and tetracyclines effectively treat the disease.

4. Serological tests are used for laboratory diagnosis.

Viral Diseases of the Cardiovascular and Lymphatic Systems

Burkitt's Lymphoma

1. Epstein-Barr (EB) virus causes Burkitt's lymphoma and nasopharyngeal cancer.

2. Burkitt's lymphoma tends to occur in patients whose immune system has been weakened, for example, by malaria or AIDS.

Infectious Mononucleosis

1. Infectious mononucleosis is caused by EB virus.

2. The virus multiplies in the parotid glands and is present in saliva. It causes the proliferation of atypical lymphocytes.

3. The disease is transmitted by ingestion of saliva from infected individuals.

4. Diagnosis is made by indirect fluorescent-antibody technique.

Classic Viral Hemorrhagic Fevers

1. Yellow fever is caused by a flavivirus (yellow fever virus). The vector is the mosquito *Aedes aegypti*.

2. Signs and symptoms include fever, chills, headache, nausea, and jaundice.

3. Diagnosis is based on the presence of virus-neutralizing antibodies in the host.

4. No treatment is available. An attenuated, live viral vaccine is available.

5. Dengue is caused by a flavivirus (dengue fever virus) and is transmitted by the mosquito *Aedes aegypti*.

6. Signs are fever, muscle and joint pain, and rash.

7. Mosquito abatement is necessary to control the disease.

8. Dengue hemorrhagic fever occurs when a person with antibodies is reinfected with the same dengue fever virus.

Emerging Viral Hemorrhagic Fevers

1. Human diseases caused by Marburg, Ebola, and Lassa fever viruses were first noticed in the late 1960s.
2. Marburg virus is found in nonhuman primates; Lassa fever viruses are found in rodents.
3. Rodents are the reservoirs for Argentine and Bolivian hemorrhagic fevers.
4. *Hantavirus* pulmonary syndrome is caused by *Hantavirus*. The virus is contracted by inhalation of dried rodent urine.

Protozoan Diseases of the Cardiovascular and Lymphatic Systems

American Trypanosomiasis (Chagas' Disease)

1. *Trypanosoma cruzi* causes Chagas' disease. The reservoir includes many wild animals. The vector is a reduviid, the "kissing bug."
2. Xenodiagnosis allows for the identification of trypanosomes in the intestinal tract of the reduviid bug, which confirms the diagnosis.

Toxoplasmosis

1. Toxoplasmosis is caused by the sporozoan *Toxoplasma gondii*.
2. *T. gondii* undergoes sexual reproduction in the intestinal tract of domestic cats, and oocysts are eliminated in cat feces.
3. In the host cell, sporozoites reproduce to form either tissue-invading tachyzoites or bradyzoites.
4. Humans contract the infection by ingesting tachyzoites or tissue cysts in undercooked meat from an infected animal, or oocysts in dried cat feces.
5. Congenital infections can occur. Signs and symptoms include severe brain damage or vision problems.
6. Toxoplasmosis can be identified by serological tests, but interpretation of the results is uncertain.

Malaria

1. The signs and symptoms of malaria are chills, fever, vomiting, and headache, which occur at intervals of 2–3 days.
2. Malaria is transmitted by *Anopheles* mosquitoes. The causative agent is any one of four species of *Plasmodium*.
3. Sporozoites reproduce in the liver and release merozoites into the bloodstream, where they infect red blood cells and produce more merozoites.
4. Laboratory diagnosis is based on microscopic observation of merozoites in red blood cells.
5. New drugs are being developed as the protozoa develop resistance to drugs such as chloroquine.

Leishmaniasis

1. *Leishmania* spp., transmitted by sandflies, cause leishmaniasis.
2. The protozoa reproduce ini the liver, spleen, and kidneys.
3. Antimony compounds are used for treatment.

Babesiosis

1. Babesiosis is caused by the protozoan *Babesia microti* and transmitted to humans by ticks.

Helminthic Diseases of the Cardiovascular and Lymphatic Systems

Schistosomiasis

1. Species of the blood fluke *Schistosoma* cause schistosomiasis.
2. Eggs eliminated with feces hatch into larvae that infect the intermediate host. Free-swimming cercariae are released and penetrate the skin of a human.
3. The adult flukes live in the veins of the liver or urinary bladder in humans.
4. Adult flukes reproduce, and eggs are excreted or remain in the host.
5. Granulomas are from the host's defense to eggs that remain in the body.
6. Observation of eggs or flukes in feces, skin tests, or indirect serological tests may be used for diagnosis.
7. Chemotherapy is used to treat the disease; sanitation and snail eradication are used to prevent the disease.

Swimmer's Itch

1. Swimmer's itch is a cutaneous allergic reaction to cercariae that penetrate the skin. The definitive host for this fluke is wildfowl.

THE LOOP

For a taxonomic approach, pages can be assigned as follows:

Answers

Review

1. Fever, decrease in blood pressure, and lymphangitis.

2. Bacteria can spread from an abscess with enzymes such as kinases and invade blood vessels.

3.

Disease	Causative Agent	Predisposing Conditions
p.s.	*S. pyogenes*	Abortion or childbirth
s.b.e.	α-hemolytic strep.	Preexisting lesions
a.b.e.	*S. aureus*	Abnormal heart valves

4. Rheumatic fever is an autoimmune disease that is precipitated by streptococcal sore throat. It is treated with anti-inflammatory drugs to relieve the symptoms. It is prevented by early diagnosis and treatment of streptococcal sore throat.

5. All are vectorborne rickettsial diseases. They differ from each other in (1) etiologic agent, (2) vector, (3) severity and mortality, and (4) incidence (e.g., epidemic, sporadic).

6.

Causative Agent	Vector	Symptoms	Treatment
Plasmodium	*Anopheles*	Recurrent fever, chills	Quinine derivative
Flavivirus	*Aedes aegypti*	Fever, nausea, jaundice	None
Flavivirus	*Aedes aegypti*	Muscle and joint pain	None
Borrelia	Soft ticks	Recurrent fever	Tetracycline
Leishmania	Sandflies	Fever, chills	Antimony

7.

Francisella tularensis	Animal reservoir, skin abrasions, ingestion, inhalation, bites	Rabbits	Small ulcer	Careful handling of animals
Brucella spp.	Animal reservoir, ingestion of milk, direct contact	Cattle	Undulant fever	Pasteurization of milk
Bacillus anthracis	Skin abrasions, inhalation, ingestion	Soil, cattle	Malignant pustule	Surveillance and vaccination of cattle
Borrelia burgdorferi	Tick bites	Deer, mice	Rash, neurologic; arthritis	Protection from ticks

8. Plague
   Causative agent—*Yersinia pestis.*
   Vector—Rat flea.
   Reservoir—Rodents.
   Prognosis—Poor if untreated; good with antibiotic treatment.
   Treatment—Tetracycline, streptomycin.
   Control—Sanitation and ratproofing buildings.

9.

Causative Agent	Transmission	Reservoir	Endemic Area
Schistosoma spp.	Penetrate skin	Aquatic snail	Asia, South America
Toxoplasma gondii	Ingestion, inhalation	Cats	United States
Trypanosoma cruzi	"Kissing bug"	Rodents	Central America

10.

	Reservoir	Etiology	Transmission	Symptoms
Cat-scratch fever	Cats	*Bartonella henselae*	Scratch; touching eyes, fleas	Swollen lymph nodes, fever, malaise
Toxoplasmosis	Cats	*Toxoplasma gondii*	Ingestion	None, congenital infections, neurologic damage

11. Gangrenous tissue is anaerobic and has suitable nutrients for *C. perfringens*.

12. Infectious mononucleosis is caused by EB virus and transmitted in oral secretions.

13. Bubonic plague—Proliferation of bacteria in lymph vessels and lymph nodes. Enlarged lymph nodes (buboes). Transmission—Flea bites.
Pneumonic plague—Growth of bacteria in the lungs. Transmission—Respiratory route.

Critical Thinking

1. Patient B has a rising antibody titer, which indicates an active infection. A therapeutic abortion could be recommended to this woman. Patient C has no immunity to *Toxoplasma* and should be advised to avoid contact with reservoirs. Patient A has antibodies against *Toxoplasma* that should provide long-lasting immunity.

2. Mosquito eradication.

3. The presence of antibodies is causing the more severe disease; this is called *antibody enhancement*. Antibodies and live viruses form immune complexes that attach to macrophages and other cells, thus increasing the number of viruses that infect a cell.

Clinical Applications

1. Tularemia. A slide agglutination test can be performed to identify the organism isolated from the lesions.

2. Incubation: March 27–29.
Prodromal: March 30.
Contacts were treated prophylactically, to prevent infection.
Causative agent: *Yersinia pestis.*
The bacterium can be identified by flourescent-antibody tests and phage typing.

3. Symptoms: due to endotoxin; fever, weakness, shock.
Gram-negative, oxidase-negative, lactose-positive, fish-eye colonies on EMB, IMViC (- - + +), glycerol-.
The inside surface of the manometer was not sterile.
Keep sterile systems closed, and sterilize internal surfaces of equipment.

4. Disease: Relapsing fever (transmitted by ticks in the cabin).
Incubation period: A little less than 2 weeks.
Lysis of the *Borrelia* by the immune response causes fever.
Cell lysis releasing gram-negative cell wall fragments can cause septic shock.

5. *Bacillus anthracis.* Animals with anthrax should not be used in manufacturing.

Learning with Technology

1. ¡Yersinia pestis*

2. *Pasteurella multocida*

Case History: A Delayed Diagnosis

Background

A 25-year-old New Mexico rancher was admitted to an El Paso Hospital on February 12 because of a two-day history of headache, chills, and fever (40°C). The day before admission, he began vomiting. The day of admission, an orange-sized swelling in the left axilla was noted. A lymph-node aspirate and a smear of peripheral blood were reported to contain gram-positive cocci, often in pairs. Under the assumption that a gram-positive organism had caused the patient's illness, he was given cefoxitin. The man was acutely ill. Within a few hours of admission, he had a cardiopulmonary arrest. During resuscitation efforts, he vomited and aspirated his vomitus; a chest X-ray showed bilateral infiltrate. Additionally, the patient bled from several body sites. The patient died within 6 hours of admission. In the 2 weeks prior to becoming ill, the patient had trapped, killed, and skinned 3 kit foxes, 4 coyotes, and 1 bobcat. The patient had cut his left hand shortly before skinning the bobcat on February 7.

After his death, biochemical testing of a gram-negative rod isolated from blood cultures identified the etiology as *Enterobacter agglomerans*.

Questions

1. Identify the following periods: incubation, prodromal, illness, decline.
2. Identify the etiologic agent of this disease. Briefly explain how you arrived at your conclusion.
3. What microbiologic tests would you perform to verify the etiology?
4. How might the patient have been treated between February 7 and 12?
5. What special precautions needed to be taken by the hospital and mortuary personnel?

The Solution

1. Incubation period: February 7–10.
   Prodromal period: February 10.
   Period of illness: February 11–12.
   Period of decline: None; the patient died.

2. *Yersinia pestis.* The organisms may have been interpreted as being cocci or diplococci because of (1) the tendency of *Y. pestis* to assume a bipolar appearance when stained and (2) the rapid division of this coccobacillary organism, which might have given the impression that the dividing organisms were in pairs. *Y. pestis* and some strains of *E. agglomerans* are relatively inactive biochemically and may be difficult to differentiate. Other characteristics, such as colonial morphology, growth characteristics in broth, and motility, will aid in the differentiation of these organisms.

3. Culture and fluorescent-antibody stains.

4. With antibiotics.

5. The patient should have been placed in isolation, and special isolation procedures should have been used in handling his body fluids. Personnel should receive antibiotic prophylaxis.

Chapter 24   *Microbial Diseases of the Respiratory System*

Learning Objectives

1. Describe how microorganisms are prevented from entering the respiratory system.
2. Characterize the normal microbiota of the upper and lower respiratory systems.
3. Differentiate among pharyngitis, laryngitis, tonsillitis, sinusitis, and epiglottitis.
4. List the causative agent, symptoms, prevention, preferred treatment, and laboratory identification tests for streptococcal pharyngitis, scarlet fever, diphtheria, cutaneous diphtheria, and otitis media.
5. List the causative agents and treatments for the common cold.
6. List the causative agent, symptoms, prevention, preferred treatment, and laboratory identification tests for pertussis and tuberculosis.
7. Compare and contrast the seven bacterial pneumonias discussed in this chapter.
8. List the causative agent, symptoms, prevention, and preferred treatment for viral pneumonia, RSV, and influenza.
9. List the causative agent, mode of transmission, preferred treatment, and laboratory identification tests for four fungal diseases of the respiratory system.

NEW IN THIS EDITION

- All morbidity and mortality data are updated to January, 2000.
- Revised tuberculosis coverage includes the latest drug therapy, new diagnostic tests, use of BCG vaccine, and a new photo of a positive skin test.
- A new Clinical Problem-Solving box follows the diagnosis of an actual case of multidrug resistant *S. pneumoniae*.
- A table lists human influenza viruses.

CHAPTER SUMMARY

Introduction

1. Infections of the upper respiratory system are the most common type of infection.
2. Pathogens that enter the respiratory system can infect other parts of the body.

Structure and Function of the Respiratory System

1. The upper respiratory system consists of the nose, pharynx, and associated structures, such as the middle ear and auditory tubes.
2. Coarse hairs in the nose filter large particles from air entering the respiratory tract.

3. The ciliated mucous membranes of the nose and throat trap airborne particles and remove them from the body.
4. Lymphoid tissue, tonsils, and adenoids provide immunity to certain infections.
5. The lower respiratory system consists of the larynx, trachea, bronchial tubes, and alveoli.
6. The ciliary escalator of the lower respiratory system helps prevent microorganisms from reaching the lungs.
7. Microbes in the lungs can be phagocytized by alveolar macrophages.
8. Respiratory mucus contains IgA antibodies.

Normal Microbiota of the Respiratory System

1. The normal microbiota of the nasal cavity and throat can include pathogenic microorganisms.
2. The lower respiratory system is usually sterile because of the action of the ciliary escalator.

MICROBIAL DISEASES OF THE UPPER RESPIRATORY SYSTEM

1. Specific areas of the upper respiratory system can become infected to produce pharyngitis, laryngitis, tonsillitis, sinusitis, and epiglottitis.
2. These infections may be caused by several bacteria and viruses, often in combination.
3. Most respiratory tract infections are self-limiting.
4. *H. influenzae* type b can cause epiglottitis.

Bacterial Diseases of the Upper Respiratory System

Streptococcal Pharyngitis (Strep Throat)

1. This infection is caused by group A β-hemolytic streptococci, the group to which *Streptococcus pyogenes* belongs.
2. Symptoms of this infection are inflammation of the mucous membrane and fever; tonsillitis and otitis media may also occur.
3. Preliminary rapid diagnosis is made by indirect agglutination tests. Definitive diagnosis is based on a rise in IgM antibodies.
4. Penicillin is used to treat streptococcal pharyngitis.
5. Immunity to streptococcal infections is type-specific.
6. Strep throat is usually transmitted by droplets but has been associated with unpasteurized milk.

Scarlet Fever

1. Strep throat caused by an erythrogenic toxin-producing *S. pyogenes* results in scarlet fever.
2. *S. pyogenes* produces erythrogenic toxin when lysogenized by a phage.
3. Symptoms include a pink rash, high fever, and a red, enlarged tongue.

Diphtheria

1.   Diphtheria is caused by exotoxin-producing *Corynebacterium diphtheriae.*
2.   Exotoxin is produced when the bacteria are lysogenized by a phage.
3.   A membrane, containing fibrin and dead human and bacterial cells, forms in the throat and can block the passage of air.
4.   The exotoxin inhibits protein synthesis, and heart, kidney, or nerve damage may result.
5.   Laboratory diagnosis is based on isolation of the bacteria and the appearance of growth on different media.
6.   Antitoxin must be administered to neutralize the toxin, and antibiotics can stop growth of the bacteria.
7.   Routine immunization in the United States includes diphtheria toxoid in the DTP vaccine.
8.   Slow-healing skin ulcerations are characteristic of cutaneous diphtheria.
9.   There is minimal dissemination of the exotoxin in the bloodstream.

Otitis Media

1.   Earache, or otitis media, can occur as a complication of nose and throat infections.
2.   Pus accumulation causes pressure on the eardrum.
3.   Bacterial causes include *Streptococcus pneumoniae,* nonencapsulated *Haemophilus influenzae, Moraxella catarrhalis, Streptococcus pyogenes,* and *Staphylococcus aureus.*

Viral Disease of the Upper Respiratory System

The Common Cold

1.   Any one of approximately 200 different viruses can cause the common cold; rhinoviruses cause about 50% of all colds.
2.   Symptoms include sneezing, nasal secretions, and congestion.
3.   Sinus infections, lower respiratory tract infections, laryngitis, and otitis media can occur as complications of a cold.
4.   Colds are most often transmitted by indirect contact.
5.   Rhinoviruses prefer temperatures slightly lower than body temperatures.
6.   The incidence of colds increases during cold weather, possibly because of increased interpersonal indoor contact or physiological changes.
7.   Antibodies are produced against the specific viruses.

MICROBIAL DISEASES OF THE LOWER RESPIRATORY SYSTEM

1.   Many of the same microorganisms that infect the upper respiratory system also infect the lower respiratory system.
2.   Diseases of the lower respiratory system include bronchitis and pneumonia.

Bacterial Diseases of the Lower Respiratory System

Pertussis (Whooping Cough)

1.   Pertussis is caused by *Bordetella pertussis.*
2.   The inital stage of pertussis resembles a cold and is called the catarrhal stage.

3. The accumulation of mucus in the trachea and bronchi causes deep coughs characteristic of the paroxysmal (second) stage.

4. The convalescence (third) stage can last for months.

5. Laboratory diagnosis is based on isolation of the bacteria on enrichment and selective media, followed by serological tests.

6. Regular immunization for children has decreased the incidence of pertussis.

Tuberculosis

1. Tuberculosis is caused by *Mycobacterium tuberculosis.*

2. Large amounts of lipids in the cell wall account for the bacterium's acid-fast characteristic as well as its resistance to drying and disinfectants.

3. *M. tuberculosis* may be ingested by alveolar macrophages; if not killed, the bacteria reproduce in the macrophages.

4. Lesions formed by *M. tuberculosis* are called tubercles; dead macrophages and bacteria form the caseous lesion that might calcify and appear in an X ray as a Ghon complex.

5. Liquefaction of the caseous lesion results in a tuberculous cavity in which *M. tuberculosis* can grow.

6. New foci of infection can develop when a caseous lesion ruptures and releases bacteria into blood or lymph vessels; this is called miliary tuberculosis.

7. Miliary tuberculosis is characterized by weight loss, coughing, and loss of vigor.

8. Chemotherapy usually involves two drugs taken for 1–2 years; multidrug-resistant *M. tuberculosis* is becoming prevalent.

9. A positive tuberculin skin test can indicate an active case of TB, or prior infection, or vaccination and immunity to the disease.

10. Laboratory diagnosis is based on the presence of acid-fast bacilli and isolation of the bacteria, which requires incubation of up to 8 weeks.

11. *Mycobacterium bovis* causes bovine tuberculosis and can be transmitted to humans by unpasteurized milk.

12. *M. bovis* infections usually affect the bones or lymphatic system.

13. BCG vaccine for tuberculosis consists of a live, avirulent culture of *M. bovis*.

14. *M. avium-intracellulare* complex infects patients in the late stages of HIV infection.

Bacterial Pneumonias

1. Typical pneumonia is caused by *S. pneumoniae* or *H. influenzae*.

2. Atypical pneumonias are caused by other microorganisms.

Pneumococcal Pneumonia

1. Pneumococcal pneumonia is caused by encapsulated *Streptococcus pneumoniae*.

2. Symptoms are fever, breathing difficulty, chest pain, and rust-colored sputum.

3. The bacteria can be identified by the production of α-hemolysins, inhibition by optochin, bile solubility, and through serological tests.

4. A vaccine consists of purified capsular material from 23 serotypes of *S. pneumoniae*.

Haemophilus influenzae Pneumonia

1. Alcoholism, poor nutrition, cancer, and diabetes are predisposing factors for *H. influenzae* pneumonia.

2. *H. influenzae* is a gram-negative coccobacillus.

Mycoplasmal Pneumonia

1. *Mycoplasma pneumoniae* causes mycoplasmal pneumonia; it is an endemic disease.

2. *M. pneumoniae* produces small "fried egg" colonies after 2 weeks' incubation on enriched media containing horse serum and yeast extract.

3. A complement-fixation test, used to diagnose the disease, is based on the rising antibody titer.

Legionellosis

1. The disease is caused by the aerobic gram-negative rod *Legionella pneumophila*.

2. The bacterium can grow in water, such as air-conditioning cooling towers, and then be disseminated in the air.

3. This pneumonia does not appear to be transmitted from person to person.

4. Bacterial culture, FA tests, and DNA probes are used for laboratory diagnosis.

Psittacosis (Ornithosis)

1. *Chlamydia psittaci* is transmitted by contact with contaminated droppings and exudates of fowl.

2. Elementary bodies allow the bacteria to survive outside a host.

3. Commercial bird handlers are most susceptible to this disease.

4. The bacteria are isolated in embryonated eggs, mice, or cell culture; identification is based on fluorescent-antibody staining.

Chlamydial Pneumonia

1. *Chlamydia pneumoniae* causes pneumonia; it is transmitted from person to person.

2. A fluorescent-antibody test is used for diagnosis.

Q Fever

1. Obligately parasitic, intracellular *Coxiella burnetii* causes Q fever.

2. The disease is usually transmitted to humans through unpasteurized milk or inhalation of aerosols in dairy barns.

3. Laboratory diagnosis is made with the culture of bacteria in embryonated eggs or cell culture.

Other Bacterial Pneumonias

1. Gram-positive bacteria that cause pneumonia include *S. aureus* and *S. pyogenes*.

2. Gram-negative bacteria that cause pneumonia include *M. catarrhalis, K. pneumoniae,* and *Pseudomonas* species.

Viral Diseases of the Lower Respiratory System

Viral Pneumonia

1. A number of viruses can cause pneumonia as a complication of infections such as influenza.

2. The etiologies are not usually identified in a clinical laboratory because of the difficulty in isolating and identifying viruses.

   Respiratory Syncytial Virus (RSV)
   1. RSV is the most common cause of pneumonia in infants.

Influenza (Flu)

1. Influenza is caused by *Influenzavirus* and is characterized by chills, fever, headache, and general muscular aches.
2. Hemagglutinin (H) and neuraminidase (N) spikes project from the outer lipid bilayer of the virus.
3. Viral strains are identified by antigenic differences in the H and N spikes; they are also divided by antigenic differences in their protein coats (A, B, and C).
4. Viral isolates are identified by hemagglutination-inhibition tests and immunofluorescence testing with monoclonal antibodies.
5. Antigenic shifts that alter the antigenic nature of the H and N spikes make natural immunity and vaccination of questionable value. Minor antigenic changes are caused by antigenic drift.
6. Deaths during an influenza epidemic are usually from secondary bacterial infections.
7. Multivalent vaccines are available for the elderly and other high-risk groups.
8. Amantadine is an effective prophylactic and curative drug against *Influenzavirus* type A.

Fungal Diseases of the Lower Respiratory System

1. Fungal spores are easily inhaled; they may germinate in the lower respiratory tract.
2. The incidence of fungal diseases has been increasing over the past 15 years or so.
3. The following mycoses can be treated with amphotericin B.

Histoplasmosis

1. *Histoplasma capsulatum* causes a subclinical respiratory infection that only occasionally progresses to a severe, generalized disease.
2. The disease is acquired by inhalation of airborne conidia.
3. Isolation of the fungus or identification of the fungus in tissue samples is necessary for diagnosis.

Coccidioidomycosis

1. Inhalation of the airborne arthrospores of *Coccidioides immitis* can result in coccidioidomycosis.
2. Most cases are subclinical, but when there are predisposing factors such as fatigue and poor nutrition, a progressive disease resembling tuberculosis can result.

Pneumocystis Pneumonia

1. *Pneumocystis carinii,* is found in healthy human lungs.
2. *Pneumocystis carinii* causes disease in immunosuppressed patients.
3. *Pneumocystis* pneumonia is currently being treated with trimethoprim or pentamidine.

Blastomycosis (North American Blastomycosis)

1. *Blastomyces dermatitidis* is the causative agent of blastomycosis.
2. The infection begins in the lungs and can spread to cause extensive abscesses.

Other Fungi Involved in Respiratory Disease

1. Opportunistic fungi can cause respiratory disease in immunosuppressed hosts, especially when large numbers of spores are inhaled.
2. Among these fungi are *Aspergillus*, *Rhizopus*, and *Mucor*.

THE LOOP

For a taxonomic approach, pages can be assigned as follows:

Answers

Review

1. <u>Droplet infection</u>. Inhalation of cells and spores; ingestion of contaminated food.

2. Coarse hairs in the nose filter dust particles from inspired air. Mucus traps dust and microorganisms, and cilia move the trapped particles toward the throat for elimination. The ciliary escalator of the lower respiratory system moves particles toward the throat. Alveolar macrophages can phagocytize microorganisms that enter the lungs. IgA antibodies are found in mucus, saliva, and tears.

3. Beta-hemolytic streptococci inhibit growth of pneumococci; faster-growing organisms can compete with pathogens.

4. Mycoplasmal pneumonia is caused by *Mycoplasma pneumoniae* bacteria. Viral pneumonia can be caused by several different viruses. Mycoplasmal pneumonia can be treated with tetracyclines, whereas viral pneumonia cannot.

5. Bacteria infecting the nose and throat can move through the eustachian tube to the inner ear. Microorganisms can enter the ear directly via swimming pool water or injury to the eardrum or skull. The bacteria that most commonly cause otitis media are *S. aureus*, *Streptococcus pneumoniae*, β-hemolytic streptococci, and *H. influenzae*. The middle ear is connected to the nose and throat.

6. **Upper Respiratory System**

Common cold	Coronaviruses	Sneezing, excessive nasal secretions, congestion

    Lower Respiratory System

Viral pneumonia	Several viruses	Fever, shortness of breath, chest pains
Influenza	*Influenzavirus*	Chills, fever, headache, muscular pains
RSV	Respiratory syncytial virus	Coughing, wheezing

    Amantadine is used to treat influenza. Ribavirin may reduce RSV symptoms.

7. | Disease | Symptoms |
    |---|---|
    | Streptococcal pharyngitis | Pharyngitis and tonsillitis |
    | Scarlet fever | Rash and fever |
    | Diphtheria | Membrane across throat |
    | Whooping cough | Paroxysmal coughing |
    | Tuberculosis | Tubercles, weight loss, and coughing |
    | Pneumococcal pneumonia | Reddish lungs, fever |
    | *H. influenzae* pneumonia | Similar to pneumococcal pneumonia |
    | Chamydial pneumonia | Low fever, cough, and headache |
    | Legionellosis | Fever and cough |
    | Psittacosis | Fever and headache |
    | Q fever | Chills and chest pain |
    | Epiglottitis | Inflamed, abscessed epiglottis |

8. Inhalation of large numbers of spores from *Aspergillus* or *Rhizopus* can cause infections in individuals with impaired immune systems, cancer, and diabetes.

9. No. Many different organisms (gram-positive bacteria, gram-negative bacteria, and viruses) can cause pneumonia. Each of these organisms is susceptible to different antimicrobial agents.

10. | Disease | Endemic Areas in the United States |
|---|---|
| Histoplasmosis | States adjoining the Mississippi and Ohio Rivers |
| Coccidioidomycosis | American Southwest |
| Blastomycosis | Mississippi |
| *Pneumocystis* | Ubiquitous |

Refer to Table 24.2.

11. In the tuberculin test, purified protein derivative (PPD) from *M. tuberculosis* is injected into the skin. Induration and reddening of the area around the injection site indicates an active infection or immunity to tuberculosis.

12. Hypothesis 1: Close indoor contact in winter promotes epidemic transmission.
Hypothesis 2: The viruses grow best at slightly cooler temperatures.
Hypothesis 3: A physiological change in humans during winter allows viral growth.

13. Gram-positive cocci

Catalase-positive	*Staphylococcus aureus*
β-hemolytic	*Streptococcus pyogenes*
α-hemolytic	*S. pneumoniae*

Gram-positive rods

Not acid-fast	*C. diphtheriae*
Acid-fast	*Mycobacterium tuberculosis*

Gram-negative cocci   *Moraxella catarrhalis*

Aerobic gram-negative rods

Coccobacilli	*B. pertussis*
Rods	*L. pneumophila*

Facultatively anaerobic gram-negative rods

Coccobacilli	*H. influenzae*
Rods	*K. pneumoniae*

Intracellular

Elementary bodies	*Chlamydia psittaci*
No elementary bodies	*Coxiella burnetii*

Wall-less   *Mycoplasma pneumoniae*

Critical Thinking

1. a. When *S. pyogenes* causing streptococcal sore throat produces an erythrogenic toxin, the infection is called scarlet fever.

   b. Diphtheroids are nonpathogenic species of corynebacteria. *C. diphtheriae*, like *S. pyogenes*, produces an exotoxin when it is lysogenized by a phage.

2. There are many strains of *Influenzavirus* because of antigenic drift and antigenic shift.

3. There are more than 200 viruses that cause colds. *Influenzavirus* changes its antigens every few months.

Clinical Applications

1. Coccidioidomycosis. Attempts to culture fungi after the initial diagnosis may have grown *C. immitis*.

2. AIDS patients were given pentamidine to prevent *Pneumocystis* pneumonia. Possible sources of infection: (1) the nurse diagnosed with tuberculosis, (2) aerosols created during pentamidine therapy, or (3) face-to-face exposure with TB patients.

3. The disease is psittacosis; transmitted by inhalation of particles from bird droppings.

4. Pneumococcal pneumonia.

5. Histoplasmosis; contracted by inhalation of conidia.

Learning with Technology

1. *Streptococcus pneumoniae*

2. *Nocardia asteroides*

Case History: Pythons in the United States

Background

On May 25, the Suffolk County, New York Health Department was informed by a hospital infection control nurse that two days earlier, a 27-year-old man had been admitted with a four-day history of fever (40°C), severe headache, chills, malaise, and vomiting. This patient had visited West Africa in April, and was an employee of an exotic bird and reptile importing company. Within one week, three other employees of that company were admitted to the same hospital with similar symptoms. All were treated with oral tetracycline; recovery was rapid and complete.

All four persons had been involved in unpacking and deticking a shipment of 500 ball pythons that were imported on May 3 from Accra, Ghana. Examination of the hemolymph of five ticks removed from these snakes indicated that all contained numerous bacteria. Three types of ticks were identified: *Amblyomma nuttalli*, *Aponomma latum*, and *A. flavomaculatum*.

Questions

1. Identify the etiologic agent of this disease.

2. What was the probable mode of transmission? The probable reservoir?

3. What information do you need to verify your conclusions?

4. Briefly explain how you arrived at your conclusions.

The Solution

1. Q fever caused by *Coxiella burnetii*.

2. Transmission could have been by aerosols or ticks.

3. A serologic test such as complement fixation or immunofluorescence.

4. Students will use clinical symptoms and association with exotic animals and ticks.

Chapter 25   *Microbial Diseases of the Digestive System*

Learning Objectives

1. Name the structures of the digestive system that contact food.

2. List examples of microbiota for each part of the gastrointestinal tract.

3. Describe the events that lead to dental caries and periodontal disease.

4. List the causative agents, suspect foods, signs and symptoms, and treatments for staphylococcal food poisoning, shigellosis, salmonellosis, typhoid fever, cholera, gastroenteritis, and peptic ulcer disease.

5. List the causative agents, modes of transmission, sites of infection, and symptoms for mumps and CMV inclusion disease.

6. Differentiate between hepatitis A, hepatitis B, hepatitis C, hepatitis D, and hepatitis E.

7. List the causative agents, mode of transmission, and symptoms of viral gastroenteritis.

8. Identify methods for preventing ergot and aflatoxin poisoning.

9. List the causative agents, modes of transmission, symptoms, and treatments for giardiasis, amoebic dysentery, cryptosporidiosis, and *Cyclospora* diarrheal infection.

10. List the causative agents, modes of transmission, symptoms, and treatments for tapeworms, hydatid disease, pinworms, hookworms, ascariasis, and trichinosis.

NEW IN THIS EDITION

- A new Clinical Problem-Solving box investigates a multistate outbreak of shigellosis.

CHAPTER SUMMARY

Introduction

1. Diseases of the digestive system are the second most common illnesses in the United States.

2. Diseases of the digestive system usually result from the ingestion of microorganisms and their toxins in food and water.

3. The fecal–oral cycle of transmission can be broken by the proper disposal of sewage, the disinfection of drinking water, and proper food preparation and storage.

Structure and Function of the Digestive System

1. The gastrointestinal (GI) tract, or alimentary canal, consists of the mouth, pharynx, esophagus, stomach, small intestine, and large intestine.

2. The teeth, tongue, salivary glands, liver, gallbladder, and pancreas are accessory structures.

3. In the GI tract, with mechanical and chemical help from the accessory structures, large food molecules are broken down into smaller molecules that can be transported by blood or lymph to cells.

4. Feces, the solids resulting from digestion, are eliminated through the anus.

Normal Microbiota of the Digestive System

1. A wide variety of bacteria colonize the mouth.

2. The stomach and small intestine have few resident microorganisms.

3. The large intestine is the habitat of *Lactobacillus, Bacteroides, E. coli, Enterobacter, Klebsiella,* and *Proteus.*

4. Bacteria in the large intestine assist in degrading food and synthesizing vitamins.

5. Up to 40% of fecal mass is microbial cells.

Bacterial Diseases of the Mouth

Dental Caries (Tooth Decay)

1. Dental caries begin when tooth enamel and dentin are eroded and the pulp is exposed to bacterial infection.

2. *Streptococcus mutans,* found in the mouth, uses sucrose to form dextran from glucose and lactic acid from fructose.

3. Bacteria adhere to teeth with a sticky dextran capsule, forming dental plaque.

4. Acid produced during carbohydrate fermentation destroys tooth enamel at the site of the plaque.

5. Gram-positive rods and filamentous bacteria can penetrate into dentin and pulp.

6. Carbohydrates such as starch, mannitol, and sorbitol are not used by cariogenic bacteria to produce dextran and do not promote tooth decay.

7. Caries are prevented by restricting ingestion of sucrose and by physical removal of plaque; a vaccine against *S. mutans* is theoretically possible.

Periodontal Disease

1. Caries of the cementum and gingivitis are caused by streptococci, actinomycetes, and anaerobic gram-negative bacteria.

2. Chronic gum disease (periodontitis) can cause bone destruction and tooth loss; periodontitis is due to inflammatory response to a variety of bacteria growing on the gums.

3. Acute necrotizing ulcerative gingivitis is caused by *Prevotella intermedia* and spirochetes.

Bacterial Diseases of the Lower Digestive System

1. A gastrointestinal infection is caused by growth of a pathogen in the intestines.

2. Incubation times, the times required for bacterial cells to grow and their products to produce symptoms, range from 12 hours to 2 weeks. Symptoms of infection generally include a fever.

3. A bacterial intoxication results from ingestion of preformed bacterial toxins.

4. Symptoms appear from 1 to 48 hours after ingestion of the toxin. Fever is not usually a symptom of intoxication.

5. Infections and intoxications cause diarrhea, dysentery, or gastroenteritis.

6. These conditions are usually treated with fluid and electrolyte replacement.

Staphylococcal Food Poisoning (Staphylococcal Enterotoxicosis)

1. Staphylococcal food poisoning is caused by ingestion of an enterotoxin produced in improperly stored foods.
2. *S. aureus* is inoculated into foods during preparation. The bacteria grow and produce enterotoxin in food stored at room temperature.
3. The exotoxin is not denatured by boiling for 30 minutes.
4. Foods with high osmotic pressure and those not cooked immediately before consumption are most often the source of staphylococcal enterotoxicosis.
5. Nausea, vomiting, and diarrhea begin 1–6 hours after eating, and the symptoms last approximately 24 hours.
6. Laboratory identification of *S. aureus* isolated from foods or the presence of thermostable nuclease in foods can confirm diagnosis.
7. Serological tests are available to detect toxins in foods.

Shigellosis (Bacillary Dysentery)

1. Shigellosis is caused by four species of *Shigella*.
2. Symptoms include blood and mucus in stools, abdominal cramps, and fever. Infections by *S. dysenteriae* result in ulceration of the intestinal mucosa.
3. Isolation and identification of the bacteria from rectal swabs are used for diagnosis.

Salmonellosis (*Salmonella* Gastroenteritis)

1. Salmonellosis, or *Salmonella* gastroenteritis, is caused by many *Salmonella* species.
2. Symptoms include nausea, abdominal pain, and diarrhea and begin 12–36 hours after eating large numbers of *Salmonella*. Septicemia can occur in infants and in the elderly.
3. Fever might be caused by endotoxin.
4. Mortality is lower than 1%, and recovery can result in a carrier state.
5. Heating food to 68°C will usually kill *Salmonella*.
6. Laboratory diagnosis is based on isolation and identification of *Salmonella* from feces and foods.

Typhoid Fever

1. *Salmonella typhi* causes typhoid fever; the bacteria are transmitted by contact with human feces.
2. Fever and malaise occur after a 2-week incubation period. Symptoms last 2–3 weeks.
3. *S. typhi* is harbored in the gallbladder of carriers.
4. Vaccines are available for high-risk people.

Cholera

1. *Vibrio cholerae* produces an exotoxin that alters membrane permeability of the intestinal mucosa; the resulting vomiting and diarrhea cause loss of body fluids.
2. The incubation period is approximately three days. The symptoms last for a few days. Untreated cholera has a 50% mortality rate.

3. Diagnosis is based on the isolation of *Vibrio* from feces.
4. *Vibrio cholerae* non-O:1 causes gastroenteritis in the United States. It is usually transmitted via contaminated seafood.

Vibrio Gastroenteritis

1. *Vibrio* gastroenteritis can be caused by the halophiles *V. parahaemolyticus* and *V. vulnificus*.
2. The onset of symptoms begins within 24 hours after eating contaminated foods. Recovery occurs within a few days.
3. The disease is contracted by eating contaminated crustaceans or contaminated mollusks.

Escherichia coli Gastroenteritis

1. *E. coli* gastroenteritis may be caused by enterotoxigenic, enteroinvasive, or enterohemorrhagic strains of *E. coli*.
2. The disease occurs as epidemic diarrhea in nurseries, as traveler's diarrhea, as endemic diarrhea in underdeveloped countries, and as hemorrhagic colitis.
3. In adults, the disease is usually self-limiting and does not require chemotherapy.
4. Enterohemorrhagic *E. coli*, such as *E. coli* O157:H7, produce Shigalike toxins that cause inflammation and bleeding of the colon.
5. Shigalike toxins can affect the kidneys to cause hemolytic uremic syndrome.

Campylobacter Gastroenteritis

1. *Campylobacter* is the second most common cause of diarrhea in the United States.
2. *Campylobacter* is transmitted in cow's milk.

Helicobacter Peptic Ulcer Disease

1. *Helicobacter pylori* produces ammonia, which neutralizes stomach acid; the bacteria colonize the stomach mucosa and cause peptic ulcer disease.
2. Bismuth and several antibiotics may be useful in treating peptic ulcer disease.

Yersinia Gastroenteritis

1. *Y. enterocolitica* and *Y. pseudotuberculosis* are transmitted in meat and milk.
2. *Yersinia* can grow at refrigeration temperatures.

Clostridium perfringens Gastroenteritis

1. A self-limiting gastroenteritis is caused by *C. perfringens*.
2. Endospores survive heating and germinate when foods (usually meats) are stored at room temperature.
3. Exotoxin produced when the bacteria grow in the intestines is responsible for the symptoms.
4. Diagnosis is based on isolation and identification of the bacteria in stool samples.

Bacillus cereus Gastroenteritis

1. Ingesting food contaminated with the soil saprophyte *Bacillus cereus* can result in diarrhea, nausea, and vomiting.

Viral Diseases of the Digestive System

Mumps

1. Mumps virus (a paramyxovirus) enters and exits the body through the respiratory tract.
2. About 16–18 days after exposure, the virus causes inflammation of the parotid glands, fever, and pain during swallowing. About 4–7 days later, orchitis may occur.
3. After onset of the symptoms, the virus is found in the blood, saliva, and urine.
4. A measles, mumps, rubella (MMR) vaccine is available.
5. Diagnosis is based on symptoms or an ELISA test is performed on viruses cultured in embryonated eggs or cell culture.

Cytomegalovirus (CMV) Inclusion Disease

1. CMV (a herpesvirus) causes intranuclear bodies and cytomegaly of host cells.
2. CMV is transmitted by saliva, urine, semen, cervical secretions, and human milk.
3. CMV inclusion disease can be asymptomatic, a mild disease, or progressive and fatal. Immunosuppressed patients may develop pneumonia.
4. If the virus crosses the placenta, it can cause congenital infection of the fetus, resulting in impaired mental development, neurological damage, and stillbirth.
5. Diagnosis is based on isolation of the virus or detection of IgG and IgM antibodies.

Hepatitis

1. Inflammation of the liver is called hepatitis. Symptoms include loss of appetite, malaise, fever, and jaundice.
2. Viral causes of hepatitis include hepatitis viruses, EB virus, and CMV.

Hepatitis A

1. Hepatitis A virus (HAV) causes infectious hepatitis A. At least 50% of all cases are subclinical.
2. HAV is ingested in contaminated food or water, grows in the cells of the intestinal mucosa, and spreads to the liver, kidneys, and spleen in the blood.
3. The virus is eliminated with feces.
4. The incubation period is 2–6 weeks; the period of disease is 2–21 days; recovery is complete within 4–6 weeks.
5. Diagnosis is based on tests for IgM antibodies.
6. Passive immunization can provide temporary protection. A vaccine is available.

Hepatitis B

1. Hepatitis B virus (HBV) causes hepatitis B, which is frequently serious.
2. HBV is transmitted by blood transfusions, contaminated syringes, saliva, sweat, breast milk, and semen.
3. Blood is tested for HB_SAg before being used in transfusions.
4. The average incubation period is 3 months; recovery is usually complete, but some patients develop a chronic infection or become carriers.
5. A vaccine against HB_SAg is available.

Hepatitis C
1. Hepatitis C virus (HCV) is transmitted via blood.
2. The incubation period is 2–22 weeks; the disease is usually mild, but some patients develop chronic hepatitis.
3. Blood is tested for HCV antibodies before being used in transfusions.

Hepatitis D (Delta Hepatitis)
1. Hepatitis D virus (HDV) has a circular strand of RNA and uses HB_SAg as a coat.

Hepatitis E (Infectious NANB Hepatitis)
1. Hepatitis E virus (HEV) is spread by the fecal–oral route.

Other Types of Hepatitis
1. There is evidence of the existence of hepatitis types F and G.

Viral Gastroenteritis
1. Viral gastroenteritis is most often caused by a rotavirus or the Norwalk agent.

Fungal Diseases of the Digestive System
1. Mycotoxins are toxins produced by some fungi.
2. Mycotoxins affect the blood, nervous system, kidney, or liver.

Ergot Poisoning
1. Ergot poisoning is caused by the mycotoxin produced by *Claviceps purpurea*.
2. Cereal grains are most often contaminated with the *Claviceps* mycotoxin.

Aflatoxin Poisoning
1. Aflatoxin is a mycotoxin produced by *Aspergillus flavus*.
2. Peanuts are most often contaminated with aflatoxin.

Protozoan Diseases of the Digestive System
Giardiasis
1. *Giardia lamblia* grows in the intestines of humans and wild animals and is transmitted in contaminated water.
2. Symptoms of giardiasis are malaise, nausea, flatulence, weakness, and abdominal cramps that persist for weeks.
3. Diagnosis is based on identification of the protozoa in the small intestine.

Cryptosporidiosis
1. *Cryptosporidium parvum* causes diarrhea; in immunosuppressed patients the disease is prolonged for months.
2. The pathogen is transmitted in contaminated water.
3. Diagnosis is based on the identification of oocysts in feces.

Cyclospora Diarrheal Infection

1. *C. cayetanensis* causes diarrhea; the protozoan was first identified in 1993.
2. It is transmitted in contaminated produce.
3. Diagnosis is based on the identification of oocysts in feces.

Amoebic Dysentery (Amoebiasis)

1. Amoebic dysentery is caused by *Entamoeba histolytica* growing in the large intestine.
2. The amoeba feeds on red blood cells and GI tract tissues. Severe infections result in abscesses.
3. Diagnosis is confirmed by observing the trophozoites in feces and by several serological tests.

Helminthic Diseases of the Digestive System

Tapeworm Infestations

1. Tapeworms are contracted by the consumption of undercooked beef, pork, or fish containing encysted larvae (cysticerci).
2. The scolex attaches to the intestinal mucosa of humans (the definitive host) and matures into an adult tapeworm.
3. Eggs are shed in the feces and must be ingested by an intermediate host.
4. Adult tapeworms can be undiagnosed in a human.
5. Diagnosis is based on the observation of proglottids and eggs in feces.
6. Neurocysticercosis in humans occurs when the pork tapeworm larvae encyst in humans.

Hydatid Disease

1. Humans infected with the tapeworm *Echinococcus granulosus* might have hydatid cysts in their lungs or other organs.
2. Dogs and wolves are usually the definitive hosts, and sheep or deer are the intermediate hosts for *E. granulosus*.

Nematode Infestations

Pinworm Infestation

1. Humans are the definitive host for pinworms, *Enterobius vermicularis*.
2. The disease is acquired by ingesting *Enterobius* eggs.

Hookworm Infestation

1. Hookworm larvae bore through skin and migrate to the intestine to mature into adults.
2. In the soil, *Necator* larvae hatch from eggs shed in feces.

Ascariasis

1. *Ascaris lumbricoides* adults live in human intestines.
2. The disease is acquired by ingesting *Ascaris* eggs.

Trichinosis

1. *Trichinella spiralis* larvae encyst in muscles of humans and other mammals to cause trichinosis.

2. The roundworm is contracted by ingesting undercooked meat containing larvae.

3. Adults mature in the intestine and lay eggs; the new larvae migrate to invade muscles.

4. Symptoms include fever, swelling around the eyes, and gastrointestinal upset.

5. Biopsy specimens and serological tests are used for diagnosis.

THE LOOP

For a taxonomic approach, pages can be assigned as follows:

The actual incidence of typhoid fever is shown in question 4 on page 433 of *Microbiology: An Introduction*. The occurence of cholera in the United States is shown in question 5 on page 433 of *Microbiology: An Introduction*.

See also "The Role of Microorganisms in Water Quality" on pp. 756–759 of *Microbiology: An Introduction*.

Answers

Review

1. Mouth—Streptococci, including *S. mutans*.
   Stomach and small intestine—None.
   Large intestine and rectum—*Lactobacillus, Bacteroides* and enterics.

2. *S. mutans* becomes established in the mouth when the teeth erupt from the gums. A sticky capsule enables the bacteria to adhere to teeth. The dextran capsule is produced when the bacteria grow on sucrose. These bacteria and others that become trapped in the dextran produce lactic acid, which erodes tooth enamel.

3. | Disease | Suspect Foods | Symptoms |
   |---------|---------------|----------|
   | Staph | Not cooked prior to eating | Vomiting and diarrhea |
   | Shigellosis | Contaminated water | Mucus and blood in stools |
   | Salmonellosis | Poultry; contaminated water | Fever, vomiting, and diarrhea |
   | Cholera | Contaminated water | Rice water stools |
   | Gastro. | Contaminated water | Vomiting and diarrhea |
   | Traveler's | Contaminated water | Vomiting and diarrhea |

   Refer to Table 25.2.

4. Both are caused by *Salmonella enterica*. However, typhoid fever is caused by a few strains of *S. enterica* that are invasive. The bacteria can cross the intestinal wall and can be disseminated throughout the body. Typhoid fever is characterized by fever and malaise without diarrhea.

5. Certain strains of *E. coli* may produce an enterotoxin or invade the epithelium of the large intestine.

6. At present there are no treatments for hepatitis. Exposed individuals can be given pooled immune globulin for hepatitis A or HBIG for passive immunity to hepatitis B. Vaccines can prevent hepatitis A and B.

7. Antibodies specific for HB_SAg are used to screen blood for HBV. A viral protein is used to test blood for antibodies against HCV.

8. Toxins produced by fungi; see pp. 705–706.

9. All four are caused by protozoa. The infections are acquired by ingesting protozoa in contaminated water. Giardiasis is a prolonged diarrhea. Amoebic dysentery is the most severe dysentery, with blood and mucus in the stools. *Cryptosporidium* and *Cyclospora* produce severe diseases in persons with immune deficiencies.

10. | Disease | Etiologic Agent | Symptoms |
    |---------|-----------------|----------|
    | Amoebic | *Entamoeba histolytica* | Blood and mucus in stools, perforation of the intestinal wall, abscesses |
    | Bacillary | *Shigella* spp. | Leukocytes in feces in addition to the symptoms listed for amoebic dysentery |

11. Fever, nausea, abdominal pain, cramps, diarrhea. The diagnosis is based on isolation of the etiologic agent from leftover food or the patient's stools.

12. **Food intoxication:** Microorganisms must be allowed to grow in food from the time of preparation to the time of ingestion. This usually occurs when foods are stored unrefrigerated or improperly canned. The etiologic agents (*Staphylococcus aureus* or *Clostridium botulinum*) must produce an exotoxin. Onset: 1 to 48 hours. Duration: A few days. Treatment: Antimicrobial agents are ineffective. The patient's symptoms may be treated.

Food infection: Viable microorganisms must be ingested with food or water. The organisms could be present during preparation and survive cooking or be inoculated during later handling. The etiologic agents are usually gram-negative organisms (*Salmonella, Shigella, Vibrio,* and *Escherichia*) that produce endotoxins. *Clostridium perfringens* is a gram-positive bacterium that causes food infection. Onset: 12 hours to 2 weeks. Duration: Longer than intoxication because the microorganisms are growing in the patient. Treatment: Rehydration.

13.

Disease	Site	Symptoms
Mumps	Parotid glands	Inflammation of the parotid glands and fever
CMV	Liver, kidneys, other organs	Liver and kidney malfunction
Infectious hepatitis	Liver	Anorexia, fever, diarrhea
Serum hepatitis	Liver	Anorexia, fever, joint pains, jaundice
Viral gastroenteritis	Lower GI tract	Nausea, diarrhea, vomiting

Refer to Table 25.2 to complete this question.

14.

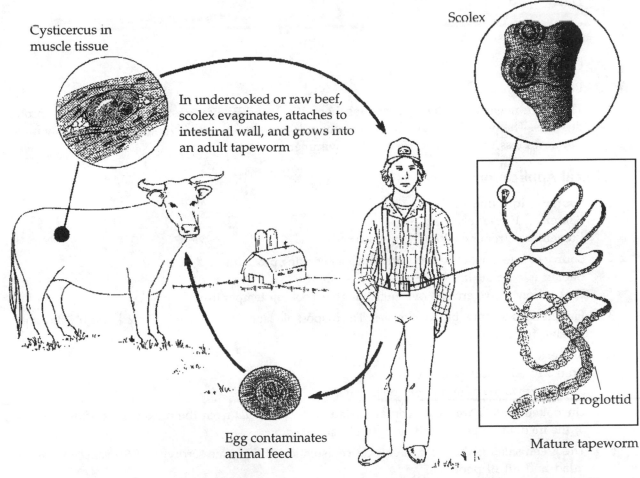

Life cycle of the beef tapeworm, *Taenia saginata.* The adult tapeworm lives in the intestine of the human, the definitive host. Tapeworm segments and eggs are eliminated with feces and are ingested by intermediate hosts, such as grazing cattle. The tapeworm eggs hatch, and cysticerci form in the animal's muscles, which are later consumed by humans. The pork tapeworm, *Taenia solium,* has a similar life cycle, except that cysticerci may also form in human tissue.

15. Refer to Figure 25.25.
16. Adequate sewage treatment and sanitary living conditions.
17. Cook meat thoroughly. Eliminate the source of contamination to cattle and pigs.

Critical Thinking

1. Humans are not usually consumed by other animals. The larval stage of *Trichinella* is encysted in humans and must be ingested to mature in the intestines of a definitive host.

2.

Disease	Conditions	Diagnosis	Prevention
Staph	Lack of refrigeration	Symptoms, presence of *S. aureus* in food	Refrigeration
Salmonellosis	Fecal contamination, inadequate heating	Isolation of *Salmonella* from stools	Sanitation, thorough heating of foods

3. Beef   c
   Deli   d
   Chicken e
   Milk   b
   Oysters a
   Pork   f

4. The gram-negative bacterial infections, hepatitis A and E, viral gastroenteritis, and the protozoan diseases. The organisms are not likely to be salt-tolerant (except *Vibrio*), and an ocean swimmer swallows less water than a freshwater swimmer.

Clinical Applications

1. Source of infection: Crabs.
   Bacteria: *Vibrio cholerae.*
   Prevention: Proper cooking temperature.

2. Source of infection: Chickens and improper cooking procedures.
   Bacteria: *Salmonella heidelberg* and *S. stanley.*
   Prevention: Refrigeration overnight; higher cooking temperature.

3. Fingersticks to draw blood samples. The disposable lancet was reused. The 2% and 0% did not receive fingersticks.

4. Bacteria: *Shigella.*
   Transmitted: Fecal–oral route.
   Source: Sewage-contaminated well water.

5. Uncooked giblets are easily contaminated with *Salmonella* from the turkey's intestines. The length of the incubation period indicates this is an infection.

6. The green salad is the most likely source. Isolation of the same serotype of *Salmonella* from the salad and all ill people.

Learning with Technology

1. *Shigella dysenteriae*
2. *Yersinia enterocolitica*

Case History: A Case of Food Poisoning

Background

Twenty-eight kindergarten children and seven adults visited a raw milk bottling plant, where they were given ice cream and raw milk. Three to six days later, nine children and three adults developed gastroenteritis. The only foods eaten by all these children (ill and well) were in the school-provided lunches. No one else in the school became sick. Stool cultures showed one bacterium in common to nine of the ill children and not present in samples from nine well children. This bacterium is a curved gram-negative rod; it neither ferments nor oxidizes glucose.

Questions

1. Identify the etiologic agent of this outbreak of food poisoning.

2. Was it food infection or intoxication?

3. How did the food get contaminated, and what item was contaminated?

4. What could be done to prevent this type of outbreak?

5. Briefly explain how you arrived at your conclusions.

The Solution

Source of infection: Raw milk

Bacteria: *Campylobacter*

Prevention: Pasteurization

Chapter 26 — *Microbial Diseases of the Urinary and Reproductive Systems*

Learning Objectives

1. List the antimicrobial features of the urinary system.
2. Identify the portals of entry for microbes into the reproductive system.
3. Describe the normal microbiota of the upper urinary tract, the male urethra, and the female urethra and vagina.
4. Describe modes of transmission for urinary and reproductive system infections.
5. List the microorganisms that cause cystitis, pyelonephritis, and leptospirosis, and name the predisposing factors for these diseases.
6. List the causative agents, symptoms, methods of diagnosis, and treatments for gonorrhea, NGU, PID, syphilis, LGV, chancroid, and bacterial vaginosis.
7. List reproductive system diseases that can cause congenital and neonatal infections, and explain how these infections can be prevented.
8. Discuss the epidemiology of genital herpes and genital warts.
9. Discuss the epidemiology of candidiasis.
10. Discuss the epidemiology of trichomoniasis.

NEW IN THIS EDITION

- A new Clinical Problem-Solving box describes a urinary tract infection among athletes.

CHAPTER SUMMARY

Introduction

1. The urinary system regulates the chemical composition of the blood and excretes nitrogenous waste.
2. The reproductive system produces gametes for reproduction and, in the female, supports the growing embryo.
3. Microbial diseases of these systems can result from infection from an outside source or from opportunistic infection by members of the normal body microbiota.

Structure and Function of the Urinary System

1. Urine is transported from the kidneys through ureters to the urinary bladder and is eliminated through the urethra.
2. Valves prevent urine from flowing back to the urinary bladder and kidneys.
3. The flushing action of urine and the acidity of normal urine have some antimicrobial value.

Structure and Function of the Reproductive System

1. The female reproductive system consists of two ovaries, two uterine tubes, the uterus, the cervix, the vagina, and the external genitals.
2. The male reproductive system consists of two testes, ducts, accessory glands, and the penis; seminal fluid leaves the male body through the urethra.

Normal Microbiota of the Urinary and Reproductive Systems

1. The urinary bladder and upper urinary tract are sterile under normal conditions.
2. Lactobacilli dominate the vaginal microbiota during the reproductive years.
3. The male urethra is normally sterile.

Diseases of the Urinary System

Bacterial Diseases of the Urinary System

1. Urethritis, cystitis, and ureteritis are terms describing the inflammations of tissues of the lower urinary tract.
2. Pyelonephritis can result from lower urinary tract infections or from systemic bacterial infections.
3. Opportunistic gram-negative bacteria from the intestines often cause urinary tract infections, especially in females.
4. Nosocomial infections following catheterization can occur in the urinary system. *E. coli* causes more than half of these infections.
5. More than 1000 bacteria of one species per milliliter of urine or 100 coliforms per milliliter of urine indicates an infection.
6. Treatment of urinary tract infections depends on the isolation and antibiotic sensitivity testing of the etiologic agents.

Cystitis
1. Inflammation of the urinary bladder, or cystitis, is common in females.
2. Microorganisms at the opening of the urethra and along the length of the urethra, careless personal hygiene, and sexual intercourse contribute to the high incidence of cystitis in females.
3. The most common etiologies are *E. coli* and *S. saprophyticus.*

Pyelonephritis
1. Inflammation of the kidneys, or pyelonephritis, is usually a complication of lower urinary tract infections.
2. About 75% of pyelonephritis cases are caused by *E. coli.*

Leptospirosis
1. The spirochete *Leptospira interrogans* is the cause of leptospirosis.
2. The disease is transmitted by urine-contaminated water.
3. Leptospirosis is characterized by chills, fever, headache, and muscle aches.
4. Diagnosis is based on isolation of the bacteria and serological identification.

Diseases of the Reproductive System

Bacterial Diseases of the Reproductive System

1. Most diseases of the genital system are sexually transmitted diseases (STDs).
2. Most STDs can be prevented by the use of condoms and are treated with antibiotics.

Gonorrhea

1. *Neisseria gonorrhoeae* causes gonorrhea.
2. Gonorrhea is a common reportable communicable disease in the United States.
3. *N. gonorrhoeae* attaches to the mucosal cells of the oral–pharyngeal area, genitals, eyes, and rectum by means of fimbriae.
4. Symptoms in males are painful urination and pus discharge. Blockage of the urethra and sterility are complications of untreated cases.
5. Females might be asymptomatic unless the infection spreads (see pelvic inflammatory disease).
6. Gonorrheal endocarditis, gonorrheal meningitis, and gonorrheal arthritis are complications that can affect both sexes if gonorrheal infections are untreated.
7. Ophthalmia neonatorum is an eye infection acquired by infants during passage through the birth canal of an infected mother.
8. Gonorrhea is diagnosed by Gram stain, ELISA, or DNA probe.

Nongonococcal Urethritis (NGU)

1. NGU, or nonspecific urethritis (NSU), is any inflammation of the urethra not caused by *N. gonorrhoeae*.
2. Most cases of NGU are caused by *Chlamydia trachomatis*.
3. *C. trachomatis* infection is the most common STD.
4. Symptoms of NGU are often mild or lacking, although salpingitis and sterility may occur.
5. *C. trachomatis* can be transmitted to newborns' eyes at birth.
6. Diagnosis is based on the detection of chlamydial DNA in urine.
7. *Ureaplasma urealyticum* and *Mycoplasma hominis* also cause NGU.

Pelvic Inflammatory Disease

1. Extensive bacterial infection of the female pelvic organs, especially of the reproductive system, is called pelvic inflammatory disease (PID).
2. PID is caused by *N. gonorrhoeae, Chlamydia,* and other bacteria that gain access to the uterine tubes. Infection of the uterine tubes is called salpingitis.
3. PID can result in blockage of the uterine tubes and sterility.

Syphilis

1. Syphilis is caused by *Treponema pallidum,* a spirochete that has not been cultured in vitro. Laboratory cultures are grown in cell cultures.
2. *T. pallidum* is transmitted by direct contact and can invade intact mucous membranes or penetrate through breaks in the skin.
3. The primary lesion is a small, hard-based chancre at the site of infection. The bacteria then invade the blood and lymphatic system, and the chancre spontaneously heals.

4. The appearance of a widely disseminated rash on the skin and mucous membranes marks the secondary stage.

5. The patient enters a latent period after the secondary lesions spontaneously heal.

6. At least 10 years after the secondary lesion, tertiary lesions called gummas can appear on many organs.

7. Congenital syphilis, resulting from *T. pallidum* crossing the placenta during the latent period, can cause neurological damage in the newborn.

8. *T. pallidum* is identifiable through darkfield microscopy of fluid from primary and seondary lesions.

9. Many serological tests, such as VDRL, RPR, and FTA–ABS, can be used to detect the presence of antibodies against *T. pallidum* during any stage of the disease.

Lymphogranuloma Venereum

1. *C. trachomatis* causes lymphogranuloma venereum, which is primarily a disease of tropical and subtropical regions.

2. The initial lesion appears on the genitals and heals without scarring.

3. The bacteria are spread in the lymph system and cause enlargement of the lymph nodes, obstruction of lymph vessels, and swelling of the external genitals.

4. The bacteria are isolated and identified from pus taken from infected lymph nodes.

Chancroid (Soft Chancre)

1. Chancroid, a swollen, painful ulcer on the mucous membranes of the genitals or mouth, is caused by *Haemophilus ducreyi*.

Gardnerella Vaginosis

1. Vaginosis is an infection without inflammation cause by *Gardnerella vaginalis*.

2. Diagnosis of *G. vaginalis* is based on increased vaginal pH, fishy odor, and the presence of clue cells.

Viral Diseases of the Reproductive System

Genital Herpes

1. Herpes simplex virus type 2 (HSV-2) causes genital herpes.

2. Symptoms of the infection are painful urination, genital irritation, and fluid-filled vesicles.

3. Neonatal herpes is contracted during fetal development or birth. It can result in neurological damage or infant fatalities.

4. The virus might enter a latent stage in nerve cells. Vesicles reappear following trauma and hormonal changes.

5. Genital herpes is associated with cervical cancer.

6. The drug acyclovir has proven effective in treating the symptoms of genital herpes, but it does not cure the disease.

Genital Warts

1. Papillomaviruses cause warts.

2. The papillomaviruses that cause genital warts have been associated with cancer of the cervix or penis.

AIDS
1. AIDS is a disease of the immune system that can be sexually transmitted (see Chapter 19).

Fungal Disease of the Reproductive System
Candidiasis
1. *Candida albicans* causes NGU in males and vulvovaginal candidiasis in females.
2. Vulvovaginal candidiasis is characterized by lesions that produce itching and irritation.
3. Predisposing factors for candidiasis are pregnancy, diabetes, tumors, and broad-spectrum antibacterial chemotherapy.
4. Diagnosis is based on observation of the fungus and its isolation from lesions.

Protozoan Disease of the Reproductive System
Trichomoniasis
1. *Trichomonas vaginalis* causes trichomoniasis when the pH of the vagina increases.
2. Diagnosis is based on observation of the protozoan in purulent discharges from the site of infection.

THE LOOP

For a taxonomic approach, pages can be assigned as follows:

Bacterial diseases	pp. 723–732
Viral diseases	pp. 732–734
Fungal diseases	pp. 734–735
Protozoan diseases	p. 735

AIDS is mentioned as a sexually transmitted disease, and the discussions on pp. 392–393 (in Chapter 13) and pp. 535–546 (in Chapter 19) are cross-referenced.

Answers

Review

1. Organs of the upper urinary tract are sterile. The resident microbiota of the urethra are *Streptococcus, Bacteroides, Mycobacterium, Neisseria,* and some enterobacteria.

2. Normal microbiota of the male genital system is the same as that of the urinary tract. During reproductive years, lactobacilli predominate in the vagina.

3. Urinary tract infections may be transmitted by improper personal hygiene and contamination during medical procedures. They are often caused by opportunistic pathogens.

4. The proximity of the anus to the urethra and the relatively short length of the urethra can allow contamination of the urinary bladder in females. Predisposing factors for cystitis in females are gastrointestinal infections and vaginal and urinary tract infections.

5. *Escherichia coli* causes about 75% of the cases. From lower urinary tract or systemic infections.

6. | Disease | Symptoms | Diagnosis |
    | --- | --- | --- |
    | *Gardnerella* | Fishy odor | Odor, pH, clue cells |
    | Gonorrhea | Painful urination | Isolation of *Neisseria* |
    | Syphilis | Chancre | FTA–ABS |
    | PID | Abdominal pain | Culture of pathogen |
    | NGU | Urethritis | Absence of *Neisseria* |
    | LGV | Lesion, lymph node enlargement | Observation of *Chlamydia* in cells |
    | Chancroid | Swollen ulcer | Isolation of *Haemophilus* |

7. Transmission—Water; enters via wounds.
    Activities—Water contact; contact with animals or rodent-infested places.
    Etiology—*Leptospira interrogans.*

8. Symptoms: Burning sensation, vesicles, painful urination.
    Etiology: Herpes simplex type 2 (sometimes type 1). When the lesions are not present, the virus is latent and noncommunicable.

9. *Candida albicans*: Severe itching; thick, yellow, cheesy discharge.
    Trichomonas vaginalis: Profuse yellow discharge with disagreeable odor.

10. | Disease | Prevention of Congenital Disease |
    | --- | --- |
    | Gonorrhea | Treatment of newborn's eyes |
    | Syphilis | Prevention and treatment of mother's disease |
    | NGU | Treatment of newborn's eyes |
    | Genital herpes | Cesarean delivery during active infection |

Critical Thinking

1. *T. pallidum pertenue* can survive on skin in the tropics because of the constant warm temperature and high humidity. In temperate regions, this type of infection (yaws) might not survive because of cooler, drier air. If *T. pallidum pertenue* successfully invaded the body, warmth and moisture are available. Sexual contact is a method of transmission that provides constant protection for the microorganisms.

2. Eliminates normal microbiota. Changes (increases) pH.

3. Nystatin will inhibit yeast without affecting bacteria. Chocolate agar and increased CO_2 provide enriched conditions but are not selective.

4. Aerobic   *Leptospira*
    Anaerobic   *Treponema*
    Coccus   *Neisseria*
    Requires X factor *Haemophilus*
    Gram-positive wall *Gardnerella*
    Parasite   *Chlamydia*
    Urease-positive   *Ureaplasma*
    Urease-negative *Mycoplasma*
    Fungus   *Candida*
    Protozoan   *Trichomonas*
    No organism   Herpes simplex

Clinical Applications

1. Her disease appears to be meningitis; however, the gram-negative cocci in the cervical culture indicate this is disseminated gonococcal infection. The *N. gonorrhoeae* was sexually transmitted.

2. Pathogen: *Neisseria gonorrhoeae*
    Treatment: Cefoxitin

    The isolates all had the same antibiotic sensitivities.

3. The diagnosis of syphilis was made on November 8. Both the woman and the infant began syphilis therapy that day. Information on the baby's father was obtained in retrospect from a correctional facility. His condition was not diagnosed or treated until after his wife was diagnosed and interviewed for contacts.

    Darkfield examination of the lesion and STS were not done on June 6 or July 1. Copies of the positive RPR results should have been forwarded to the local STD control office to ensure treatment and epidemiologic case management. The regular laboratory clerk was on leave at the time, and her replacement inadvertently forwarded all copies of the laboratory reports to the hospital. Thus, the STD Control Program was unaware that the serologic results were reactive. The hospital's Infection Control Nurse, who was responsible for reviewing STS results, also was away on leave; both patients' results were filed without being brought to the attention of their attending physicians.

Learning with Technology

1. *Escherichia coli*
2. *Proteus vulgaris*

Case History: An Epidemic

Background

A pregnant 18-year-old woman came to the Ford County urgent-care clinic with a low-grade fever, malaise, and headache. She was sent home with a diagnosis of influenza. She again sought treatment 7 days later with a macular rash on her trunk, arms, hands, and feet. Further questioning of the patient when serology results were known revealed that 1 month previously, she had a painless ulcer on her vagina that healed spontaneously.

The same day, patient #2 sought medical treatment for a penile ulcer.

In a routine examination, patient #3, a pregnant female, had positive serologic tests for this disease but was asymptomatic.

Patient #4 was tested because of her sexual contact with patient #2. She had no symptoms and a positive serologic test.

Patient #5, a contact of patients #3 and #6, was also serologically positive. He frequently traveled to a neighboring county, which reported a 290% increase in this disease over the preceding year.

Patient #6, a female, had a rash and also tested positive. Patients 1 and 2 were in drug-abuse rehabilitation; these two were the only two who reported use of crack cocaine.

Questions

1. What bacterial diseases can cause rashes?
2. What serologic tests are used to diagnose these infections?
3. What is the disease? How did six residents of Ford County get this disease?
4. What are the consequences of not treating this infection?

The Solution

1. *Rickettsia rickettsii* (spotted fever), *Salmonella typhi*, *Borrelia burgdorferi*, *Staphylococcus aureus* (toxic shock syndrome), *Streptococcus pyogenes* (scarlet fever), and *Treponema pallidum*.
2. VDRL, rapid plasma reagin, and FTA–ABS for syphilis; agglutination tests for Lyme disease, *S. pyogenes*, and spotted fever. *S. aureus* is usually diagnosed from cultures (gram-positive, coagulase positive).
3. Syphilis. Patient #5 apparently brought the disease to Ford County and gave it to patients #3 and #6 during sexual intercourse; they could have given it to the others.
4. Tertiary syphilis may develop in the adults; the fetus can contract congenital syphilis.

Chapter 27   *Environmental Microbiology*

NEW IN THIS EDITION

- Revised and expanded discussions of biofilms and bioremediation.

CHAPTER SUMMARY

Metabolic Diversity

1. Microorganisms live in a wide variety of habitats because of their metabolic diversity, their ability to use a variety of carbon and energy sources and grow under different physical conditions.

Habitat Variety

1. Extremophiles live in extreme conditions of temperature, acidity, alkalinity, or salinity.

Symbiosis

1. Symbiosis is a relationship between two different organisms or populations.
2. Parasitism is a type of symbiosis in which one organism gets its nutrients and reproductive capability from another organism.
3. Mutualism is a type of symbiosis in which both partners benefit.
4. Symbiotic fungi called mycorrhizae live in and on plant roots; they increase the surface area and nutrient absorption of the plant.

Soil Microbiology and Biogeochemical Cycles

1. In biogeochemical cycles, certain chemical elements are recycled.
2. Microorganisms in the soil decompose organic matter and transform carbon-, nitrogen-, and sulfur-containing compounds into usable forms.
3. Microbes are essential to the continuation of biogeochemical cycles.
4. Elements are oxidized and reduced by microorganisms during these cycles.

The Carbon Cycle

1. Carbon dioxide is incorporated, or fixed, into organic compounds by photoautotrophs and chemoautotrophs.
2. These organic compounds provide nutrients for chemoheterotrophs.
3. Chemoheterotrophs release CO_2 that is then used by photoautotrophs.
4. Carbon is removed from the cycle when it is in $CaCO_3$ and fossil fuels.

The Nitrogen Cycle

1. Microorganisms decompose proteins from dead cells and release amino acids.
2. Ammonia is liberated by microbial ammonification of the amino acids.
3. The nitrogen in ammonia is oxidized to produce nitrates for energy by nitrifying bacteria.
4. Denitrifying bacteria reduce nitrates to molecular nitrogen (N_2).
5. N_2 is converted into ammonia by nitrogen-fixing bacteria.
6. Nitrogen-fixing bacteria include free-living genera such as *Azotobacter*, cyanobacteria, and the symbiotic bacteria *Rhizobium* and *Frankia*.
7. Ammonium and nitrate are used by bacteria and plants to synthesize amino acids that are assembled into proteins.

The Sulfur Cycle

1. Hydrogen sulfide (H_2S) is used by autotrophic bacteria; the sulfur is oxidized to form S^0 or SO_4^{2-}.
2. Winogradsky discovered that *Beggiatoa* bacteria oxidize sulfur (H_2S and S^0) for energy.
3. Plants and other microorganisms can reduce SO_4^{2-} to make certain amino acids; these amino acids are in turn used by animals.

4. H_2S is released by decay or dissimilation of these amino acids.
5. Sulfur dioxide produced by combustion of fossil fuels combines with water to form sulfurous acid.

The Phosphorus Cycle

1. Phosphorus (as PO_4^{3-}) is found in rocks and bird guano.
2. When solubilized by microbial acids, the PO_4^{3-} is available for plants and microorganisms.
3. Endolithic bacteria live in solid rock; these autotrophic bacteria use hydrogen as an energy source.

The Degradation of Synthetic Chemicals in Soil and Water

1. Many synthetic chemicals, such as pesticides, are resistant to degradation by microbes.

Bioremediation
1. The use of microorganisms to remove pollutants is called bioremediation.
2. The growth of oil-degrading bacteria can be enhanced by the addition of nitrogen and phosphorous fertilizer.

Solid Municipal Waste
1. Municipal landfills prevent decomposition of solid wastes because they are dry and anaerobic.
2. In some landfills, methane produced by methanogens can be recovered for an energy source.
3. Composting can be used to promote biodegradation of organic matter.

Aquatic Microbiology and Sewage Treatment

Biofilms
1. Microbes adhere to surfaces and accumulate as biofilms on solid surfaces in contact with water.

Aquatic Microorganisms

1. The study of microorganisms and their activities in natural waters is called aquatic microbiology.
2. Natural waters include lakes, ponds, streams, rivers, estuaries, and oceans.
3. The concentration of bacteria in water is proportional to the amount of organic material in the water.
4. Most aquatic bacteria tend to grow on surfaces rather than in a free-floating state.

Freshwater Microbiota
1. Numbers and locations of freshwater microbiota depend on the availability of oxygen and light.
2. Photosynthetic algae are the primary producers of a lake; they are found in the limnetic zone.
3. Pseudomonads, *Cytophaga*, *Caulobacter*, and *Hyphomicrobium* are found in the limnetic zone, where oxygen is abundant.

4. Microbes in stagnant water use available oxygen and can cause odors and the death of fish.

5. The amount of dissolved oxygen is increased by wave action.

6. Purple and green sulfur bacteria are found in the profundal zone, which contains light and H_2S but not oxygen.

7. *Desulfovibrio* reduces SO_4^{2-} to H_2S in benthic mud.

8. Methane-producing bacteria are also found in the benthic zone.

Seawater Microbiota
1. Phytoplankton, consisting mainly of diatoms, are the primary producers of the open ocean.

2. Some algae and bacteria are bioluminescent. They possess the enzyme luciferase, which can emit light.

The Role of Microorganisms in Water Quality

Water Pollution
1. Microorganisms are filtered from water that percolates into groundwater supplies.

2. Some pathogenic microorganisms are transmitted to humans in drinking and recreational waters.

3. Resistant chemical pollutants may be concentrated in animals in an aquatic food chain.

4. Mercury is metabolized by certain bacteria into a soluble compound that is concentrated in animals.

5. Nutrients such as phosphates cause algal blooms, which can lead to eutrophication of aquatic ecosystems.

6. Eutrophication, meaning well nourished, is the result of the addition of pollutants or natural nutrients.

7. *T. ferrooxidans* produces sulfuric acid at coal-mining sites.

Water Purity Tests
1. Tests for the bacteriological quality of water are based on the presence of indicator organisms, the most common of which are coliforms.

2. Coliforms are aerobic or facultatively anaerobic, gram-negative, non–endospore-forming rods that ferment lactose with the production of acid and gas within 48 hours of being placed in a medium at 35°C.

3. Fecal coliforms, predominantly *E. coli*, are used to indicate the presence of human feces.

Water Treatment
1. Drinking water is held in a holding reservoir long enough that suspended matter settles.

2. Flocculation treatment uses a chemical such as alum to coalesce and then settle colloidal material.

3. Filtration removes protozoan cysts and other microorganisms.

4. Drinking water is disinfected with chlorine to kill remaining pathogenic bacteria.

Sewage Treatment
1. Domestic waste water is called sewage; it includes household water, toilet wastes, industrial wastes, and rainwater.

Primary Sewage Treatment
1. Primary sewage treatment is the removal of solid matter called sludge.
2. Biological activity is not very important in primary treatment.

Biochemical Oxygen Demand
1. Biochemical oxygen demand (BOD) is a measure of the biologically degradable organic matter in water.
2. Primary treatment removes about 25–35% of the BOD of sewage.
3. BOD is determined by measuring the amount of oxygen bacteria require to degrade the organic matter.

Secondary Sewage Treatment
1. Secondary sewage treatment is the biological degradation of organic matter in sewage after primary treatment.
2. Activated sludge systems, trickling filters, and rotating biological contactors are methods of secondary treatment.
3. Microorganisms degrade the organic matter aerobically.
4. Secondary treatment removes up to 95% of the BOD.

Disinfection and Release
1. Treated sewage is disinfected, usually by chlorination, before discharge onto land or into water.

Sludge Digestion
1. Sludge is placed in an anaerobic sludge digester; bacteria degrade organic matter and produce simpler organic compounds, methane, and CO_2.
2. The methane produced in the digester is used to heat the digester and operate other equipment.
3. Excess sludge is periodically removed from the digester, dried, and disposed of (as landfill or as soil conditioner) or incinerated.

Septic Tanks
1. Septic tanks can be used in rural areas to provide primary treatment of sewage.
2. They require a large leaching field for the effluent.

Oxidation Ponds
1. Small communities can use oxidation ponds for secondary treatment.
2. These require a large area in which to build an artificial lake.

Tertiary Sewage Treatment
1. Tertiary sewage treatment uses physical filtration and chemical precipitation to remove all the BOD, nitrogen, and phosphorus from water.
2. Tertiary treatment provides drinkable water, whereas secondary treatment provides water usable only for irrigation.

THE LOOP

Bioremediation, "Bacterial Banqueters Attend Oil Spill"   p. 35
Symbiosis   pp. 161, 407–409

Answers

Review

1. Extremophiles include thermophiles such as *Thermus aquaticus*, acidophiles such as *Thiobacillus*, halophiles such as *Halobacterium*, and endoliths.

2. The koala should have an organ housing a large population of cellulose-degrading microorganisms.

3. *Penicillium* might make penicillin to reduce competition from faster-growing bacteria.

4.

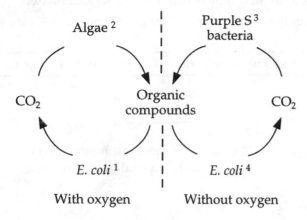

1—Any chemoheterotroph using aerobic respiration
2—Any aerobic autotroph
3—Any anaerobic autotroph
4—Any chemoheterotroph producing CO_2 via fermentation

5. Amino acids; SO_4^{2-}; plants and bacteria; H_2S; carbohydrates; S^0.

6. Phosphorus must be available for all organisms.

7.

Process	Reactions	Microorganisms
Ammonification	$-NH_2 \rightarrow NH_3$	Proteolytic bacteria
Nitrification	$NH_3 \rightarrow NO_2^-$	*Nitrosomonas*
	$NO_2 \rightarrow NO_3^-$	*Nitrobacter*
Denitrification	$NO_3 \rightarrow N_2$	*Bacillus*
N fixation	$N_2 \rightarrow NH_3$	*Rhizobium*

8. Cyanobacteria: With fungi, cyanobacteria act as the photoautotrophic partner in a lichen; they may also fix nitrogen in the lichen. With *Azolla*, they fix nitrogen.

   Mycorrhizae: Fungi that grow in and on the roots of higher plants; increase absorption of nutrients.

   Rhizobium: In root nodules of legumes; fix nitrogen.

   Frankia: In root nodules of alders, roses, and other plants; fix nitrogen.

9. Settling
   Flocculation treatment
   Sand filtration (or activated charcoal filtration)
   Chlorination
   The amount of treatment prior to chlorination depends on the amount of inorganic and organic matter in the water.

10. A coliform count is used to determine the bacteriologic quality of water; that is, the presence of human pathogens or evidence of fecal contamination.

11.
_b_	Leaching field
_a_	Removal of solids
_b_	Biological degradation
_b_	Activated sludge
_c_	Chemical precipitation of phosphorus
_b_	Trickling filter
_c_	Results in drinking water

12. Activated sludge is an aerobic process that can result in complete oxidation of organic matter.

13. Both require large areas of land and can result in the pollution of surface or groundwater if they are overloaded.

14.

	BOD	Rate of Eutrophication	Dissolved Oxygen
Untreated	3+	3+	+
Primary	2+	2+	2+
Secondary	+	+	3+

Accumulation of BOD and loss of dissolved oxygen would be much less in a fast-moving river. Continual aeration caused by the river's movement would result in rapid oxidation of organic matter.

15. Biodegradation of sewage, herbicides, oil, or PCBs.

Critical Thinking

1. The straight chain is readily degraded by beta oxidation (refer to "Lipid Catabolism," page 124).

2.

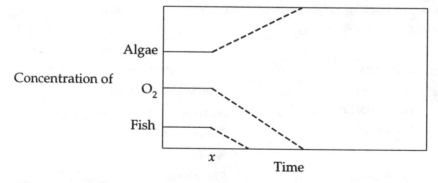

Clinical Applications

1. The source of his recurrent infection was a biofilm on his pacemaker. The infection was cured with antibiotics because freed cells were sensitive to antibiotics; however, antibiotics couldn't penetrate the biofilm to eliminate the source.

2. Nitrates, phosphates, oxygen, or water may not be present in sufficient amounts for bacterial growth. The naturally occurring bacteria may be able to degrade the hydrocarbons.

3. Sewage infiltrated the municipal water supply via a pipe that was damaged by the flooding. Flow through the damaged pipe must be stopped; drinking water must be hyperchlorinated; infected individuals should be treated.

Learning with Technology

1.   *Vibrio cholerae*

2.   *Arthrobacter globiformis*

Case History: An Epidemic of Food Poisoning

Background

An outbreak of food poisoning occurred the last week of June, affecting two-thirds of the 1700 persons who attended a dinner in Port Allen, Louisiana. A questionnaire to obtain information concerning the illness was administered to 122 persons. Of this sample, 82 (67.2%) reported illness. Physicians treated 32 patients (26.2%); 9 (7.4%) required hospitalization.

Laboratory analysis yielded positive cultures of the etiological agents from the leftover shrimp, potatoes, corn, and hogshead cheese, and from 7 of 15 stool samples from patients. The person who collected the food stored all of it in one container.

Potatoes were purchased from a farmer. They had been stored in a root cellar and were transported loose in a pickup truck. The morning of the dinner the potatoes were washed, peeled, and boiled until tender. The pots were covered with aluminum foil and transported unrefrigerated 50 miles to the dinner.

Corn was purchased from a produce market. The night before the dinner, the corn was shucked and kept in the wooden boxes in which it had been purchased. The day of the dinner, the corn was boiled. For transport, it was wrapped in aluminum foil.

Hogshead cheese was purchased the morning of the dinner along with the salt meat. The headcheese and salt meat were kept refrigerated by the delicatessen. After purchase, they were stored unrefrigerated until dinner. (Read labels in the deli section of your grocery store to ascertain the ingredients in headcheese.)

Raw shrimp was purchased at one location and shipped to a second location in standard wooden seafood boxes. It was boiled on the morning of the dinner and placed back into the boxes. After being covered with aluminum foil, it was transported 40 miles in an unrefrigerated truck to the dinner. An inspection of the wholesale seafood establishment was undertaken six days later. The investigation revealed that the shrimp had been boiled in 300-lb. batches in the following manner: A batch was placed in a container until the water came to a rolling boil; then the gas was turned off, and the shrimp kept in the hot water for 15 minutes.

Questions

1.   On one page, identify the etiological agent of this outbreak of food poisoning.

2.   Was it food infection or intoxication?

3.   How did the food get contaminated, and which item was contaminated?

4.   Briefly explain how you arrived at your conclusions. Why did you eliminate the other major causes of food poisoning?

Data

Case	Age	Sex	D	C	W	N	C	H	F	V	1	2	3	4	5	6	7	Day	Hour	Duration (Days)
			Symptoms								Foods Eaten							Onset		
1	13	F	+	+	+	+	+	+	+	+	+	+	+	+	+	+	+	S	11p	1
2	21	F	+	+	+	+	+	+	+	+	+		+			+	+	M	12n	4
3	64	F	+	+	+	+	+		+		+	+	+	+	+	+	+	M	1p	3
4	32	M	+	+	+	+	+	+	+	+	+	+	+	+	+		+	M	5p	2
5	41	F	+	+	+	+	+	+		+	+	+	+		+		+	T	8a	4
6	42	M	+	+	+	+	+	+	+	+	+	+	+	+	+		+	M	8a	6
7	30	M	+	+	+	+					+		+	+		+	+	S	12m	7
8	50	M						+				+	+		+	+	+	W	8a	
9	55	M	+	+	+	+		+	+		+	+	+		+		+	S	11p	1
10	60	F	+								+	+	+	+	+		+	T	7p	3
11	61	M	+		+		+	+	+		+	+	+	+	+			M	12n	4
12	19	M	+	+			+	+	+		+		+			+	+	S	10p	4
13	22	F									+	+		+	+	+				
14	70	F	+	+	+		+	+	+		+	+	+	+	+		+	M	1p	6
15	69	F	+	+	+		+	+	+		+	+	+	+	+		+	M	9a	6
16	68	M	+	+	+		+					+	+	+	+	+	+	M	12n	5
17	34	F	+	+	+		+				+	+	+		+		+	M	1p	5
18	40	F			+	+	+					+	+		+	+	+	M	pm	7
19	41	F	+	+	+	+	+	+	+		+	+		+	+	+	+	S	pm	5
20	59	M										+		+	+	+				
21	78	M	+	+	+	+	+	+	+		+			+		+	+	T	am	5
22	73	F	+	+	+		+				+		+	+			+	M	3p	4
23	46	F	+	+	+		+				+		+		+		+	M	4a	3
24	57	M	+	+	+						+	+	+	+	+		+	M	9a	2
25	34	M	+	+	+	+					+		+			+	+	M	9p	8
26	72	M	+	+	+						+		+	+		+		T	1a	8+
27	32	M	+	+	+	+	+	+	+	+	+		+				+	T	am	7
28	61	F	+	+	+	+	+	+	+	+	+		+		+			S	12m	1
29	36	F	+	+	+	+	+	+	+	+	+		+	+		+		M	1a	2
30	59	M	+	+	+	+	+	+	+		+		+		+		+	M	pm	2
31	74	M	+	+	+	+	+	+	+		+		+		+		+	M	10a	4
32	50	M										+	+	+		+	+	+			
33	54	M	+	+		+						+	+		+	+	+	M	am	5
34	?	M	+	+	+	+					+		+	+		+		T	am	6
35	39	M	+	+	+	+					+	+	+		+		+	S	pm	8+
36	40	M	+	+	+	+		+			+		+	+		+		M	10a	1
37	78	F	+	+	+	+						+	+	+	+		+	M	9a	8+
38	57	F	+		+	+						+	+	+		+	+	M	8a	7
39	14	M	+		+	+						+	+		+	+	+	M	pm	4
40	15	M		+	+	+						+	+		+	+		M	10a	2

Case	Age	Sex	D	C	W	N	C	H	F	V	1	2	3	4	5	6	7	Day	Hour	Duration (Days)
41	60	F	+	+	+	+	+	+	+		+		+	+		+	+	M	11a	4
42	63	M	+	+	+	+	+	+	+		+		+		+	+		S	11p	5
43	81	F	+	+	+	+	+	+	+		+		+	+	+	+	+	T	11a	6
44	57	F	+	+	+	+	+	+	+		+		+			+	+	T	12n	7
45	32	M	+	+	+	+	+	+	+		+		+	+				M	12n	6
46	?	F	+	+	+	+	+	+	+		+			+	+	+	+	M	8a	7
47	20	F	+	+	+	+	+	+	+		+			+	+		+	M	9a	2
48	25	M	+	+	+	+						+				+		S	10p	2
49	32	F	+	+	+	+					+	+		+		+	+	S	12n	3
50	38	M	+	+	+	+					+	+			+	+		M	10a	4
51	40	F	+	+	+	+					+	+	+	+		+	+	M	12n	5
52	45	M	+	+	+	+					+		+		+		+	T	12n	5
53	18	F	+	+	+	+					+		+	+		+	+	M	12n	3
54	10	M	+	+	+	+	+	+	+		+		+	+	+	+	+	M	11a	2
55	20	F	+	+	+	+	+	+	+		+		+	+				T	1a	2
56	60	F	+	+	+	+	+	+	+		+		+		+	+	+	M	11a	3
57	63	M	+	+	+	+						+			+	+	+	T	11a	4
58	57	M	+	+	+						+		+		+			S	pm	4
59	46	F									+		+	+		+				
60	45	M	+	+	+						+		+		+	+		M	am	3
61	60	M	+	+	+		+	+	+		+	+		+	+	+	+	M	8a	1
62	14	M	+	+	+						+	+	+		+	+	+	M	6p	5
63	39	M	+	+	+								+	+	+	+	+	M	5a	3
64	48	M	+	+	+						+	+	+		+	+	+	T	1a	2
65	33	F	+	+	+	+	+	+	+	+	+	+	+	+		+	+	T	11p	3
66	50	F	+	+	+						+	+	+		+	+		M	3p	7
67	17	F	+	+	+	+	+	+	+	+	+	+	+	+	+		+	M	2a	3
68	59	M	+	+	+	+	+	+	+	+	+		+		+		+	M	11a	5
69	46	F	+	+	+	+	+	+	+	+	+		+	+		+	+	S	10p	6
70	34	F	+	+	+						+		+	+	+		+	T	3a	1
71	35	M	+	+	+						+		+		+		+	T	4a	2
72	50	F	+	+	+								+	+		+	+	M	9a	8
73	56	M	+	+	+	+					+		+			+		M	9p	8
74	45	M	+	+	+	+					+		+		+		+	M	1a	5
75	?	F	+	+	+	+	+	+	+	+	+		+					T	2a	7
76	63	M	+	+	+	+	+	+	+	+	+				+	+	+	S	12m	3
77	54	F	+	+	+								+	+	+		+	S	11p	4
78	35	F	+	+	+	+	+	+	+		+		+		+		+	M	12n	4
79	38	M	+	+	+	+	+	+	+		+			+			+	M	4a	5
80	33	M											+	+		+				
81	32	F									+	+		+		+				
82	47	F	+	+	+	+	+	+	+		+		+	+		+	+	M	4a	7
83	49	M				+					+		+		+		+	S	9p	
84	38	F	+	+	+	+	+	+	+		+		+	+			+	M	3p	1
85	42	F	+	+	+	+	+	+	+		+	+			+		+	M	5a	5

Case	Age	Sex	D	C	W	N	C	H	F	V	1	2	3	4	5	6	7	Day	Hour	Duration (Days)
86	45	M	+	+	+	+	+	+	+		+			+			+	M	2a	5
87	49	M	+	+	+	+	+	+	+		+			+	+	+		T	3a	6
88	71	F	+	+	+	+	+	+	+	+	+				+	+	+	M	10a	3
89	75	F	+	+	+	+	+	+	+		+				+		+	M	2p	8
90	39	M	+	+	+	+	+	+	+		+			+			+	S	11p	2
91	70	F	+	+	+										+	+	+	M	5a	4
92	48	F	+	+	+	+	+	+	+		+			+		+	+	M	6p	6
93	21	F	+	+		+					+	+		+	+		+	M	2a	5
94	24	F	+			+					+	+	+	+	+	+	+	M	11p	7
95	25	M	+		+	+					+				+			M	9a	5
96	40	M	+	+	+	+	+	+			+	+		+			+	M	8a	2
97	36	F		+	+	+	+				+						+	T	2a	4
98	58	M	+	+	+						+	+	+	+	+	+	+	S	10p	1
99	?	F	+	+	+	+	+	+	+		+	+		+		+	+	T	1a	7
100	41	F	+	+	+	+	+	+	+		+	+	+	+			+	M	3p	5
101	55	F	+	+	+	+	+	+	+		+	+	+		+		+	S	9p	3
102	42	M	+	+	+	+	+	+	+		+	+	+			+	+	T	2p	4
103	27	M	+	+	+	+	+	+	+		+		+	+		+		M	3a	3
104	25	F	+	+	+	+	+	+	+		+			+			+	M	11a	6
105	20	M												+	+	+	+			
106	18	F									+		+	+	+					
107	43	M	+	+	+	+						+		+			+	M	2a	5
108	33	F	+	+							+		+		+	+	+	M	5p	6
109	65	M										+		+		+				
110	14	F	+	+	+	+	+	+	+		+	+				+	+	M	6a	1
111	13	M	+	+	+	+						+			+	+		M	3p	7
112	50	F	+	+	+	+	+	+	+		+	+	+	+		+	+	S	1p	7
113	?	F	+	+	+	+						+		+			+	M	2a	5
114	12	F	+	+	+	+	+	+	+		+			+		+	+	T	3a	4
115	18	M	+	+	+	+						+	+	+				S	10p	6
116	70	M	+	+	+	+	+	+	+	+	+			+		+		T	1a	3
117	33	M	+	+	+	+						+			+	+		M	11p	2
118	60	F	+	+	+	+	+				+	+	+			+	+	M	6a	5
119	50	M	+	+	+	+	+				+			+			+	T	2a	5
120	48	F	+	+	+	+	+		+		+				+	+	+	M	3p	7
121	13	F	+	+	+	+	+									+		M	4a	4
122	19	F	+	+	+	+	+		+		+	+		+		+	+	S	10p	6

Legend:

Symptoms: D = Diarrhea, C = Cramps, W = Weakness, N = Nausea, C = Chills, H = Headache, F = Fever, V = Vomiting.

Foods Eaten: 1-Boiled shrimp, 2-Headcheese, 3-Boiled potatoes, 4-Boiled corn, 5-Boiled salt meat, 6-Bread and butter, 7-Watermelon.

Dinner was at 6 P.M. Sunday.

Hints

1. Make a summary table of the persons not ill.

2. Make a table of the onset of symptoms following the dinner.

The Solution

1. *Vibrio parahaemolyticus.*

2. Infection.

3. The shrimp became contaminated while feeding in polluted coastal waters.

4. Students should find a correlation between eating the shrimp and illness. Unrefrigerated transport of the shrimp allowed bacterial growth, and low-temperature cooking did not kill the bacteria.

Chapter 28 · *Applied and Industrial Microbiology*

Learning Objectives

1. Describe thermophilic anaerobic spoilage and flat sour spoilage by mesophilic bacteria.
2. Compare and contrast food preservation by industrial food canning, aseptic packaging, and radiation.
3. Name four beneficial activities of microorganisms in food production.
4. Define industrial fermentation and bioreactor.
5. Differentiate between primary and secondary metabolites.
6. Describe the role of microorganisms in the production of industrial chemicals and pharmaceuticals.
7. Define bioconversion, and list its advantages.

NEW IN THIS EDITION

- Expanded discussion of irradiation of food.
- Revised discussion of microbiological processes in (copper) ore extraction.

CHAPTER SUMMARY

Food Microbiology

1. The earliest methods of preserving foods were drying, the addition of salt or sugar, and fermentation.

Industrial Food Canning

1. Commercial sterilization of food is accomplished by steam under pressure in a retort.
2. Commercial sterilization heats canned foods to the minimum temperature necessary to destroy *Clostridium botulinum* endospores while minimizing alteration of the food.
3. The commercial sterilization process uses sufficient heat to reduce a population of *C. botulinum* by 12 logarithmic cycles (12D treatment).
4. Endospores of thermophiles can survive commercial sterilization.
5. Canned foods stored above 45°C can be spoiled by thermophilic anaerobes.
6. Thermophilic anaerobic spoilage is sometimes accompanied by gas production; if no gas is formed, the spoilage is called flat sour spoilage.
7. Spoilage by mesophilic bacteria is usually from improper heating procedures or leakage.

8. Acidic foods can be preserved by heat of 100°C because microorganisms that survive are not capable of growth in a low pH.

9. *Byssochlamys*, *Aspergillus*, and *Bacillus coagulans* are acid-tolerant and heat-resistant microbes that can spoil acidic foods.

Aseptic Packaging

1. Presterilized materials are assembled into packages and aseptically filled with heat-sterilized liquid foods.

Radiation and Industrial Food Preservation

1. Gamma radiation can be used to sterilize food, kill insects and parasitic worms, and prevent the sprouting of fruits and vegetables.

The Role of Microorganisms in Food Production

Cheese

1. The milk protein casein curdles because of the action by lactic acid bacteria or the enzyme rennin or chymosin.

2. Cheese is the curd separated from the liquid portion of milk, called whey.

3. Hard cheeses are produced by lactic acid bacteria growing in the interior of the curd.

4. The growth of microorganisms in cheeses is called ripening.

5. Semisoft cheeses are ripened by bacteria growing on the surface; soft cheeses are ripened by *Penicillium* growing on the surface.

Other Dairy Products

1. Old-fashioned buttermilk was produced by lactic acid bacteria growing during the butter-making process.

2. Commerical buttermilk is made by letting lactic acid bacteria grow in skim milk for 12 hours.

3. Sour cream, yogurt, kefir, and kumiss are produced by lactobacilli, streptococci, or yeasts growing in low-fat milk.

Nondairy Fermentations

1. Sugars in bread dough are fermented by yeast to ethanol and CO_2; the CO_2 causes the bread to rise.

2. Sauerkraut, pickles, olives, and soy sauce are the products of microbial fermentations.

Alcoholic Beverages and Vinegar

1. Carbohydrates obtained from grains, potatoes, or molasses are fermented by yeasts to produce ethanol in the production of beer, ale, and distilled spirits.

2. The sugars in fruits such as grapes are fermented by yeasts to produce wines.

3. In wine-making, lactic acid bacteria convert malic acid into lactic acid in malolactic fermentation.

4. *Acetobacter* and *Gluconobacter* oxidize ethanol in wine to acetic acid (vinegar).

Industrial Microbiology

1. Microorganisms produce alchohols and acetone that are used in industrial processes.
2. Industrial microbiology has been revolutionized by the ability of genetically engineered cells to make many new products.
3. Biotechnology is a way of making commercial products by using living organisms.

Fermentation Technology

1. The growth of cells on a large scale is called industrial fermentation.
2. Industrial fermentation is carried out in bioreactors, which control aeration, pH, and temperature.
3. Primary metabolites such as ethanol are formed as the cells grow (during the trophophase).
4. Secondary metabolites such as penicillin are produced during the stationary phase (idiophase).
5. Mutant strains that produce a desired product can be selected.

Immobilized Enzymes and Microorganisms
1. Enzymes or whole cells can be bound to solid spheres or fibers. When substrate passes over the surface, enzymatic reactions change the substrate to the desired product.
2. They are used to make paper, textiles, and leather and are environmentally safe.

Industrial Products

1. Most amino acids used in foods and medicine are produced by bacteria.
2. Microbial production of amino acids can be used to produce L-isomers; chemical production results in both D- and L-isomers.
3. Lysine and glutamic acid are produced by *Corynebacterium glutamicum*.
4. Citric acid, used in foods, is produced by *Aspergillus niger*.
5. Enzymes used in manufacturing foods, medicines, and other goods are produced by microbes.
6. Some vitamins used as food supplements are made by microorganisms.
7. Vaccines, antibiotics, and steroids are products of microbial growth.
8. The metabolic activities of *T. ferrooxidans* can be used to recover uranium and copper ores.
9. Yeasts are grown for wine- and bread-making; other microbes (*Rhizobium, Bradyrhizobium,* and *Bacillus thuringiensis*) are grown for agricultural use.

Alternative Energy Sources Using Microorganisms

1. Organic waste, called biomass, can be converted by microorganisms into alternative fuels, a process called bioconversion.
2. Fuels produced by microbial fermentation are methane and ethanol.

Industrial Microbiology and the Future

1. Genetic engineering will enhance the ability of industrial microbiology to produce medicines and other useful products.

THE LOOP ○━━━━▭

The topics in Chapter 28 can be studied along with general principles discussed in earlier chapters. These topics are cross-referenced to the principles in *Microbiology: An Introduction*.

Answers

Review

1. Industrial microbiology is the science of using microorganisms to produce products or accomplish a process. Industrial microbiology provides (1) chemicals such as antibiotics that would not otherwise be available, (2) processes to remove or detoxify pollutants, (3) fermented foods that have desirable flavors or enhanced shelf life, and (4) enzymes for manufacturing a variety of goods.

2. The goal of commercial sterilization is to eliminate spoilage and disease-causing organisms. The goal of hospital sterilization is complete sterilization.

3. The acid in the berries will prevent the growth of some microbes.

4. A presterilized package is aseptically filled with presterilized food.

5. Milk $\xrightarrow{\text{Lactic acid bacteria}}$ Curd   +   Whey

    $\hspace{5.9cm}$ Rennin

    $\hspace{5cm}\downarrow\hspace{1.5cm}\downarrow$

    $\hspace{4.7cm}$ Cheese $\hspace{0.7cm}$ Waste

    Hard cheese is ripened by lactic acid bacteria growing anaerobically in the interior of the curd. Soft cheese is ripened by molds growing aerobically on the outside of the curd.

6. Fruit juice $\xrightarrow{\text{Yeast}}$ Ethyl alcohol + CO_2

7. Nutrients must be dissolved in water; water is also needed for hydrolysis. Malt is the carbon and energy source that the yeast will ferment to make alcohol. Malt contains glucose and maltose from the action of amylases on starch in seeds (barley).

8. A primary metabolite is produced during trophophase; a secondary metabolite, during idiophase.

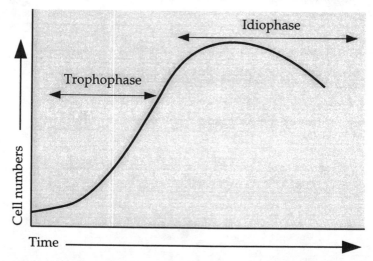

9. A bioreactor provides the following advantages over simple flask containers:
    *   Larger culture volumes can be grown.
    *   Process instrumentation for monitoring and controlling critical environmental conditions such as pH, temperature, dissolved oxygen, and aeration can be used.
    *   Sterilization and cleaning systems are designed in place.
    *   Aseptic sampling and harvest systems for in-process sampling exist.

- Improved aeration and mixing characteristics result in improved cell growth and high final cell densities.
- A high degree of automation is possible.
- Process reproducibility is improved.

10. (1) Enzymes don't produce hazardous wastes; (2) Enzymes work under reasonable conditions, e.g., they don't require high temperatures or acidity; (3) Eliminates the need to use petroleum in chemical syntheses of solvents such as alcohol and acetone; (4) Enzymes are biodegradable; (5) Enzymes are not toxic.

11. The production of ethyl alcohol from corn; or methane from sewage. Alcohols and hydrogen are produced by fermentation; methane is produced by anaerobic respiration.

Critical Thinking

1. Lactic acid bacteria. Lactic acid bacteria are found in milk and are responsible for fermenting it. Lactic acid bacteria, likewise, could contaminate any food made with milk and cause fermentation (e.g., sourdough bread). Some lactic acid bacteria are found on plants and are responsible for their fermentation (e.g., sauerkraut and the malolactic fermentation of wine).

2. See page 146. Amination of α-ketoglutaric acid will make glutamic acid.

3. Cellulases hydrolyze the fibers to soften the fabric. The cellulases are extracted from the fungus, *Trichoderma*. See page 251.

Clinical Applications

1. a. New medium added will neutralize pH. Some old medium must flow out as new medium is added, so some acid will leave.

   b. During stationary phase

   c. During log phase

2. The *E. coli* grew well in the cider at room temperature (25°C). Refrigeration and sorbate both slowed the growth but did not completely stop it or kill the bacteria. None of these treatments are sufficient to prevent transmission of *E. coli* by cider.

3. a. Production is best after logarithmic growth stops; lipids and oxygen are necessary. It is a secondary metabolite.

   b. Glucose is used first.

   c. The organic molecules supply carbon and energy; NaCl is used to maintain osmotic pressure; the elements N, S, K, and P are provided by the inorganic salts; and the KH_2PO_4 and Na_2HPO_4 are buffers.

   d. *Nocardia* are soil bacteria that resemble actinomycetes; they may produce short filaments and be acid-fast.

Learning with Technology

1. *Leuconostic lactis*

2. *Bacillus subtilis*

Case History: Making a Product

Background

You have isolated a bacterium that produces a bacteriocin against *Salmonella*. The bacterium grows well in Medium 1 but doesn't produce its bacteriocin. Testing one component at a time, you determine that iron interferes with bacteriocin production. When you try Medium 1 without iron, the bacteria don't grow. Bacteriocin production is best in Medium 2.

Medium 1		Medium 2	
Glucose	0.5%	Glycerol	3.5%
Citric acid	0.1%	Glucose	0.5%
20 amino acids, each	0.04%	Citric acid	0.1%
K_2HPO_4	0.15%	20 amino acids, each	0.04%
$MgSO_4$	0.12%	K_2HPO_4	0.15%
$FeSO_4$	$1.52 \times 10^{-5}\%$	$MgSO_4$	0.12%

Data

Medium	Generation time (min)	Bacteriocin (mg/l)
1	27	0.02
2	300	0.20

Questions

1. Design a procedure to maximize the biomass (total cells) and the yield of the bacteriocin.
2. What are bacteriocins?
3. How might a bacteriocin against *Salmonella* be used?

The Solution

1. Use a two-stage production. Grow the cells in Medium 1 until stationary phase, then transfer them to Medium 2 to maximize bacteriocin production.
2. Bacteriocins are toxins produced by bacteria that kill other bacteria.
3. Bacteriocins against *Salmonella* could be useful in food production or storage to prevent *Salmonella* infections.

Answers to Multiple-Choice Questions

Chapter 1	Chapter 2	Chapter 3	Chapter 4	Chapter 5	Chapter 6	Chapter 7
1. a	1. c	1. e	1. e	1. a	1. c	1. d
2. c	2. b	2. c	2. d	2. d	2. a	2. b
3. d	3. b	3. b	3. b	3. b	3. c	3. d
4. c	4. e	4. a	4. a	4. c	4. a	4. d
5. b	5. b	5. a	5. d	5. c	5. c	5. b
6. e	6. c	6. e	6. e	6. b	6. d	6. b
7. c	7. a	7. d	7. b	7. b	7. e	7. b
8. a	8. a	8. b	8. e	8. a	8. a	8. a
9. c	9. b	9. a	9. a	9. c	9. b	9. a
10. a	10. c	10. c	10. b	10. b	10. b	10. b

Chapter 8	Chapter 9	Chapter 10	Chapter 11	Chapter 12	Chapter 13	Chapter 14
1. c	1. b	1. b	1. d	1. c	1. e	1. a
2. d	2. b	2. e	2. b	2. b	2. c	2. b
3. c	3. b	3. d	3. e	3. b	3. b	3. a
4. d	4. b	4. b	4. a	4. a	4. c	4. d
5. c	5. c	5. e	5. b	5. d	5. b	5. b
6. b	6. d	6. a	6. c	6. b	6. e	6. c
7. a	7. c	7. a	7. e	7. a	7. c	7. d
8. c	8. b	8. e	8. b	8. c	8. d	8. a
9. d	9. e	9. a	9. b	9. a	9. d	9. c
10. a	10. a	10. b	10. a	10. d	10. c	10. b

Chapter 15	Chapter 16	Chapter 17	Chapter 18	Chapter 19	Chapter 20	Chapter 21
1. e	1. a	1. d	1. c	1. b	1. b	1. c
2. c	2. d	2. e	2. d	2. b	2. a	2. d
3. d	3. c	3. b	3. b	3. b	3. a	3. b
4. d	4. d	4. d	4. a	4. b	4. b	4. c
5. c	5. b	5. e	5. a	5. d	5. a	5. d
6. a	6. a	6. c	6. b	6. e	6. d	6. d
7. b	7. c	7. d	7. c	7. a	7. e	7. e
8. a	8. b	8. d	8. a	8. d	8. b	8. d
9. d	9. d	9. a	9. b	9. c	9. c	9. a
10. c	10. e	10. d	10. c	10. b	10. d	10. d

Chapter 22	Chapter 23	Chapter 24	Chapter 25	Chapter 26	Chapter 27	Chapter 28
1. a	1. e	1. a	1. d	1. b	1. a	1. c
2. c	2. b	2. c	2. e	2. e	2. b	2. b
3. a	3. d	3. e	3. e	3. a	3. b	3. e
4. b	4. c	4. a	4. c	4. c	4. b	4. c
5. a	5. a	5. c	5. e	5. d	5. c	5. b
6. c	6. e	6. b	6. b	6. c	6. c	6. c
7. b	7. a	7. a	7. b	7. c	7. b	7. a
8. a	8. c	8. e	8. e	8. e	8. b	8. a
9. c	9. c	9. b	9. a	9. b	9. e	9. b
10. a	10. c	10. d	10. d	10. a	10. c	10. a

CHAPTER 1   The Microbial World and You

OBJECTIVE QUESTIONS

1) Which of the following pairs is mismatched?

   A) Ehrlich — chemotherapy

   B) Koch — aseptic surgery

   C) Pasteur — proof of biogenesis

   D) Jenner — vaccination

   Answer: B
   Skill: Recall

2) Which of the following is a scientific name?

   A) IGAS

   B) Flesh–eating bacteria

   C) Group A streptococcus

   D) *Streptococcus pyogenes*

   E) Streptococci

   Answer: D
   Skill: Analysis

3) Which of the following is *not* a kingdom in the five–kingdom system?

   A) Virus   B) Monera   C) Fungi   D) Plant   E) Animal

   Answer: A
   Skill: Recall

4) Which of the following is true about fungi?

   A) All are prokaryotic.

   B) All are multicellular.

   C) All require organic material for growth.

   D) All grow using sunlight and carbon dioxide.

   E) All are plants.

   Answer: C
   Skill: Recall

5) Which of the following is *not* true about protozoa?

   A) They have rigid cell walls.

   B) They are classified by their method of locomotion.

   C) All are unicellular.

   D) All have complex cells.

   E) All are eukaryotic.

   Answer: A
   Skill: Recall

6) Which of the following is true about viruses?

   A) They are not composed of cells.

   B) They cannot metabolize nutrients.

   C) They cannot reproduce themselves.

   D) They have DNA or RNA.

   E) All of the above.

   Answer: E
   Skill: Recall

7) Which of the following is probably true about all the experiments that proved spontaneous generation?

   A) Air was lacking.

   B) Too much heat was applied.

   C) The food source could not support life.

   D) Microorganisms were already present.

   E) All of the above.

   Answer: D
   Skill: Analysis

8) Regarding Pasteur's experiments with the S–neck flask, which of the following statements is true?

   A) There was air involved.

   B) There was a food source involved.

   C) Any possibility of contamination was removed.

   D) All microorganisms were killed before beginning.

   E) All of the above.

   Answer: E
   Skill: Understanding

9) Which of the following is true about insect control by microorganisms?

   A) The insects develop resistance to the microorganisms.

   B) The microorganisms are permanent in the environment.

   C) The microorganisms are specific for the insect pest.

   D) The microorganisms may cause disease in other animals.

   E) This technique is just as dangerous as the use of chemical pesticides.

   Answer: C
   Skill: Analysis

10) Which of the following pairs is mismatched?

    A) Hooke — cell theory

    B) van Leeuwenhoek — germ theory

    C) Lister — aseptic surgery

    D) Pasteur — fermentation

    E) None of the above

    Answer: B
    Skill: Recall

11) Who disproved the theory of spontaneous generation?

    A) van Leeuwenhoek

    B) Hooke

    C) Pasteur

    D) Koch

    E) None of the above

    Answer: C
    Skill: Recall

12) Who observed cells in plant material?

    A) van Leeuwenhoek

    B) Hooke

    C) Pasteur

    D) Koch

    E) None of the above

    Answer: B
    Skill: Recall

13) Who was the first to observe microorganisms with a microscope?

    A) van Leeuwenhoek

    B) Hooke

    C) Pasteur

    D) Koch

    E) None of the above

    Answer: A
    Skill: Recall

14) Who proved that microorganisms cause disease?

    A) van Leeuwenhoek

    B) Hooke

    C) Pasteur

    D) Koch

    E) None of the above

Answer: D
Skill: Recall

15) Which of the following is a scientific name?

    A) *Mycobacterium leprae*   B) Hansen's bacillus

Answer: A
Skill: Recall

16) Which of the following is *not* a kingdom in the five-kingdom system?

    A) Monera   B) Algae   C) Protista   D) Plant   E) Animal

Answer: B
Skill: Analysis

17) Which of the following statements is untrue?

    A) All bacteria lack nuclear membranes.

    B) All fungi are multicellular.

    C) All protozoa are unicellular.

    D) All viruses are parasites.

    E) All fungi have nuclear membranes.

Answer: B
Skill: Analysis

18) Which of the following statements is true?

    A) Viruses cannot reproduce outside of a host cell.

    B) Bacteria cannot move.

    C) Fungi are plants.

    D) Protozoa have rigid cell walls.

    E) Algae are parasites.

Answer: A
Skill: Recall

19) Which of the following findings was essential for Jenner's vaccination process?

    A) A weakened microorganism may produce immunity.

    B) A weakened microorganism will not cause disease.

    C) Someone who recovers from a disease will not acquire that disease again.

    D) Disease is caused by viruses.

    E) Vaccination provides immunity.

    Answer: A
    Skill: Analysis

20) Which of the following requirements was necessary for Pasteur to disprove spontaneous generation?

    A) Providing a food source that would support growth

    B) Supplying air

    C) Keeping microorganisms out

    D) Removing microorganisms that were initially present

    E) All of the above

    Answer: E
    Skill: Analysis

21) Which of the following pairs is mismatched?

    A) Immunologist — studies ecology of *Legionella pneumophila*

    B) Virologist — studies human immunodeficiency virus

    C) Microbial ecologist — studies bacteria that degrade oil

    D) Microbial physiologist — studies fermentation of sourdough bread

    E) Molecular biologist — studies recombinant DNA

    Answer: A
    Skill: Analysis

22) Which of the following pairs is mismatched?

    A) Chemotherapy — treatment of disease

    B) Pathogen — disease causing

    C) Vaccine — a preparation of microorganisms

    D) Penicillin — antibiotic

    E) Normal microbiota — harmful

    Answer: E
    Skill: Recall

23) Which of the following is *not* part of the study of microbiology?

    A) Bacteria   B) Fungi   C) Viruses   D) Insects   E) Helminths

    Answer: D
    Skill: Recall

24) Sourdough bread differs from conventional bread during leavening because
A) Yeasts produce carbon dioxide and ethyl alcohol.
B) Yeasts produce acid.
C) *Lactobacillus* produces acids.
D) Acids are added to it during rising.
E) Of the temperature and humidity.

Answer: C
Skill: Recall

25) Which of the following is *not* an example of biotechnology?
A) Bacterial production of French bread
B) Bacterial degradation of a dead animal
C) Bacterial production of yogurt
D) Bacterial production of vinegar
E) None of the above

Answer: E
Skill: Analysis

26) Genetic engineering can be used to make all of the following *except*
A) Vaccines.
B) Human hormones.
C) Drugs.
D) Life.
E) None of the above.

Answer: D
Skill: Understanding

27) The best definition of biotechnology is
A) The development of genetic engineering.
B) The use of living organisms to make desired products.
C) Curing diseases.
D) The use of microorganisms in sewage treatment.
E) All of the above.

Answer: B
Skill: Analysis

28) You are observing a cell through a microscope and note that it has no apparent nucleus. You conclude that it most likely

   A) Has a peptidoglycan cell wall.

   B) Has a cellulose cell wall.

   C) Moves by pseudopods.

   D) Is part of a multicellular animal.

   E) None of the above.

   Answer: A
   Skill: Analysis

29) A nucleated, green cell that moves by means of flagella is _____.

   A) Alga   B) Bacterium   C) Fungus   D) Helminth   E) Virus

   Answer: A
   Skill: Recall

30) An agent that reproduces in cells but is not composed of cells and contains RNA as its genetic material is _____.

   A) Alga   B) Bacterium   C) Fungus   D) Helminth   E) Virus

   Answer: E
   Skill: Recall

31) A multicellular organism that has chitin cell walls and absorbs organic material is _____.

   A) Alga   B) Bacterium   C) Fungus   D) Helminth   E) Virus

   Answer: C
   Skill: Recall

32) A multicellular organism that has a mouth and lives in an animal host is _____.

   A) Alga   B) Bacterium   C) Fungus   D) Helminth   E) Virus

   Answer: D
   Skill: Recall

33) In the name *Escherichia coli*, *coli* is the

   A) Kingdom.

   B) Family.

   C) Genus.

   D) Specific epithet.

   E) None of the above

   Answer: D
   Skill: Analysis

34) Which of the following pairs is mismatched?

   A) Lancefield — immunology

   B) Weizmann — virology

   C) Dubos — antibiotics

   D) Jacob and Monod — microbial genetics

   E) Winogradsky — microbial ecology

   Answer: B
   Skill: Recall

35) Which of the following does *not* belong with the others?

   A) Recycling elements

   B) Human diseases

   C) Sewage treatment

   D) Bioremediation

   E) Insect control

   Answer: B
   Skill: Analysis

36) You are looking at a white cottony growth on a culture medium. Microscopic examination reveals it is multicellular. You can conclude all of the following about this organism *except* that it

   A) Has cell walls.

   B) Has DNA enclosed in a nucleus.

   C) Is eucaryotic.

   D) Is a bacterium.

   E) Absorbs organic nutrients.

   Answer: D
   Skill: Analysis

37) All members of the following groups contain DNA *except*

   A) Bacteria.   B) Fungi.   C) Helminths.   D) Protozoa.   E) Viruses.

   Answer: E
   Skill: Recall

38) Which one of the following does *not* belong with the others?

   A) Cellulose   B) Chitin   C) Nucleus   D) Peptidoglycan

   Answer: C
   Skill: Understanding

39) All of the following have cell walls *except*

A) Animalia.

B) Bacteria.

C) Fungi.

D) Plantae.

E) None of the above.

Answer: A
Skill: Recall

40) Which of the following does *not* belong with the others?

A) Animalia   B) Fungi   C) Helminth   D) Plantae   E) Protista

Answer: C
Skill: Understanding

41) Fungi differ from bacteria because fungi

A) Have cell walls.

B) Have DNA.

C) Have a nucleus.

D) Spoil food.

E) None of the above.

Answer: C
Skill: Analysis

42) Archaea differ from eubacteria because archaea

A) Lack peptidoglycan.

B) Lack nuclei.

C) Use organic compounds for food.

D) Reproduce by binary fission.

E) None of the above.

Answer: A
Skill: Analysis

43) Bacteria differ from viruses because bacteria

A) Have DNA and RNA.   B) Have cells.

C) Can live without a host.   D) All of the above.

Answer: D
Skill: Analysis

44) Which of the following lack a nucleus?

A) Animalia

B) Bacteria

C) Fungi

D) Protozoa

E) None of the above

Answer: B
Skill: Recall

45) Which one of the following does *not* belong with the others?

A) Archaea   B) Bacteria   C) Eukarya   D) Fungi

Answer: D
Skill: Understanding

ESSAY QUESTIONS

1) In 1835 Bassi showed that a silkworm disease was caused by a fungus, and in 1865 Pasteur found another silkworm disease was caused by a protozoan. Why do we use Koch's postulates instead of "Bassi's" or "Pasteur's" postulates?

Answer:

Skill:

2) List two examples of biotechnology that involve genetic engineering and two examples that do not.

Answer:

Skill:

3) Paul Berg received the Nobel Prize for developing the procedure for incorporating fragments of animal DNA into bacteria. List some reasons why his work was a major contribution to science.

Answer:

Skill:

CHAPTER 2   Chemical Principles

OBJECTIVE QUESTIONS

1) Which of the following statements is *not* true about the atom $^{12}_{6}C$?

    A) It has 6 protons in its nucleus.

    B) It has 12 neutrons in its nucleus.

    C) It has 6 electrons orbiting the nucleus.

    D) Its atomic number is 6.

    E) Its atomic weight is 12.

Answer: B
Skill: Understanding

Figure 2.1

$^{16}_{8}O$

$^{12}_{6}C$

$^{1}_{1}H$

2) Calculate the molecular weight of ethyl alcohol, C_2H_5OH, using the information in Figure 2.1.

    A) 96   B) 46   C) 34   D) 33   E) Can't tell

Answer: B
Skill: Analysis

3) Which of the following is *not* true about enzymes?

    A) Enzymes are made of proteins.

    B) Enzymes lower the activation energy of a reaction.

    C) Enzymes increase the number of collisions in a chemical reaction.

    D) Enzymes are not used up in a reaction.

    E) None of the above.

Answer: C
Skill: Recall

4) Which of the following is *not* true?

   A) Salts readily dissolve in water.

   B) Water molecules are formed by hydrolysis.

   C) Water freezes from the top down.

   D) Water is a part of a dehydration reaction.

   E) Water is a polar molecule.

   Answer: B
   Skill: Recall

5) Which of the following is the type of bond holding K^+ and I^- ions in KI?

   A) Ionic bond   B) Covalent bond   C) Hydrogen bond

   Answer: A
   Skill: Analysis

6) Which of the following is the type of bond between molecules of water in a beaker of water?

   A) Ionic bond   B) Covalent bond   C) Hydrogen bond

   Answer: C
   Skill: Analysis

7) What are the type of bonds holding hydrogen and oxygen atoms in the molecule of H_2O?

   A) Ionic bond   B) Covalent bond   C) Hydrogen bond

   Answer: B
   Skill: Analysis

8) Identify the following reaction: Glucose + fructose → sucrose + water

   A) Dehydration synthesis reaction

   B) Hydrolysis reaction

   C) Exchange reaction

   D) Reversible reaction

   E) Ionic reaction

   Answer: A
   Skill: Analysis

9) Identify the following reaction: Lactose + H_2O → glucose + galactose

   A) Dehydration synthesis reaction

   B) Hydrolysis reaction

   C) Exchange reaction

   D) Reversible reaction

   E) Ionic reaction

   Answer: B
   Skill: Analysis

10) Identify the following reaction: $HCl + NaHCO_3 \rightarrow NaCl + H_2CO_3$

    A) Dehydration synthesis reaction

    B) Hydrolysis reaction

    C) Exchange reaction

    D) Reversible reaction

    E) Ionic reaction

Answer: C
Skill: Analysis

11) Identify the following reaction: $NH_4OH \Longleftrightarrow NH_3 + H_2O$

    A) Dehydration synthesis reaction

    B) Hydrolysis reaction

    C) Exchange reaction

    D) Reversible reaction

    E) Ionic reaction

Answer: D
Skill: Analysis

12) Which type of molecule contains the alcohol glycerol?

    A) Carbohydrates

    B) Lipids

    C) Nucleic acids

    D) Proteins

    E) None of the above

Answer: B
Skill: Recall

13) Which type of molecule is composed of (CH_2O) units?

    A) Carbohydrates

    B) Lipids

    C) Nucleic acids

    D) Proteins

    E) None of the above

Answer: A
Skill: Recall

14) Which type of molecule contains $-NH_2$ groups?

   A) Carbohydrates

   B) Lipids

   C) Nucleic acids

   D) Proteins

   E) None of the above

   Answer: D
   Skill: Recall

15) Which type of molecule never contains a phosphate group?

   A) Carbohydrates

   B) Lipids

   C) Nucleic acids

   D) Proteins

   E) None of the above

   Answer: A
   Skill: Recall

16) Which of the following statements is *not* true about the atom $^{16}_{8}O$?

   A) It has 8 protons in its nucleus.

   B) It has 8 electrons in its nucleus.

   C) It has 8 neutrons in its nucleus.

   D) Its atomic number is 8.

   E) Its atomic weight is 16.

   Answer: B
   Skill: Understanding

Figure 2.1

$^{16}_{8}O$

$^{12}_{6}C$

$^{1}_{1}H$

17) Calculate the number of moles in 92 grams of ethyl alcohol, C_2H_5OH, using the information in Figure 2.1.

   A) 1   B) 2   C) 3   D) 4   E) Can't tell

   Answer: B
   Skill: Analysis

18) Which of the following statements is *not* true?

    A) Enzymes are made of proteins.

    B) Enzymes are used up in a chemical reaction.

    C) Enzymes lower the activation energy required for a reaction.

    D) Enzymes increase the probability of a reaction.

    E) None of the above.

Answer: B
Skill: Recall

19) Which of the following pairs is mismatched?

    A) $NaOH \rightleftharpoons Na^+ + OH^-$ — base

    B) $HF \rightleftharpoons H^+ + F^-$ — acid

    C) $MgSO_4 \rightleftharpoons Mg^{2+} + SO_4^{2-}$ — salt

    D) $KH_2PO_4 \rightleftharpoons K^+ + H_2PO_4^-$ — acid

    E) $H_2SO_4 \rightleftharpoons 2H^+ + SO_4^{2-}$ — acid

Answer: D
Skill: Analysis

Table 2.20

Refer to these reactions to answer the question below.

    $NaOH \rightleftharpoons Na^+ + OH^-$ — base

    $HF \rightleftharpoons H^+ + F^-$ — acid

    $MgSO_4 \rightleftharpoons Mg^{2+} + SO_4^{2-}$ — salt

    $KH_2PO_4 \rightleftharpoons K^+ H_2PO_4^-$ — acid

    $H_2SO_4 \rightleftharpoons 2H^+ + SO_4^{2-}$ — salt

20) Which of the following is *not* true about the reactions listed above?

    A) They are exchange reactions.

    B) They are ionization reactions.

    C) They occur when the reactants are dissolved in water.

    D) They are dissociation reactions.

    E) They are reversible reactions.

Answer: A
Skill: Understanding

21) What is the type of bond between the hydrogen of one molecule and the nitrogen of another molecule?

    A) Ionic bond   B) Covalent bond   C) Hydrogen bond

Answer: C
Skill: Recall

22) What is the type of bond between carbon, hydrogen, and oxygen atoms in organic molecules?

    A) Ionic bond   B) Covalent bond   C) Hydrogen bond

Answer: B
Skill: Recall

23) What type of bond is between ions in salt?

    A) Ionic bond   B) Covalent bond   C) Hydrogen bond

Answer: A
Skill: Recall

24) Identify the following reaction: $H_2O + CO_2 <==> H_2CO_3$

    A) Dehydration synthesis reaction

    B) Hydrolysis reaction

    C) Exchange reaction

    D) Reversible reaction

    E) Covalent reaction

Answer: C
Skill: Analysis

25) Identify the following reaction: Glycine + lysine → dipeptide + H_2O

    A) Dehydration synthesis reaction

    B) Hydrolysis reaction

    C) Exchange reaction

    D) Reversible reaction

    E) Covalent reaction

Answer: E
Skill: Analysis

26) Identify the following reaction: Sucrose + H_2O --------> glucose + fructose

A) Dehydration synthesis reaction

B) Hydrolysis reaction

C) Exchange reaction

D) Reversible reaction

E) Covalent reaction

Answer: B
Skill: Analysis

27) Structurally, ATP is most like this type of molecule.

A) Carbohydrates

B) Lipids

C) Proteins

D) Nucleic acids

E) None of the above

Answer: D
Skill: Recall

28) Which molecule has chemicals in genes?

A) Carbohydrates

B) Lipids

C) Proteins

D) Nucleic acids

E) None of the above

Answer: D
Skill: Recall

29) Which molecule is composed of a chain of amino acids?

A) Carbohydrates

B) Lipids

C) Proteins

D) Nucleic acids

E) None of the above

Answer: C
Skill: Recall

30) These are the primary molecules making up plasma membranes in cells.

   A) Carbohydrates

   B) Lipids

   C) Proteins

   D) Nucleic acids

   E) None of the above

   Answer: B
   Skill: Recall

31) Oil–degrading bacteria are naturally present in the environment but cannot degrade an oil spill fast enough to avoid ecological damage. The actions of these bacteria can be speeded up by

   A) Providing oil for them.

   B) Providing sugar as a carbon source.

   C) Providing nitrogen and phosphorus.

   D) Adding water.

   E) All of the above.

   Answer: C
   Skill: Recall

Figure 2.2

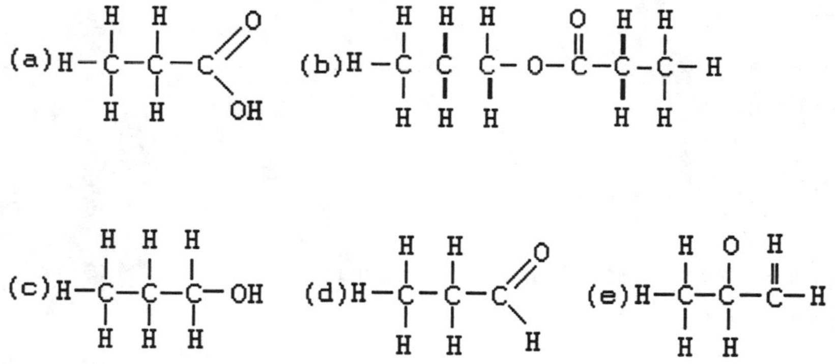

32) In Figure 2.2, which is an alcohol?

   A) a   B) b   C) c   D) d   E) e

   Answer: C
   Skill: Analysis

33) In Figure 2.2, which is an ester?

   A) a   B) b   C) c   D) d   E) e

   Answer: B
   Skill: Analysis

34) In Figure 2.2, which is an organic acid?

   A) a   B) b   C) c   D) d   E) e

Answer: A
Skill: Analysis

Figure 2.3

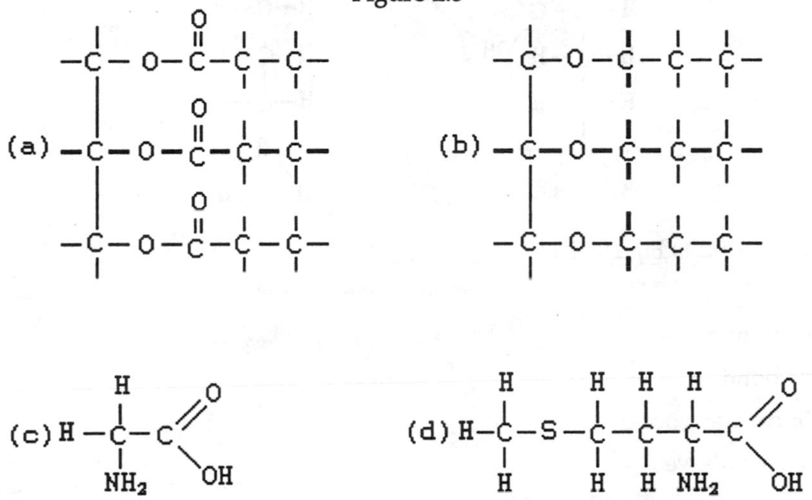

35) Use Figure 2.3 to answer the following question. Archaea differ from eubacteria in the composition of the cell membrane lipids. Archaea have ether–bonded lipids shown in __1__ above and eubacteria have ester–bonded lipids shown in __2__ above.

   A) 1–b; 2–a

   B) 1–b; 2–c

   C) 1–a; 2–c

   D) 1–c; 2–d

   E) None of the above

Answer: A
Skill: Analysis

19

Figure 2.4

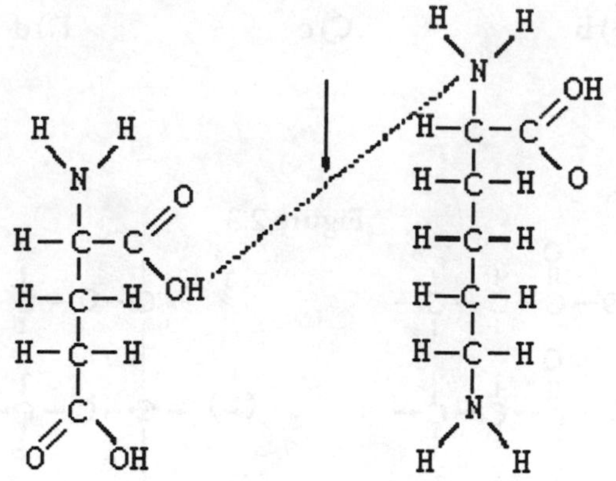

36) What kind of bond is Figure 2.4?

    A) Ioinic bond

    B) Hydrogen bond

    C) Peptide bond

    D) Double covalent bond

    E) None of the above

Answer: B
Skill: Analysis

Figure 2.5

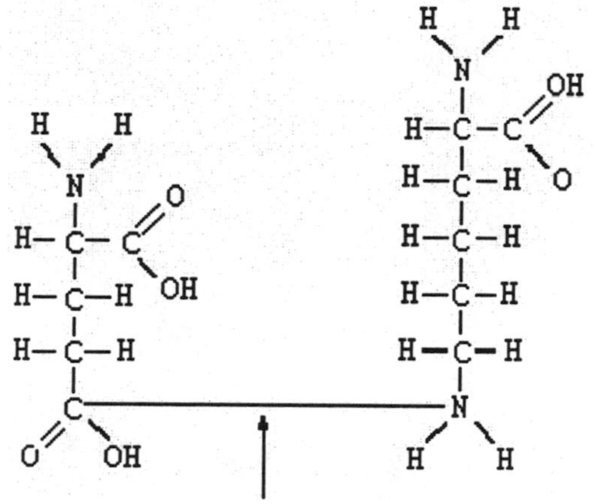

37) What kind of bond is Figure 2.5?

    A) Ionic bond

    B) Hydrogen bond

    C) Peptide bond

    D) Double covalent bond

    E) None of the above

Answer: C
Skill: Analysis

38) An *E. coli* culture that has been growing at 37 °C is moved to 25 °C. Which of the following changes must be made in its plasma membrane?

A) Increase the number of phosphate groups.

B) Increase the viscosity.

C) Increase the number of saturated chains.

D) Increase the number of unsaturated chains.

E) No changes are necessary.

Answer: D
Skill: Understanding

39) Assume *Saccharomyces cerevisiae* is grown in a nutrient medium containing the radioisotope ^{35}S. After 48 hr incubation, the ^{35}S would most likely be found in the *S. cerevisiae's*

A) Carbohydrates.

B) Nucleic acids.

C) Water.

D) Lipids.

E) Proteins.

Answer: E
Skill: Understanding

40) Assume *Saccharomyces cerevisiae* is grown in a nutrient medium containing the radioisotope ^{32}P. After 48 hr incubation, the ^{32}P would most likely be found in the *S. cerevisiae's*

A) Plasma membrane.

B) Cell wall.

C) Water.

D) Proteins.

E) None of the above.

Answer: A
Skill: Understanding

41) Starch, dextran, glycogen, and cellulose are polymers of

A) Amino acids.

B) Glucose.

C) Fatty acids.

D) Nucleic acids.

E) None of the above.

Answer: B
Skill: Recall

42) Which of the following is a base?

A) $C_2H_5OCOOH \rightarrow H^+ + C_2H_5OCOO^-$

B) C_2H_5OH

C) $NaOH \rightarrow Na^+ + OH^-$

D) $H_2O \rightarrow H^+ + OH^-$

E) All of the above

Answer: C
Skill: Analysis

43) Two glucose molecules are combined to make a maltose molecule. The chemical formula for maltose is

A) $C_3H_6O_3$ B) $C_6H_{12}O_6$ C) $C_{12}H_{24}O_{12}$ D) $C_{12}H_{22}O_{11}$ E) $C_{12}H_{23}O_{10}$

Answer: D
Skill: Analysis

44) *Desulfovibrio* bacteria can perform the following reaction: $S^{6-} \rightarrow S^{2-}$ These bacteria are

A) synthesizing sulfur.   B) reducing sulfur.

C) hydrolyzing sulfur.   D) oxidizing sulfur.

Answer: D
Skill: Understanding

45) The antimicrobial drug imidazole inhibits sterol synthesis. This would most likely interfere with

A) Bacterial cell walls.

B) Fungal cell walls.

C) Eucaryotic plasma membranes.

D) Procaryotic plasma membranes.

E) Genes.

Answer: C
Skill: Analysis

ESSAY QUESTIONS

1) Describe how the properties of phospholipids make these molecules well–suited for plasma membranes.

Answer:

Skill:

Figure 2.6

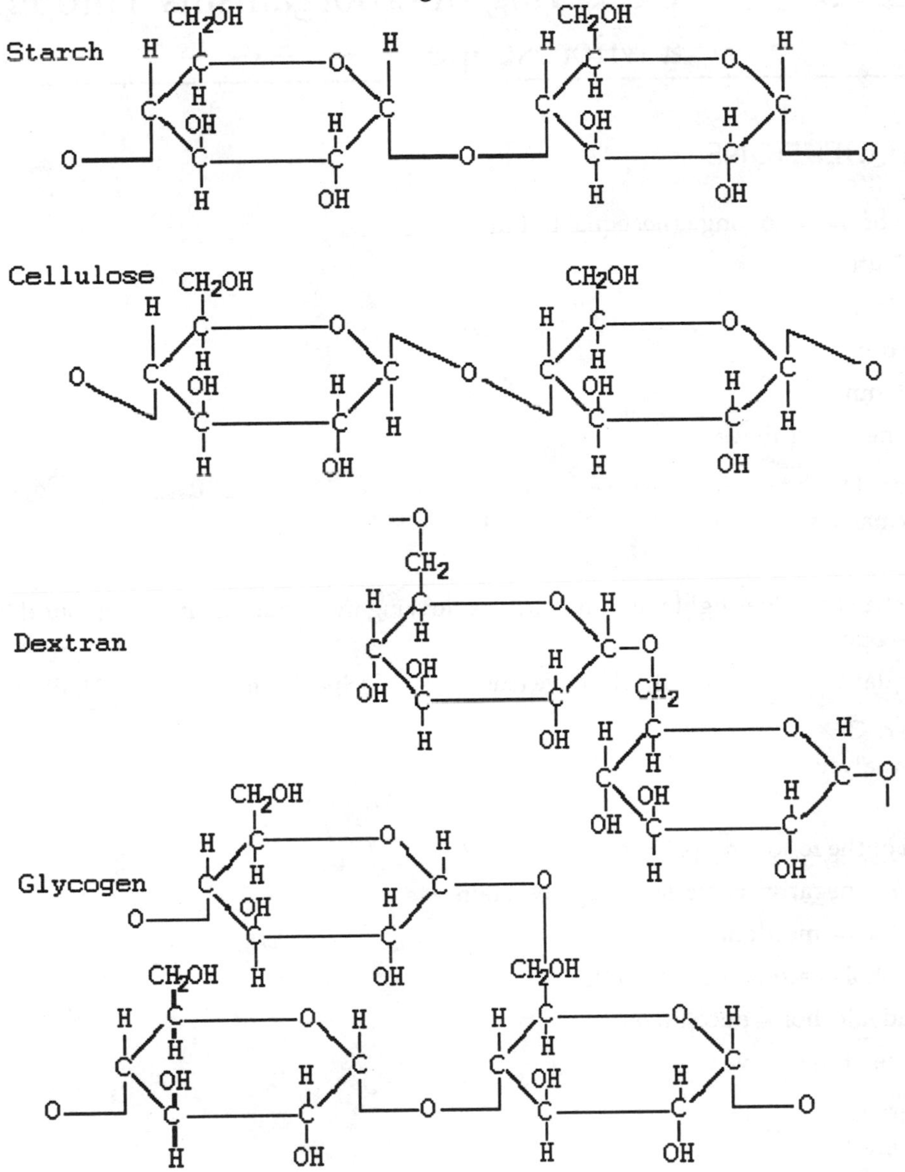

2) Use Figure 2.6 to answer the following question. Starch, cellulose, dextran, and glycogen are polysaccharides. How are they similar? To what are their different properties due? Why can't an enzyme that hydrolyzes starch degrade cellulose?

Answer:

Skill:

CHAPTER 3   Observing Microorganisms Through a Microscope

OBJECTIVE QUESTIONS

1) Which of the following is *not* equal to 1 m?
   A) 10^6 μm
   B) 10^9 nm
   C) 10 dm
   D) 100 mm
   E) None of the above

   Answer: D
   Skill: Analysis

2) What structure does light pass through after leaving the condenser in a compound light microscope?
   A) Ocular lens   B) Objective lens   C) Specimen   D) Illuminator

   Answer: C
   Skill: Analysis

3) Which of the following pairs is mismatched?
   A) Gram-negative bacteria — negative stain
   B) Iodine — mordant
   C) Alcohol — acetone-decolorizer
   D) Acid-alcohol — decolorizer
   E) None of the above

   Answer: A
   Skill: Recall

4) Place the steps of the Gram stain in the correct order:
   1-Alcohol-acetone; 2-Crystal violet; 3-Safranin; 4-Iodine.
   A) 1-2-3-4   B) 2-1-4-3   C) 2-4-1-3   D) 4-3-2-1   E) 1-3-2-4

   Answer: C
   Skill: Recall

5) Which of the following pairs is mismatched?

   A) Alcohol–acetone—decolorizer

   B) Crystal violet–basic dye

   C) Safranin—acid dye

   D) Iodine—mordant

   E) None of the above

   Answer: C
   Skill: Recall

6) The counterstain in the acid–fast stain is

   A) A basic dye.

   B) An acid dye.

   C) A negative stain.

   D) A mordant.

   E) Necessary to determine acid–fast cells.

   Answer: A
   Skill: Recall

7) The purpose of a mordant in the Gram stain is

   A) To remove the simple stain.

   B) To make the bacterial cells larger.

   C) To make the flagella visible.

   D) To prevent the crystal violet from leaving the cells.

   E) None of the above.

   Answer: D
   Skill: Recall

8) Place the following steps in the correct sequence:
   1–Staining; 2–Making a smear; 3–Fixing.

   A) 1–2–3

   B) 3–2–1

   C) 2–3–1

   D) 1–3–2

   E) The order doesn't matter

   Answer: C
   Skill: Recall

9) The best use of a negative stain is
   A) To determine cell size.
   B) To determine cell shape.
   C) To determine Gram reaction.
   D) To see endospores.
   E) a and b

   Answer: E
   Skill: Recall

10) Simple staining is often necessary to improve contrast in this microscope.
    A) Compound light microscope
    B) Phase-contrast microscope
    C) Darkfield microscope
    D) Fluorescence microscope
    E) Electron microscope

    Answer: E
    Skill: Recall

11) This microscope is used to see internal structures of cells in a natural state.
    A) Compound light microscope
    B) Phase-contrast microscope
    C) Darkfield microscope
    D) Fluorescence microscope
    E) Electron microscope

    Answer: B
    Skill: Recall

12) This microscope uses an ultraviolet light source.
    A) Compound light microscope
    B) Phase-contrast microscope
    C) Darkfield microscope
    D) Fluorescence microscope
    E) Electron microscope

    Answer: D
    Skill: Recall

13) This microscope achieves the highest magnification and greatest resolution.
   A) Compound light microscope
   B) Phase–contrast microscope
   C) Darkfield microscope
   D) Fluorescence microscope
   E) Electron microscope

   Answer: E
   Skill: Recall

14) In this microscope, the observer does not look at an image through a lens.
   A) Compound light microscope
   B) Phase–contrast microscope
   C) Darkfield microscope
   D) Fluorescence microscope
   E) Electron microscope

   Answer: E
   Skill: Recall

15) This microscope produces an image of a light cell against a dark background; internal structures are not visible.
   A) Compound light microscope
   B) Phase–contrast microscope
   C) Darkfield microscope
   D) Fluorescence microscope
   E) Electron microscope

   Answer: C
   Skill: Recall

16) Which of the following is *not* correct?
   A) $1\ \mu m = 10^{-6}\ m$
   B) $1\ nm = 10^{-9}\ m$
   C) $1\ \mu m = 10^{3}\ nm$
   D) $1\ \mu m = 10^{-3}\ mm$
   E) $1\ nm = 10^{-6}\ \mu m$

   Answer: E
   Skill: Analysis

17) The counterstain in the Gram stain is
    A) A negative stain.
    B) A mordant.
    C) A basic dye.
    D) An acid dye.
    E) Necessary to determine the Gram reaction.

    Answer: C
    Skill: Understanding

Figure 3.1

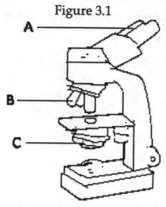

18) In Figure 3.1 line A points to the microscope's
    A) Illuminator.   B) Condenser.   C) Ocular lens.   D) Objective lens.

    Answer: C
    Skill: Recall

19) In Figure 3.1 line B points to the microscope's
    A) Illuminator.   B) Condenser.   C) Ocular lens.   D) Objective lens.

    Answer: D
    Skill: Recall

20) In Figure 3.1 line C points to the microscope's
    A) Illuminator.   B) Condenser.   C) Ocular lens.   D) Objective lens.

    Answer: B
    Skill: Recall

21) The light that hits the specimen is scattered and does not come directly from the light source in this microscope.

A) Compound light microscope

B) Phase–contrast microscope

C) Darkfield microscope

D) Fluorescence microscope

E) Electron microscope

Answer: C
Skill: Recall

22) This microscope is used to observe a specimen that emits light when illuminated with an ultra–violet light.

A) Compound light microscope

B) Phase–contrast microscope

C) Darkfield microscope

D) Fluorescence microscope

E) Electron microscope

Answer: D
Skill: Recall

23) This microscope does not use a light.

A) Compound light microscope

B) Phase–contrast microscope

C) Darkfield microscope

D) Fluorescence microscope

E) Electron microscope

Answer: E
Skill: Recall

24) This microscope takes advantage of differences in the refractive indexes of cell structures.

A) Compound light microscope

B) Phase–contrast microscope

C) Darkfield microscope

D) Fluorescence microscope

E) Electron microscope

Answer: B
Skill: Recall

25) The appearance of gram-positive bacteria after addition of the first dye in the Gram stain.
   A) Purple
   B) Red
   C) Colorless
   D) Brown
   E) None of the above

   Answer: A
   Skill: Analysis

26) The appearance of gram-negative bacteria after addition of the mordant in the Gram stain.
   A) Purple
   B) Red
   C) Colorless
   D) Brown
   E) None of the above

   Answer: A
   Skill: Analysis

27) The appearance of gram-negative bacteria after addition of the decolorizing agent in the Gram stain.
   A) Purple
   B) Red
   C) Colorless
   D) Brown
   E) None of the above

   Answer: C
   Skill: Analysis

28) The appearance of gram-positive bacteria after adding the counterstain in the Gram stain.
   A) Purple
   B) Red
   C) Colorless
   D) Brown
   E) None of the above

   Answer: A
   Skill: Analysis

29) The appearance of gram–negative bacteria after completing the Gram stain.

   A) Purple

   B) Red

   C) Colorless

   D) Brown

   E) None of the above

   Answer: B
   Skill: Analysis

30) *Bdellovibrio* are unusual bacteria because they

   A) Phagocytize other bacteria.

   B) Live inside another bacterium as a parasite.

   C) Kill nearby bacteria.

   D) Enter and digest other bacteria.

   E) Release their cellular contents to the outside.

   Answer: D
   Skill: Recall

31) What is the total magnification of a chloroplast viewed with a 10× ocular lens and a 45× objective lens?

   A) 10×

   B) 45×

   C) 100×

   D) 450×

   E) None of the above

   Answer: D
   Skill: Analysis

32) You suspect a 100 nm structure is present in a cell. Which of the following provides the lowest magnification that you can use to see this structure?

   A) Brightfield microscope

   B) Darkfield microscope

   C) Transmission electron microscope

   D) Phase–contrast microscope

   E) Scanning electron microscope

   Answer: E
   Skill: Analysis

33) This microscope uses two beams of light to produce a three–dimensional, color image.
   A) Fluorescence microscope
   B) Phase–contrast microscope
   C) Darkfield microscope
   D) DIC microscope
   E) None of the above

   Answer: D
   Skill: Recall

34) This microscope is used to see intracellular detail.
   A) Fluorescence microscope
   B) Phase–contrast microscope
   C) Darkfield microscope
   D) DIC microscope
   E) None of the above

   Answer: B
   Skill: Recall

35) Image looks like a negative stain in this microscope:
   A) Fluorescence microscope
   B) Phase–contrast microscope
   C) Darkfield microscope
   D) DIC microscope
   E) None of the above

   Answer: A
   Skill: Recall

36) This microscope is used to see detail of a 300 nm virus.
   A) Fluorescence microscope
   B) Phase–contrast microscope
   C) Darkfield microscope
   D) DIC microscope
   E) None of the above

   Answer: E
   Skill: Recall

37) Assume you stain *Bacillus* by applying malachite green with heat and then counterstaining with safranin. Through the microscope, the green structures are

   A) Cell walls.

   B) Capsules.

   C) Endospores.

   D) Flagella.

   E) Can't tell.

   Answer: C
   Skill: Analysis

38) Cells are differentiated after which step in the Gram stain?

   A) Safranin   B) Alcohol–acetone   C) Iodine   D) Crystal violet

   Answer: B
   Skill: Understanding

39) You find colorless areas in cells in a Gram-stained smear. What should you do next?

   A) An acid–fast stain

   B) A flagella stain

   C) A capsule stain

   D) An endospore stain

   E) A simple stain

   Answer: D
   Skill: Analysis

40) What Gram reaction do you expect from acid-fast bacteria?

   A) Gram–positive   B) Gram–negative

   C) Both gram–positive and gram–negative   D) Can't tell

   Answer: A
   Skill: Analysis

41) Bacterial smears are fixed before staining to

   A) Kill the bacteria.

   B) Affix the cells to the slide.

   C) Make their walls permeable.

   D) a and b.

   E) All of the above.

   Answer: D
   Skill: Recall

42) The resolution of a microscope can be improved by changing the
   A) Condenser.
   B) Fine adjustment.
   C) Wavelength of light.
   D) Diaphragm.
   E) Coarse adjustment.

   Answer: C
   Skill: Recall

43) Van Leeuwenhoek's microscope was a(n)
   A) Electron microscope.
   B) Phase–contrast microscope.
   C) Simple microscope.
   D) Confocal microscope.
   E) None of the above.

   Answer: C
   Skill: Recall

44) The purpose of the ocular lens is to
   A) Improve resolution.
   B) Magnify the image from the objective lens.
   C) Decrease the refractive index.
   D) Increase the light.
   E) None of the above.

   Answer: B
   Skill: Recall

45) Which of the following pairs is mismatched?
   A) Fluorescence microscope–uses fluorescent light source
   B) Brightfield microscope–used to view stained specimens
   C) Confocal microscope–produces a three–dimensional image
   D) Scanning electron microscope–used to view surface of specimen
   E) Scanning tunneling microscope–used to visualize DNA

   Answer: A
   Skill: Analysis

ESSAY QUESTIONS

1) In 1877, Robert Koch thought preparing permanently stained slides would be valuable. Why was his assessment correct?
   Answer:
   Skill:

2) In 1884, Hans Christian Gram described a method of staining bacterial cells while not staining surrounding animal tissues. However, he thought that the staining method he developed was faulty because all bacteria did not stain. In a letter to the editor of the journal in which Gram published his findings, write your response to Gram's concern.

Answer:

Skill:

CHAPTER 4   Functional Anatomy of Prokaryotic and Eukaryotic Cells

OBJECTIVE QUESTIONS

1) Which of the following is *not* a distinguishing characteristic of prokaryotic cells?

    A) Their DNA is not enclosed within a membrane.

    B) They lack membrane–enclosed organelles.

    C) They have cell walls containing peptidoglycan.

    D) Their DNA is not associated with histones.

    E) None of the above.

    Answer: E
    Skill: Recall

2) Which of the following is *not* true about a gram–positive cell wall?

    A) It maintains the shape of the cell.

    B) It is sensitive to lysozyme.

    C) It protects the cell in a hypertonic environment.

    D) It contains teichoic acids.

    E) None of the above.

    Answer: C
    Skill: Analysis

3) Which of the following best describes what happens when a bacterial cell is placed in a solution containing 5% NaCl?

    A) Sucrose will move into the cell from a higher to a lower concentration.

    B) The cell will undergo osmotic lysis.

    C) Water will move out of the cell.

    D) Water will move into the cell.

    E) No change will result; the solution is isotonic.

    Answer: C
    Skill: Understanding

4) The best definition of osmotic pressure is
   A) The movement of solute molecules from a higher to a lower concentration.
   B) The force with which a solvent moves across a semi-permeable membrane from a higher to a lower concentration.
   C) The movement of a substance across a semi-permeable membrane from a higher to a lower concentration.
   D) The active transport of a substance out of a cell to maintain equilibrium.
   E) The movement of solute molecules from a lower to a higher concentration across a semi-permeable membrane.

   Answer: B
   Skill: Understanding

5) By which of the following mechanisms can a cell transport a substance from a lower to a higher concentration?
   A) Simple diffusion
   B) Facilitated diffusion
   C) Active transport
   D) Extracellular enzymes
   E) Any of the above

   Answer: C
   Skill: Analysis

6) Which of the following is *not* a characteristic of the plasma membrane?
   A) Maintains cell shape
   B) Composed of a phospholipid bilayer
   C) Contains proteins
   D) The site of cell wall formation
   E) Selectively permeable

   Answer: A
   Skill: Analysis

7) All of the following are lacking a cell wall *except*
   A) Protoplasts.
   B) Fungi.
   C) L-forms.
   D) Mycoplasmas.
   E) Animal cells.

   Answer: B
   Skill: Recall

8) Which of the following statements is true?

   A) Endospores are for reproduction.

   B) Endospores allow a cell to survive environmental changes.

   C) Endospores are easily stained in a Gram stain.

   D) A cell produces one endospore and keeps growing.

   E) A cell can produce many endospores.

   Answer: B
   Skill: Recall

9) Which of the following pairs is mismatched?

   A) Endoplasmic reticulum — internal transport

   B) Golgi complex — secretion

   C) Mitochondria — ATP production

   D) Centrosome — food storage

   E) Lysosome — digestive enzymes

   Answer: D
   Skill: Recall

10) Which of the following organelles most closely resembles a procaryotic cell?

   A) Nucleus

   B) Mitochondrion

   C) Golgi complex

   D) Vacuole

   E) Cell wall

   Answer: B
   Skill: Analysis

Figure 4.1

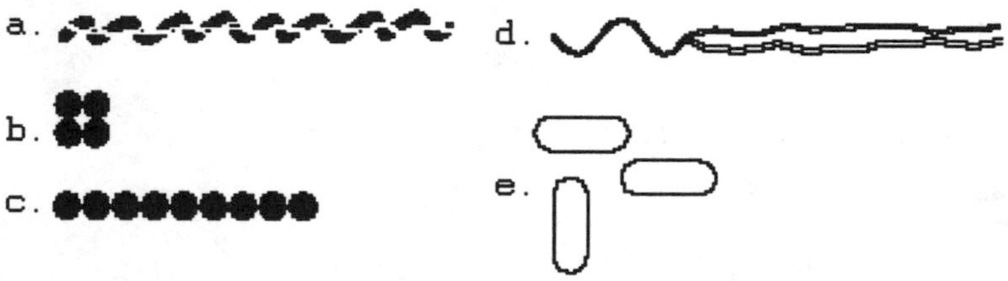

11) In Figure 4.1, which drawing is a tetrad?

   A) drawing a   B) drawing b   C) drawing c   D) drawing d   E) drawing e

   Answer: B
   Skill: Recall

12) In Figure 4.1, which drawing possesses an axial filament?

    A) drawing a   B) drawing b   C) drawing c   D) drawing d   E) drawing e

Answer: A
Skill: Recall

13) In Figure 4.1, which drawing is Streptococci?

    A) drawing a   B) drawing b   C) drawing c   D) drawing d   E) drawing e

Answer: C
Skill: Recall

14) In Figure 4.1, which drawing is Bacilli?

    A) drawing a   B) drawing b   C) drawing c   D) drawing d   E) drawing e

Answer: E
Skill: Recall

15) Which of the following is generally not true of procaryotic cells?

    A) They have a semirigid cell wall.

    B) They are motile by means of flagella.

    C) They possess 80S ribosomes.

    D) They reproduce by binary fission.

    E) None of the above.

Answer: C
Skill: Recall

16) Which of the following is *not* true about gram–negative cell walls?

    A) They protect the cell in a hypotonic environment.

    B) They have an extra outer layer composed of lipoproteins, lipopolysaccharides, and phospholipids.

    C) They are toxic to humans.

    D) They are sensitive to penicillin.

    E) Their Gram reaction is due to the outer membrane.

Answer: D
Skill: Analysis

17) Which of the following is *not* a structure found in procaryotic cells?

    A) Flagella

    B) Axial filament

    C) Cilia

    D) Pili

    E) Peritrichous flagella

Answer: C
Skill: Recall

18) Which of the following is *not* true about the glycocalyx?

   A) It may be composed of polysaccharide.

   B) It may be composed of polypeptide.

   C) It may be responsible for virulence.

   D) It is used to adhere to surfaces.

   E) None of the above.

   Answer: E
   Skill: Analysis

19) Which of the following is *not* a chemical component of a bacterial cell wall?

   A) Cellulose

   B) Peptidoglycan

   C) Teichoic acids

   D) Peptide chains

   E) N–acetylmuramic acid

   Answer: A
   Skill: Analysis

20) Which of the following is *not* part of the active transport process?

   A) Plasma membrane

   B) Transporter proteins

   C) ATP

   D) Cell wall

   E) None of the above

   Answer: D
   Skill: Analysis

Figure 4.2

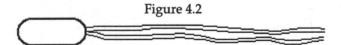

21) Which of the following terms best describes the cell in Figure 4.2?

   A) Peritrichous flagella

   B) Amphitrichous flagella

   C) Lophotrichous flagella

   D) Monotrichous flagella

   E) Axial filament

   Answer: C
   Skill: Recall

22) In bacteria, photosynthetic pigments are found in
    A) Chloroplasts.
    B) Cytoplasm.
    C) Chromatophores.
    D) Mesosomes.
    E) None of the above.

    Answer: C
    Skill: Recall

23) The difference between simple diffusion and facilitated diffusion is that facilitated diffusion
    A) Can move materials from a higher to a lower concentration.
    B) Can move materials from a lower to a higher concentration.
    C) Requires ATP.
    D) Requires transporter proteins.
    E) Doesn't require ATP.

    Answer: D
    Skill: Analysis

24) Possible functions of magnetosomes include all of the following *except*
    A) Get cells to the North Pole.
    B) Protect cells from hydrogen peroxide accumulation.
    C) Synthesize ATP.
    D) Locate suitable environments.
    E) None of the above.

    Answer: A
    Skill: Recall

25) Which of the following cell structures has a role in the initiation of disease?
    A) Gram-positive cell wall
    B) Lipid A
    C) Cell membrane
    D) Fimbriae
    E) All of the above

    Answer: D
    Skill: Recall

26) Fimbriae and pili differ in that pili

    A) Are composed of pilin.

    B) Are composed of flagellin.

    C) Are used to transfer DNA.

    D) Are used for asexual reproduction.

    E) Are used for attachment.

Answer: C
Skill: Recall

Figure 4.3

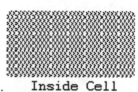

27) In Figure 4.3, which diagram of a cell wall is a gram-negative cell wall?

    A) Figure a

    B) Figure b

    C) Both Figure a and Figure b

    D) Neither Figure a nor Figure b

    E) Can't tell

Answer: A
Skill: Analysis

28) In Figure 4.3, which diagram of a cell wall is a toxic cell wall?

    A) Figure a

    B) Figure b

    C) Both Figure a and Figure b

    D) Neither Figure a nor Figure b

    E) Can't tell

Answer: A
Skill: Analysis

29) In Figure 4.3, which diagram of a cell wall has a wall that protects against osmotic lysis?

A) Figure a

B) Figure b

C) Both Figure a and Figure b

D) Neither Figure a nor Figure b

E) Can't tell

Answer: C
Skill: Analysis

30) In Figure 4.3, which diagram of a cell wall is decolorized by acetone–alcohol?

A) Figure a

B) Figure b

C) Both Figure a and Figure b

D) Neither Figure a nor Figure b

E) Can't tell

Answer: A
Skill: Analysis

31) In Figure 4.3, which diagram of a cell wall is resistant to many antibiotics (e.g., penicillin)?

A) Figure a

B) Figure b

C) Both Figure a and Figure b

D) Neither Figure a nor Figure b

E) Can't tell

Answer: A
Skill: Analysis

32) In Figure 4.3, which diagram of a cell wall contains teichoic acids?

A) Figure a

B) Figure b

C) Both Figure a and Figure b

D) Neither Figure a nor Figure b

E) Can't tell

Answer: B
Skill: Analysis

33) In Figure 4.3, which diagram of a cell wall contains porins?

    A) Figure a
    B) Figure b
    C) Both Figure a and Figure b
    D) Neither Figure a nor Figure b
    E) Can't tell

    Answer: A
    Skill: Analysis

34) Where are phospholipids most likely found in a prokaryotic cell?

    A) Flagella
    B) Around organelles
    C) Plasma membrane
    D) Ribosomes
    E) b and c

    Answer: C
    Skill: Understanding

35) Where are phospholipids most likely found in a eukaryotic cell?

    A) Flagella
    B) Around organelles
    C) Plasma membrane
    D) Ribosomes
    E) b and c

    Answer: E
    Skill: Understanding

36) Found in gram–positive bacteria.

    A) Pseudomurein
    B) Sterol–rich cell membranes
    C) Peptidoglycan
    D) Nucleus
    E) a and c

    Answer: C
    Skill: Understanding

37) Found in archaea.
A) Pseudomurein
B) Sterol–rich cell membranes
C) Peptidoglycan
D) Nucleus
E) a and c

Answer: A
Skill: Understanding

38) Found in mycoplasmas.
A) Pseudomurein
B) Sterol–rich cell membranes
C) Peptidoglycan
D) Nucleus
E) a and c

Answer: B
Skill: Understanding

39) You have isolated a motile, gram–positive cell with no visible nucleus. You can safely assume that the cell
A) Has 9 pairs + 2 flagella.
B) Has a mitochondrion.
C) Has a cell wall.
D) Lives in an extreme environment.
E) Has a nucleus.

Answer: C
Skill: Analysis

40) What will happen if a bacterial cell is placed in distilled water with lysozyme?
A) The cell will plasmolyze.
B) The cell will undergo osmotic lysis.
C) Water will leave the cell.
D) Lysozyme will diffuse into the cell.
E) No change will result; the solution is isotonic.

Answer: B
Skill: Analysis

41) What will happen if a bacterial cell is placed in 10% NaCl with penicillin?

    A) The cell will plasmolyze.

    B) The cell will undergo osmotic lysis.

    C) Water will enter the cell.

    D) Penicillin will diffuse into the cell.

    E) No change will result; the solution is isotonic.

Answer: A
Skill: Analysis

42) Which one of the following pairs is *not* correctly matched?

    A) Metachromatic granules — phosphate storage

    B) Lipid inclusions — energy reserve

    C) Ribosomes — protein storage

    D) Sulfur granules — energy reserve

    E) Gas vacuoles — flotation

Answer: C
Skill: Analysis

43) All of the following are energy reserves *except*

    A) Carboxysomes.

    B) Polysaccharide granules.

    C) Lipid inclusions.

    D) Sulfur granules.

    E) Metachromatic granules.

Answer: A
Skill: Recall

44) Which one of the following is *not* a functionally analogous pair?

    A) Nucleus — nuclear region

    B) Mitochondria — procaryotic plasma membrane

    C) Chloroplasts — thylakoids

    D) Cilia — pili

    E) 9+2 flagella — bacterial flagella

Answer: D
Skill: Understanding

45) All of the following can be found in mitochondria and procaryotes *except*

    A) Circular chromosome.

    B) 70S ribosomes.

    C) Cell wall.

    D) Binary fission.

    E) ATP–generating mechanism.

Answer: C
Skill: Analysis

ESSAY QUESTIONS

1) What is the importance of a phospholipid bilayer to living cells? Where is it in prokaryotic cells?

Answer:

Skill:

2) Provide evidence to substantiate the hypothesis that eucaryotic cells evolved from prokaryotic cells.

Answer:

Skill:

3) Compare and contrast gram–positive and gram–negative cell walls with regard to (a) sensitivity to antimicrobial agents, (b) resistance to phagocytosis, (c) chemical composition, and (d) decolorization by alcohol.

Answer:

Skill:

4) Group A, β hemolytic streptococci ("the flesh–eating bactera") rapidly spreads through the human body by digesting the hyaluronic acid between ceslls. Label the plasma membrane shown below to illustrate digestion of the polysaccharide, hyaluronic acid.

Add these labels to the diagram in the appropriate places:
hyaluronic acid (a polysaccharaide); GluUA–GlcNAc (a disaccharide);
glucose (a monosaccharide); hyaluronidase (an exoenzyme);
hyaluronic acid permease.

Answer:

Skill:

CHAPTER 5   Microbial Metabolism

OBJECTIVE QUESTIONS

1) Which of the following compounds is *not* an enzyme?

   A) Dehydrogenase

   B) Cellulase

   C) Coenzyme A

   D) β–galactosidase

   E) None of the above

Answer: C
Skill: Recall

Figure 5.1

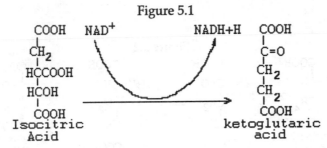

2) Which compound is being reduced in the reaction in Figure 5.1?

   A) Isocitric acid and α–ketoglutaric acid

   B) α–ketoglutaric acid and NAD^+

   C) NAD^+

   D) NADH

   E) NADH and isocitric acid

Answer: C
Skill: Analysis

3) Which organism is *not* correctly matched to its energy source?

   A) Photoheterotroph — light

   B) Photoautotroph — CO_2

   C) Chemoautotroph — Fe^{2+}

   D) Chemoheterotroph — glucose

   E) Chemoautotroph — NH_3

Answer: B
Skill: Recall

4) How many molecules of ATP can be generated from the complete oxidation of glucose to CO_2 and H_2O?

   A) 2   B) 4   C) 34   D) 38   E) 76

   Answer: D
   Skill: Recall

5) Which of the following is *not* true about anaerobic respiration?

   A) It involves glycolysis only.

   B) It involves the Krebs cycle.

   C) It involves the reduction of nitrate.

   D) It generates ATP.

   E) It requires cytochromes.

   Answer: A
   Skill: Recall

Figure 5.2

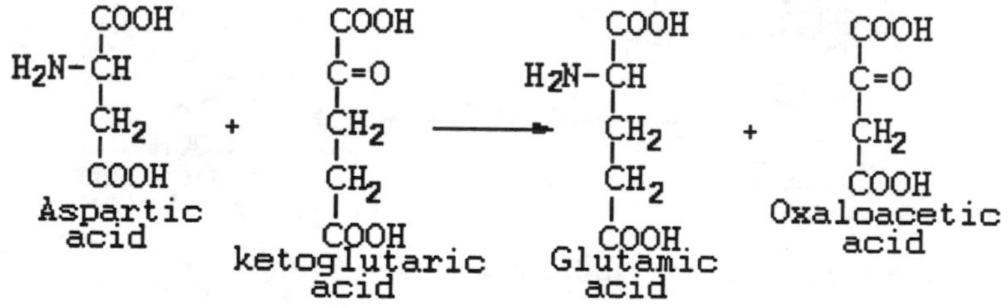

6) What type of reaction is in Figure 5.2?

   A) Decarboxylation

   B) Transamination

   C) Dehydrogenation

   D) Oxidation

   E) Reduction

   Answer: B
   Skill: Analysis

7) What is the fate of pyruvic acid in an organism that uses aerobic respiration?

   A) It is reduced to lactic acid.

   B) It is oxidized in the Krebs cycle.

   C) It is oxidized in the electron transport chain.

   D) It is catabolized in glycolysis.

   E) It is reduced in the Krebs cycle.

   Answer: B
   Skill: Understanding

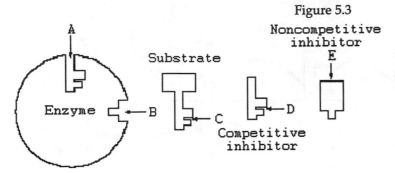

Figure 5.3

8) How would a noncompetitive inhibitor interfere with a reaction involving the enzyme in Figure 5.3?

A) It would bind to A.

B) It would bind to B.

C) It would bind to C.

D) It would bind to D.

E) Can't tell.

Answer: B
Skill: Analysis

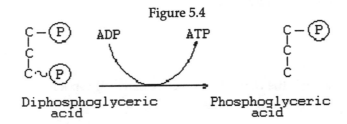

Figure 5.4

9) How is ATP generated in the reaction in figure 5.4?

A) Glycolysis

B) Fermentation

C) Photophosphorylation

D) Oxidative phosphorylation

E) Substrate–level phosphorylation

Answer: E
Skill: Analysis

10) Fatty acids are catabolized in

A) The Krebs cycle.

B) The electron transport chain.

C) Glycolysis.

D) The pentose phosphate pathway.

E) The Entner Doudoroff pathway.

Answer: A
Skill: Recall

Figure 5.5

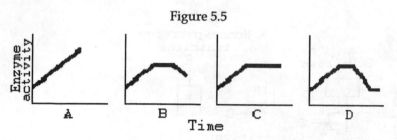

Enzyme activity

A   B   C   D

Time

11) Which of the graphs in Figure 5.5 best illustrates the activity of an enzyme that is saturated with substrate?

   A) A   B) B   C) C   D) D   E) A or C

   Answer: C
   Skill: Analysis

12) According to the chemiosmotic mechanism, ATP is generated when

   A) Electrons are transferred between carrier molecules.

   B) A high–energy phosphate group is transferred from an intermediate metabolite to ADP.

   C) Chlorophyll liberates an electron.

   D) Protons are moved across a membrane.

   E) Cells lyse in a hypotonic environment.

   Answer: D
   Skill: Recall

13) Which of the following is the best definition of oxidative phosphorylation?

   A) Electrons are passed through a series of carriers to O_2.

   B) The energy released as carrier molecules are oxidized is used to generate ATP.

   C) The energy released in the reduction of carrier molecules is used to generate ATP.

   D) The transfer of a high–energy phosphate group to ADP.

   Answer: B
   Skill: Analysis

14) All the following are true about substrate–level phosphorylation *except*

   A) It involves the direct transfer of a high–energy phosphate group from an intermediate metabolic compound to ADP.

   B) No final electron acceptor is required.

   C) It occurs in glycolysis.

   D) The oxidation of intermediate metabolic compounds releases energy that is used to generate ATP.

   E) All of the above are true.

   Answer: D
   Skill: Analysis

15) Which of the following is *not* true about photophosphorylation?

    A) Light liberates an electron from chlorophyll.

    B) The oxidation of carrier molecules releases energy.

    C) Energy from oxidation reactions is used to generate ATP from ADP.

    D) It requires CO_2.

    E) It occurs in photosynthesizing cells.

Answer: D
Skill: Analysis

16) Which of the following is *not* an end-product of fermentation?

    A) Lactic acid

    B) Ethyl alcohol

    C) Glycerol

    D) Pyruvic acid

    E) Acetone

Answer: D
Skill: Analysis

17) A strictly fermentative bacterium produces energy

    A) By glycolysis only.

    B) By aerobic respiration only.

    C) By fermentation or aerobic respiration.

    D) Only in the absence of oxygen.

    E) Only in the presence of oxygen.

Answer: A
Skill: Understanding

18) The advantage of the pentose phosphate pathway is that it produces all of the following *except*

    A) Precursors for nucleic acids.

    B) Precursors for the synthesis of glucose.

    C) Three ATPs.

    D) NADPH.

    E) Precursors for the synthesis of amino acids.

Answer: C
Skill: Recall

19) What percent of the total ATP produced from the complete catabolism of glucose is produced by aerobic respiration?

    A) 5%   B) 11%   C) 50%   D) 89%   E) 95%

Answer: D
Skill: Analysis

20) Which of the following is *not* true about beta oxidation?

    A) It is a method of catabolizing fatty acids.

    B) It involves the formation of 2–carbon units.

    C) It involves the formation of acetyl–CoA.

    D) It is a step in glycolysis.

    E) None of the above.

Answer: D
Skill: Analysis

21) Which of the following reactions generates ATP?

    A) Glucose → glucose–6–phosphate

    B) Phosphoenolpyruvic acid → pyruvic acid

    C) Glucose–6–phosphate → glucose–1,6–diphosphate

    D) Glyceraldehyde–3–phosphate → 1,3–diphosphoglyceric acid

    E) NADH → NAD$^+$

Answer: B
Skill: Understanding

22) Which of the following is the best definition of fermentation?

    A) The reduction of glucose to pyruvic acid.

    B) The oxidation of glucose with organic molecules serving as electron acceptors.

    C) The complete catabolism of glucose to CO_2 and H_2O.

    D) The production of energy by substrate–level phosphorylation.

    E) The production of ethyl alcohol from glucose.

Answer: B
Skill: Analysis

23) Which of the following is *not* necessary for respiration?

    A) Cytochromes

    B) Flavoproteins

    C) A source of electrons

    D) Oxygen

    E) Quinones

Answer: D
Skill: Understanding

24) Aerobic respiration differs from anaerobic respiration in which of the following respects?

   A) Anaerobic respiration is glycolysis.

   B) The final electron acceptors are different.

   C) Aerobic respiration requires the electron transport chain.

   D) Aerobic respiration gets electrons from the Krebs cycle.

   E) Aerobic respiration produces more ATP.

Answer: B
Skill: Analysis

25) Which one of the following would you predict is an allosteric inhimbitor of the Krebs cycle enzyme, ketoglutarate dehydrogenase?

   A) Citric acid

   B) α–ketoglutaric acid

   C) NAD$^+$

   D) NADH

   E) All of the above

Answer: D
Skill: Understanding

26) When oxygen is unavailable, *Halobacterium* produces ATP by

   A) Fermentation.

   B) Photophosphorylation.

   C) Oxidative phosphorylation.

   D) Substrate–level phosphorylation.

   E) The Krebs cycle.

Answer: B
Skill: Recall

27) To a microbiologist, fermentation is best defined as

   A) The state of being in high activity or commotion.

   B) Any process that produces alcoholic beverages.

   C) Any spoilage of food by microorganisms.

   D) Any large–scale microbial process.

   E) All metabolic processes that release energy from a sugar or other organic molecule, do not require oxygen or an electron transport system, and use an organic molecule as the final electron acceptor.

Answer: E
Skill: Recall

28) Uses CO_2 for carbon and H_2 for energy.

   A) Chemoautotroph

   B) Chemoheterotroph

   C) Photoautotroph

   D) Photoheterotroph

   E) None of the above

   Answer: A
   Skill: Analysis

29) Uses glucose for carbon and energy.

   A) Chemoautotroph

   B) Chemoheterotroph

   C) Photoautotroph

   D) Photoheterotroph

   E) None of the above

   Answer: B
   Skill: Analysis

30) Has bacteriochlorophylls and uses alcohols for carbon.

   A) Chemoautotroph

   B) Chemoheterotroph

   C) Photoautotroph

   D) Photoheterotroph

   E) None of the above

   Answer: D
   Skill: Analysis

31) Cyanobacteria are an example of this type.

   A) Chemoautotroph

   B) Chemoheterotroph

   C) Photoautotroph

   D) Photoheterotroph

   E) None of the above

   Answer: C
   Skill: Analysis

32) Which statements are true?
1—Electron carriers are located at ribosomes.
2—ATP is a common intermediate between catabolic and anabolic pathways.
3—ATP is used for the long–term storage of energy and so is often found in storage granules.
4—Anaerobic organisms are capable of respiration.
5—ATP is generated by the flow of protons across the cell membrane.

A) 2, 4, 5   B) 1, 3, 4   C) 2, 3, 5   D) 1, 2, 3   E) All

Answer: A
Skill: Analysis

33) C_2H_5OH $\xrightarrow{\textit{Acetobacter}}$ C_2H_3OOH
Ethyl alcohol   Acetic acid

A) The process requires O_2.   B) The process occurs anaerobically.

Answer: A
Skill: Understanding

34) $C_5H_{12}O_6$ $\xrightarrow{\textit{Saccharomyces}}$ $2C_2H_5OH + 2CO_2$
Glucose   Ethyl alcohol

A) The process requires O_2.   B) The process occurs anaerobically.

Answer: B
Skill: Understanding

35) $NO_3^- + 2H^+$ $\xrightarrow{\textit{Pseudomonas}}$ $NO_2^- + H_2O$

Nitrate ion   Nitrite ion

A) The process requires O_2.   B) The process occurs anaerobically.

Answer: B
Skill: Understanding

36) $2H^+$ $\xrightarrow{\textit{E. coli}}$ H_2O
Hydrogen ions   Water

A) The process requires O_2.   B) The process occurs anaerobically.

Answer: A
Skill: Understanding

37) Assume you are working for a chemical company and you are responsible for growing a yeast culture that produces ethyl alcohol. The yeasts are growing well on the maltose medium but are not producing alcohol. The most likely explanation is

A) The maltose is toxic.

B) O_2 is in the medium.

C) Not enough protein is provided.

D) Yeasts don't produce ethyl alcohol.

E) None of the above.

Answer: B
Skill: Understanding

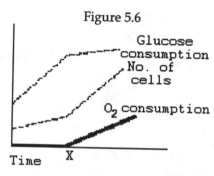

Figure 5.6

38) The rates of O_2 and glucose consumption by a bacterial culture are shown in Figure 5.6. Assume a bacterial culture was grown in a glucose medium without O_2. Then O_2 was added at the time marked X. The data indicate that

A) These bacteria don't use O_2.

B) These bacteria get more energy anaerobically.

C) Aerobic metabolism is more efficient than fermentation.

D) These bacteria can't grow anaerobically.

Answer: C
Skill: Analysis

39) An enzyme, citrate synthase, in the Krebs cycle is inhibited by ATP; this is an example of

A) Allosteric inhibition.

B) Competitive inhibition.

C) Feedback inhibition.

D) Noncompetitive inhibition.

E) Turnover rate.

Answer: A
Skill: Understanding

40) A shipping company employee notices that the inside of ships' hulls where ballast water is stored are deteriorating. The hull paint contained cyanide to prevent microbial growth. Since bacteria were growing on the hulls, you can conclude that the

A) Bacteria were using aerobic respiration.

B) Bacteria were using anaerobic respiration.

C) Bacteria were growing by fermentation.

D) Bacteria were using cytochromes.

E) None of the above.

Answer: C
Skill: Analysis

41) *Beggiatoa* bacteria get energy by oxidizing S^{2-} to S^{6+}. This means they take __(1)__ for their __(2)__.

A) 1–electrons; 2–electron transport chain

B) 1–electrons; 2–fermentation

C) 1–protons; 2–NAD^+

D) 1–sulfur; 2–photophosphorylation

E) 1–glucose; 2–glycolysis

Answer: A
Skill: Analysis

Figure 5.7

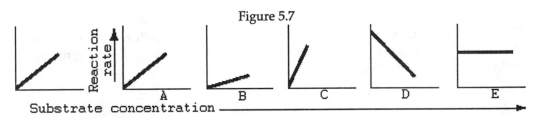

42) The graph at the left of Figure 5.7 shows the reaction rate for an enzyme at its optimum temperature. Which graph shows enzyme activity at a higher temperature?

A) Graph A   B) Graph B   C) Graph C   D) Graph D   E) Graph E

Answer: B
Skill: Analysis

43) A bacterial culture grown in a glucose–peptide medium causes the pH to increase. The bacteria are most likely

A) Fermenting the glucose.   B) Oxidizing the glucose.

C) Using the peptides.   D) Not growing.

Answer: C
Skill: Analysis

44) *Gallionella* bacteria can get energy from the reaction: $Fe^{2+} \rightarrow Fe^{3+}$. This reaction is an example of

   A) Oxidation.

   B) Reduction.

   C) Fermentation.

   D) Photophosphorylation.

   E) The Calvin–Benson cycle.

   Answer: A
   Skill: Analysis

45) Assume you are growing bacteria on a lipid medium that started at pH 7. The action of bacterial lipases should cause the pH of the medium to

   A) Increase.   B) Decrease.   C) Stay the same.

   Answer: B
   Skill: Understanding

ESSAY QUESTIONS

1) Compare and contrast photophosphorylation and oxidative phosphorylation.

   Answer:

   Skill:

2) *Rhodopseudomonas* is an anaerobic photoautotroph that uses organic compounds as an electron donor. It is also capable of chemoheterotrophic metabolism. Diagram the metabolic pathways of this bacterium.

   Answer:

   Skill:

3) Identify the catabolic pathways used by the following bacteria:

Pseudomonas	Oxidizes glucose
Lactobacillus	Ferments glucose
Alcaligenes	Neither oxidizes nor ferments glucose
Escherichia	Oxidizes and ferments glucose

   Answer:

   Skill:

4) Differentiate between the following two laboratory tests: starch hydrolysis and starch fermentation.

   Answer:

   Skill:

5) *Streptococcus* lacks an electron transport chain. How does this bacterium reoxidize NADH? Where is the NADH formed?

Answer:

Skill:

6) You look in the refrigerator at some orange drink you had forgotten was there. The drink now has an off taste and bubbles. What is the most likely explanation for the changes in the drink?

Answer:

Skill:

7) Why is *Clostridium perfringens* likely to grow in gangrenous wounds?

Answer:

Skill:

CHAPTER 6   Microbial Growth

OBJECTIVE QUESTIONS

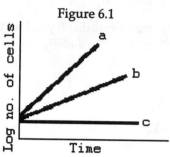

Figure 6.1

1) In Figure 6.1, which line best depicts a facultative anaerobe in the absence of O_2?

    A) line a   B) line b   C) line c

Answer: B
Skill: Understanding

2) In Figure 6.1, which line best illustrates a mesophile at 5°C above its optimum temperature?

    A) line a   B) line b   C) line c

Answer: B
Skill: Understanding

3) In Figure 6.1 , which line best depicts an obligate anaerobe in the presence of O_2?

    A) line a   B) line b   C) line c

Answer: C
Skill: Understanding

4) In Figure 6.1, which line best depicts a mesophile with an optimum temperature of 35°C incubated at 40°C?

    A) line a   B) line b   C) line c

Answer: B
Skill: Understanding

5) In Figure 6.1, which line shows the growth of an obligate aerobe incubated anaerobically?

    A) line a   B) line b   C) line c

Answer: C
Skill: Understanding

6) In Figure 6.1, which line best illustrates the growth of a facultative anaerobe incubated aerobically?

    A) line a   B) line b   C) line c

Answer: A
Skill: Understanding

7) In Figure 6.1, which line best depicts a catalase–negative cell incubated aerobically?

    A) line a   B) line b   C) line c

Answer: C
Skill: Understanding

8) Micrococci are facultative halophiles. In Figure 6.1, which line best depicts the growth of *M. luteus* in a nutrient medium containing 7.5% NaCl?

    A) line a   B) line b   C) line c

Answer: B
Skill: Understanding

9) In Figure 6.1, which line best depicts a psychrophile incubated at room temperature?

    A) line a   B) line b   C) line c

Answer: C
Skill: Understanding

10) In Figure 6.1, which line best depicts a psychrotroph incubated at 0°C?

    A) line a   B) line b   C) line c

Answer: B
Skill: Understanding

11) In Figure 6.1, which line best depicts *Neisseria gonorrhoeae* when growing inside the human body?

    A) line a   B) line b   C) line c

Answer: A
Skill: Understanding

12) The addition of which of the following to a culture medium will neutralize acids?

    A) Buffers   B) Sugars   C) pH   D) Heat   E) Carbon

Answer: A
Skill: Recall

13) Salts and sugars work to preserve foods by creating a
    A) Depletion of nutrients.
    B) Hypotonic environment.
    C) Lower osmotic pressure.
    D) Hypertonic environment.
    E) Lower pH.

    Answer: D
    Skill: Analysis

14) The term facultative anaerobe refers to an organism that
    A) Doesn't use oxygen but tolerates it.
    B) Is killed by oxygen.
    C) Uses oxygen or grows without oxygen.
    D) Requires less oxygen than is present in air.
    E) Prefers to grow without oxygen.

    Answer: C
    Skill: Analysis

15) Which of the following is *not* a disadvantage of the standard plate count?
    A) Cells may form aggregates.
    B) Requires incubation time.
    C) Determines viable cells.
    D) Chemical and physical requirements are determined by media and incubation.
    E) None of the above.

    Answer: C
    Skill: Recall

16) Which of the following is *not* a disadvantage of the direct microscopic count?
    A) Some organisms are motile.
    B) Enumerates dead cells.
    C) No incubation time.
    D) Sample volume is unknown.
    E) Large number of cells is required.

    Answer: D
    Skill: Recall

17) Which of the following is *not* used to determine metabolic activity?

    A) Acid production from fermentation.

    B) CO_2 produced from the Krebs cycle.

    C) NO_2^- produced from the electron transport chain.

    D) Decreased dissolved oxygen.

    E) Turbidity.

Answer: E
Skill: Understanding

18) Thirty–six colonies grew in nutrient agar from 1.0 ml of undiluted sample in a standard plate count. How many cells were in the original sample?

    A) 4   B) 9   C) 18   D) 36   E) 72

Answer: D
Skill: Understanding

Figure 6.2

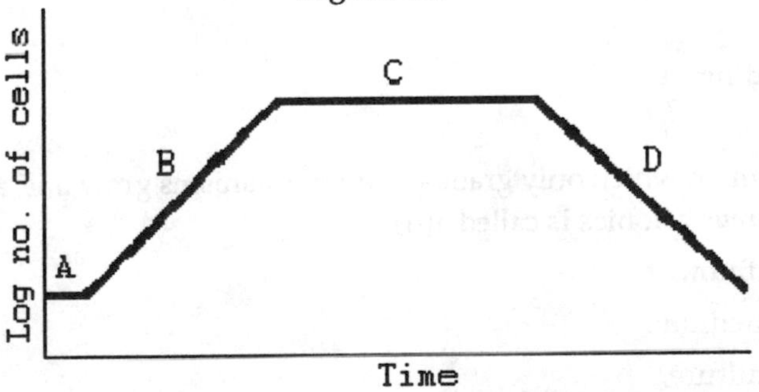

19) In Figure 6.2, which section shows a growth phase where the number of cells dying equals the number of cells dividing?

    A) A   B) B   C) C   D) D   E) A and C

Answer: C
Skill: Recall

20) In Figure 6.2, which sections of the graph illustrate a logarithmic change in cell numbers?

    A) A and C   B) B and D   C) A and B   D) C and D   E) B only

Answer: B
Skill: Analysis

21) Most bacteria grow best at pH

    A) 1.   B) 5.   C) 7.   D) 9.   E) 14.

Answer: C
Skill: Recall

Figure 6.3

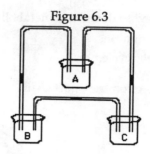

22) Figure 6.3 shows three containers of water connected by tubes. A selectively permeable membrane divides each tube. Solutes are added to each container to give final concentrations of 5% NaCl in A; 10% NaCl in B; and 5% sucrose in C. When the experiment is first set up, the initial movement of water will be

A) A to B; B to C; C to A.

B) A to B; C to B; C to A.

C) A to C; B to C; C to A.

D) A to C; C to B; C to A.

E) B to A; B to C; C to A.

Answer: B
Skill: Understanding

23) A culture medium on which only gram–positive organisms grow and a yellow halo surrounds *Staphylococcus aureus* colonies is called a(n)

A) Selective medium.

B) Differential medium.

C) Enrichment culture.

D) a and b.

E) b and c.

Answer: D
Skill: Analysis

24) A culture medium consisting of agar, human blood, and beef heart is a

A) Chemically defined medium.

B) Complex medium.

C) Selective medium.

D) Differential medium.

E) Reducing medium.

Answer: B
Skill: Analysis

25) Which of the following pairs is mismatched?

   A) Psychrotroph — growth at 0°C

   B) Thermophile — growth at 37°C

   C) Mesophile — growth at 25°C

   D) Psychrophile — growth at 15°C

   E) None of the above

Answer: B
Skill: Recall

26) During which growth phase will gram-positive bacteria be most susceptible to penicillin?

   A) Lag phase

   B) Log phase

   C) Death phase

   D) Stationary phase

   E) The culture is equally susceptible during all phases.

Answer: B
Skill: Understanding

27) Which of the following is the best definition of generation time?

   A) The length of time it takes for lag phase.

   B) The length of time it takes for a cell to divide.

   C) The minimum rate of doubling.

   D) The duration of log phase.

   E) The time it takes for nuclear division.

Answer: B
Skill: Recall

28) All of the following are direct methods to measure microbial growth *except*

   A) Direct microscopic count.

   B) Standard plate count.

   C) Filtration.

   D) Metabolic activity.

   E) MPN.

Answer: D
Skill: Analysis

29) Which group of microorganisms is most likely to spoil a freshwater trout preserved with salt?
   A) Psychrophiles
   B) Halophiles
   C) Anaerobes
   D) Thermophiles
   E) None of the above

   Answer: B
   Skill: Recall

30) Which of the following is an organic growth factor?
   A) Glucose
   B) NAD^+
   C) Peptone
   D) $NH_4H_2PO_4$
   E) All of the above

   Answer: B
   Skill: Analysis

31) Which of the following is an example of a metabolic activity that could be used to measure microbial growth?
   A) Standard plate count
   B) Glucose consumption
   C) Direct microscopic count
   D) Turbidity
   E) MPN

   Answer: B
   Skill: Understanding

32) An experiment began with 4 cells and ended with 128 cells. How many generations did the cells go through?
   A) 64   B) 32   C) 6   D) 5   E) 4

   Answer: D
   Skill: Analysis

33) Three cells with generation times of 30 minutes are inoculated into a culture medium. How many cells are there after 5 hours?
   A) 3×2^{10} B) 1024   C) 243   D) 48   E) 16

   Answer: A
   Skill: Analysis

34) This organism produces catalase and superoxide dismutase.

   A) Aerobe   B) Aerotolerant anaerobe   C) Obligate anaerobe

   Answer: A
   Skill: Understanding

35) This organism is killed by atmospheric O_2.

   A) Aerobe   B) Aerotolerant anaerobe   C) Obligate anaerobe

   Answer: C
   Skill: Understanding

36) Producers in the hydrothermal vents on the ocean floor use CO_2 for their carbon source and what for energy?

   A) Light
   B) Sulfide
   C) Organic molecules
   D) Carbon dioxide
   E) None of the above

   Answer: B
   Skill: Recall

Table 6.1

Below are three different culture media.

Medium A	Medium B	Medium C
Na_2HPO_4	Tide detergent	Glucose
KH_2PO_4	Na_2HPO_4	Peptone
$MgSO_4$	KH_2PO_4	$(NH_4)_2SO_4$
$CaCl_2$	$MgSO_4$	KH_2PO_2
$NaHCO_3$	$(NH_4)_2SO_4$	Na_2HPO_4

37) In Table 6.1, which medium (media) is (are) chemically defined?

   A) A
   B) B
   C) A and B
   D) A and C
   E) None of the above

   Answer: A
   Skill: Analysis

38) In Table, 6.1, in which medium would an autotroph grow?
   A) A
   B) B
   C) A and B
   D) A and C
   E) None of the above

Answer: A
Skill: Analysis

39) Which of the following should *not* be included in a medium used to select for a nitrogen–fixing chemoheterotroph?
   A) KH_2PO_4 B) $(NH_4)_2SO_4$ C) Glucose   D) $MgSO_4$ E) Na_2HPO_4

Answer: B
Skill: Analysis

40) Assume you inoculated 100 cells into 100 ml nutrient broth. You then inoculated 100 cells of the same species into 200 ml nutrient broth. After incubation for 24 hr, you should have
   A) More cells in the 100 ml.
   B) More cells in the 200 ml.
   C) The same number of cells in both.

Answer: C
Skill: Understanding

41) The source of nutrients in nutrient agar is
   A) Agar.
   B) Nutrient.
   C) Peptone and beef extract.
   D) Peptone and NaCl.
   E) All of the above.

Answer: C
Skill: Analysis

42) Catalyzes the reaction: $O_2^- \cdot + O_2^- \cdot + 2H^+ \rightarrow H_2O_2 + O_2$

   A) Catalase
   B) Oxidase
   C) Peroxidase
   D) Superoxide dismutase
   E) None of the above

Answer: D
Skill: Recall

43) Catalyzes the reaction: $2H_2O_2 \rightarrow 2H_2O + O_2$

   A) Catalase

   B) Oxidase

   C) Peroxidase

   D) Superoxide dismutase

   E) None of the above

   Answer: A
   Skill: Recall

44) Catalyzes the reaction: $H_2O_2 + 2H^+ \rightarrow 2H_2O$

   A) Catalase

   B) Oxidase

   C) Peroxidase

   D) Superoxide dismutase

   E) None of the above

   Answer: C
   Skill: Recall

Table 6.2
The following data show growth of two bacteria on different media.

Amount of Growth

	Staphylococcus aureus	Streptococcus pyogenes
Nutrient agar	++	++
Nutrient agar + 7.5% NaCl	+	–

45) The data in Table 6.2 indicate that *S. aureus* is a(n)

   A) Mesophile.

   B) Facultative anaerobe.

   C) Facultative halophile.

   D) Aerobe.

   E) Halophile.

   Answer: C
   Skill: Analysis

ESSAY QUESTIONS

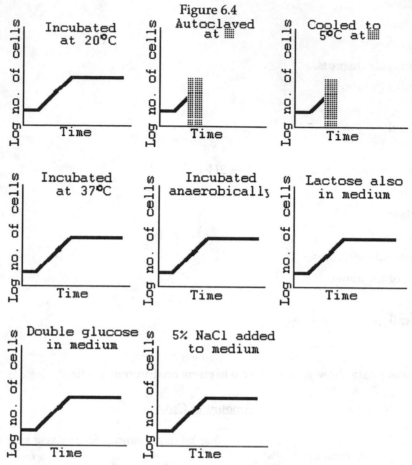

Figure 6.4

1) In each graph in Figure 6.4, the culture was incubated at 30°C in glucose minimal salts broth, aerobically. The bacterium is a facultative anaerobe with an optimum temperature of 37°C, and can metabolize glucose and lactose. Indicate how each growth curve would change under the conditions indicated in each graph. Draw the new graphs or write "no change."

Answer:

Skill:

Table 6.3

Bacterial generation times were calculated in the media shown below. All media were prepared with tap water and incubated aerobically in the light.

Medium	Generation times (min.)			
	Escherichia coli	*Pseudomonas aeruginosa*	*Lactobacillus*	*Nitrobacter*
NaCl, NO_3^-, $MgSO_4$	0	0	0	80
Glucose	100	0	0	0
Glucose, NaCl	82	0	0	0
Glucose, NaCl, PO_4^{3-}	56	200	0	0
Glucose, NaCl, PO_4^{3-}, $MgSO_4$	43	100	0	0
Glucose, NaCl, PO_4^{3-}, $MgSO_4$, 8 amino acids	28	40	0	0
Glucose, NaCl, PO_4^{3-}, $MgSO_4$, 19 amino acids	25	25	80	0

2) Compare and contrast the growth requirements of the bacteria. Which of the media, if any, are chemically defined?

Answer:

Skill:

73

CHAPTER 7   The Control of Microbial Growth

OBJECTIVE QUESTIONS

1) Which of the following is best to sterilize heat–labile solutions?
   A) Dry heat
   B) Autoclave
   C) Membrane filtration
   D) Pasteurization
   E) None of the above

   Answer: C
   Skill: Recall

2) Which of the following best describes the pattern of microbial death?
   A) The cells in a population die at a constant rate.
   B) All the cells in a culture die at once.
   C) All of the cells in a culture are never killed.
   D) The pattern varies depending on the antimicrobial agent.
   E) The pattern varies depending on the species.

   Answer: A
   Skill: Recall

3) Bacterial death will result from damage to which of the following structures?
   A) Cell wall
   B) Plasma membrane
   C) Proteins
   D) Nucleic acids
   E) All of the above

   Answer: E
   Skill: Analysis

4) Which of the following substances can sterilize?
   A) Alcohol
   B) Phenolics
   C) Ethylene oxide
   D) Chlorine
   E) Soap

   Answer: C
   Skill: Recall

5) Which of the following is used for surgical hand scrubs?

 A) Phenol

 B) Chlorine bleach

 C) Chlorhexidine

 D) Soap

 E) Glutaraldehyde

Answer: C
Skill: Recall

6) Which of the following is *not* a heavy metal?

 A) Silver nitrate

 B) Mercurochrome

 C) Merthiolate

 D) Copper sulfate

 E) Chlorine

Answer: E
Skill: Recall

7) Place the following surfactants in order from the most effective to the least effective antimicrobial activity:
1–Soap; 2–Anionic detergent; 3–Cationic detergent.

 A) 1, 2, 3   B) 1, 3, 2   C) 2, 1, 3   D) 3, 2, 1   E) 3, 1, 2

Answer: D
Skill: Analysis

8) The antimicrobial activity of chlorine is due to which of the following?

 A) The formation of hypochlorous acid

 B) The formation of hydrochloric acid

 C) The formation of ozone

 D) The formation of free O

 E) Disruption of the plasma membrane

Answer: A
Skill: Recall

9) Iodophors differ from iodine (I_2) in that iodophors

 A) Don't stain.

 B) Are less irritating.

 C) Are longer lasting.

 D) Are combined with a nonionic detergent.

 E) All of the above.

Answer: E
Skill: Analysis

10) Phenolics differ from phenol in that phenolics are all of the following *except*

    A) Nonirritating.

    B) Relatively odorless.

    C) More effective antibacterial agents.

    D) More toxic to human cells.

    E) All of the above.

Answer: D
Skill: Recall

11) Which of the following does *not* achieve sterilization?

    A) Dry heat

    B) Pasteurization

    C) Autoclave

    D) Formaldehyde

    E) Ethylene oxide

Answer: B
Skill: Recall

12) Which of the following is a limitation of the autoclave?

    A) Length of time

    B) Ability to inactivate viruses

    C) Ability to kill endospores

    D) Use with heat–labile materials

    E) Use with glassware

Answer: D
Skill: Analysis

13) Which of the following affects the elimination of bacteria from an object?

    A) Number of bacteria present

    B) Temperature

    C) pH

    D) Presence of organic matter

    E) All of the above

Answer: E
Skill: Recall

14) Which of the following is *not* a direct result of heat?

   A) Breaking hydrogen bonds

   B) Breaking sulfhydryl bonds

   C) Denaturing enzymes

   D) Cell lysis

   E) None of the above

   Answer: D
   Skill: Analysis

15) Which of the following substances is the least effective antimicrobial agent?

   A) Soap

   B) Cationic detergents

   C) Phenolics

   D) Iodine

   E) Alcohol

   Answer: A
   Skill: Recall

16) Which of these disinfectants acts by denaturing proteins?

   A) Phenolics

   B) Aldehydes

   C) Halogens

   D) Alcohols

   E) All of the above

   Answer: E
   Skill: Analysis

17) Which of these is *not* an oxidizing agent?

   A) Ozone

   B) Hydrogen peroxide

   C) Iodine

   D) Chlorine

   E) None of the above

   Answer: E
   Skill: Analysis

18) Which of the following is *not* used for the disinfection of water?

 A) Ozone

 B) Gamma radiation

 C) Chlorine

 D) Copper sulfate

 E) None of the above

Answer: E
Skill: Analysis

19) Which of the following is *not* effective against nonenveloped viruses?

 A) Alcohol

 B) Chlorine

 C) Ethylene oxide

 D) Ozone

 E) All are equally effective

Answer: A
Skill: Understanding

20) Glutaraldehyde is considered one of the most effective disinfectants for hospital use. Which of the following would *not* be true about glutaraldehyde?

 A) Stains and corrodes.

 B) Safe to transport.

 C) Acts rapidly.

 D) Not hampered by organic material.

 E) Attacks all microorganisms.

Answer: A
Skill: Understanding

21) Which concentration of ethyl alcohol is the most effective bactericide?

 A) 100%   B) 70%   C) 50%   D) 40%   E) 30%

Answer: B
Skill: Recall

22) All of the following contribute to hospital-acquired infections *except*

 A) Some bacteria metabolize disinfectants.

 B) Gram-negative bacteria are often resistant to disinfectants.

 C) Invasive procedures can provide a portal of entry for bacteria.

 D) Bacteria may be present in commercial products such as mouthwash.

 E) None of the above.

Answer: E
Skill: Analysis

23) Which of the following treatments is the most effective for controlling microbial growth?

    A) 63°C for 30 min.

    B) 72°C for 15 sec.

    C) 140°C for < 1 sec.

    D) They are equivalent treatments.

    E) None are effective.

Answer: D
Skill: Recall

24) Which of the following could be used to sterilize plastic Petri plates in a plastic wrapper?

    A) Microwaves

    B) Ultraviolet radiation

    C) Gamma radiation

    D) Sunlight

    E) None of the above

Answer: C
Skill: Analysis

25) Which of the following treatments does *not* yield a sterile product?

    A) 0.45 µm filtration

    B) Autoclaving

    C) Gamma radiation

    D) Ionizing radiation

    E) 170°C for 2 hr

Answer: A
Skill: Analysis

Figure 7.1

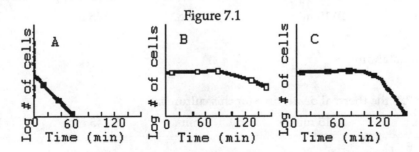

26) In Figure 7.1, what is the thermal death time for culture A?

    A) 150°C   B) 60 min.   C) 120 min.   D) 100°C   E) Can't tell

Answer: B
Skill: Analysis

27) Assume that one culture is a freshly opened package of dried yeast with 0.1% moisture, another culture is the same yeast with 7.5% moisture, and the third culture is a package of yeast mixed with water. Each culture was exposed to 130°C in a hot–air oven. Which graph in Figure 7.1 most likely shows the yeast with 0.1% moisture?

A) A   B) B   C) C   D) A and B   E) B and C

Answer: C
Skill: Understanding

Figure 7.2

A suspension of 10^6 *Bacillus cereus* endospores was put in a hot–air oven at 170°C. Plate counts were used ot determine the number of endospores surviving at the time intervals shown.

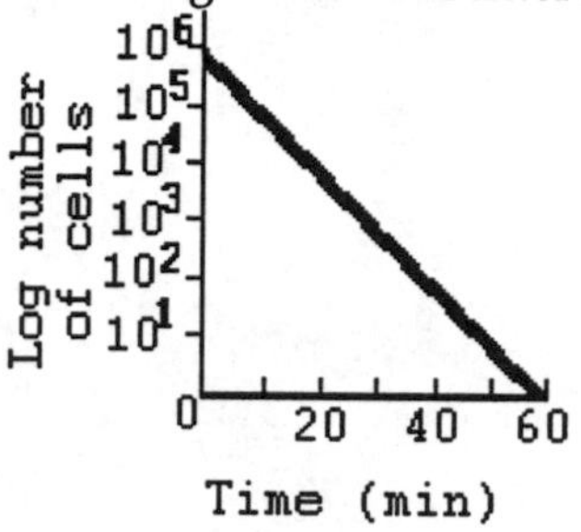

28) In Figure 7.2, the thermal death point for this culture is
A) 15 min.
B) 50°C.
C) 30 min.
D) 170°C.
E) Can't tell from the data provided.

Answer: E
Skill: Understanding

29) In Figure 7.2, the decimal reduction time (D value) for the culture is approximately
A) 0 min.   B) 15 min.   C) 30 min.   D) 45 min.   E) 60 min.

Answer: B
Skill: Understanding

30) In Figure 7.2, the thermal death time for this culture is
A) 0 min.   B) 15 min.   C) 30 min.   D) 45 min.   E) 60 min.

Answer: E
Skill: Understanding

31) Which of the following pairs is *not* correctly matched?

    A) $AgNO_3$ — newborns' eyes

    B) Merthiolate — open wounds

    C) $CuSO_4$ — algicide

    D) H_2O_2 — open wounds

    E) None of the above

Answer: B
Skill: Recall

Figure 7.3

Assume 10^9 *E. coli* cells/ml are in a flask.

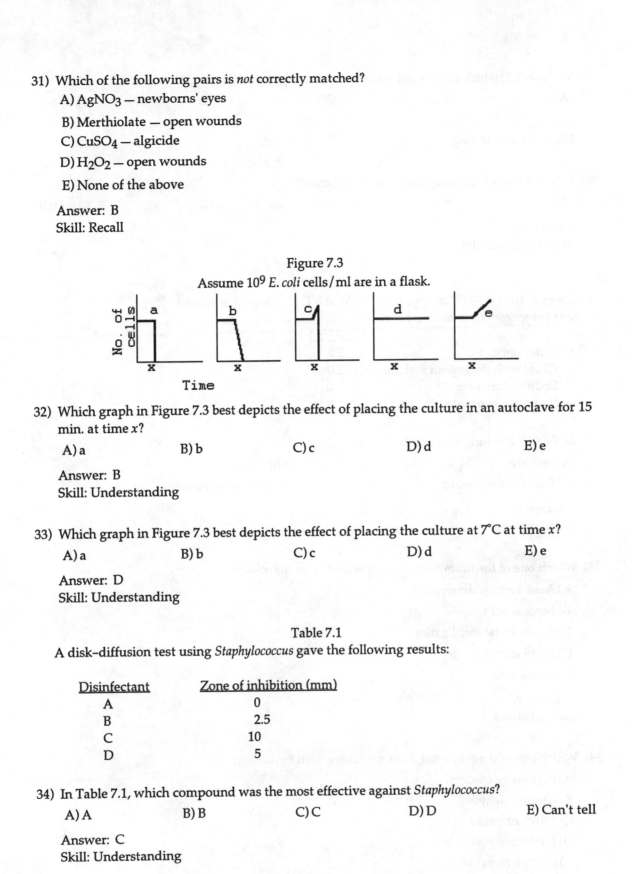

32) Which graph in Figure 7.3 best depicts the effect of placing the culture in an autoclave for 15 min. at time *x*?

    A) a   B) b   C) c   D) d   E) e

Answer: B
Skill: Understanding

33) Which graph in Figure 7.3 best depicts the effect of placing the culture at $7°C$ at time *x*?

    A) a   B) b   C) c   D) d   E) e

Answer: D
Skill: Understanding

Table 7.1

A disk–diffusion test using *Staphylococcus* gave the following results:

Disinfectant	Zone of inhibition (mm)
A	0
B	2.5
C	10
D	5

34) In Table 7.1, which compound was the most effective against *Staphylococcus*?

    A) A   B) B   C) C   D) D   E) Can't tell

Answer: C
Skill: Understanding

35) In Table 7.1, which compound was the most effective against *E. coli*?

A) A   B) B   C) C   D) D   E) Can't tell

Answer: E
Skill: Understanding

36) In Table 7.1, which compound was bactericidal?

A) A   B) B   C) C   D) D   E) Can't tell

Answer: E
Skill: Understanding

Table 7.2

The fate of *E. coli* O157:H7 in apple cider held at 8°C for 2 weeks, with and without preservatives, is shown below:

	Bacteria / ml
Cider only	2.2
Cider with Potassium sorbate	2.0
Sodium benzoate	0
Sorbate + benzoate	0

37) In Table 7.2, which preservative is most effective?

A) Sorbate   B) Benzoate

C) Sorbate + benzoate   D) No preservative

Answer: B
Skill: Analysis

38) Which one of the following does *not* belong with the others?

A) Acid–anionic detergents

B) Benzoic acid

C) Commercial sterilization

D) Pasteurization

E) Sorbic acid

Answer: A
Skill: Analysis

39) Which one of the following does *not* belong with the others?

A) Beta–propiolactone

B) Glutaraldehyde

C) Ethylene oxide

D) Hydrogen peroxide

E) Propylene oxide

Answer: D
Skill: Analysis

40) Foods are preserved with all of the following *except*

   A) Biguanides
   B) Nisin
   C) Potassium sorbate
   D) Sodium nitrite
   E) Sodium propionate

   Answer: A
   Skill: Recall

41) Which one of the following is *most* useful for disinfecting medical instruments?

   A) Benzoic acid
   B) Bisphenols
   C) Chlorine
   D) Phenol
   E) Quats

   Answer: E
   Skill: Recall

42) Which one of the following is the best advertisement for a disinfectant?

   A) Kills *E. coli.*
   B) Kills *Staphylococcus aureus.*
   C) Kills *Pseudomonas.*
   D) Kills lipophilic viruses.
   E) All are equal.

   Answer: C
   Skill: Analysis

Table 7.3

The following data were obtained by incubating the bacteria in nutrient medium + disinfectant for 24 hr, then transferring one loopful to nutrient medium (subculturing).

| | Doom | | | K.O. | |
Dilution	Initial	Subculture		Initial	Subculture
1:16	–	+		+	+
1:32	–	+		+	+
1:64	–	+		+	+
1:128	+	+		+	+

| | Mortum | | | Sterl | |
Dilution	Initial	Subculture		Initial	Subculture
1:16	–	–		–	+
1:32	–	+		+	+
1:64	+	+		+	+
1:128	+	+		+	+

43) In Table 7.3, which disinfectant is the most effective at stopping bacterial growth?

    A) Doom   B) K.O.   C) Mortum   D) Sterl   E) Can't tell

Answer: A
Skill: Understanding

44) In Table 7.3, which disinfectant was bactericidal?

    A) Doom   B) K.O.   C) Mortum   D) Sterl   E) Can't tell

Answer: C
Skill: Understanding

45) In Table 7.3, which disinfectant was most effective against *Salmonella*?

    A) Doom   B) K.O.   C) Mortum   D) Sterl   E) Can't tell

Answer: E
Skill: Understanding

ESSAY QUESTIONS

Table 7.4

The results below were obtained from a use–dilution test of two disinfectants. Cultures were inoculated into tubes with varying concentrations of disinfectants and incubated for 24 hr at 20°C. (+ = growth; – = no growth.)

Concentration	Disinfectant 1		Disinfectant 2	
	Initial	Subculture	Initial	Subculture
1: 10	–	+	–	–
1: 90	–	+	–	–
1: 900	+	+	–	–
1: 90,000	+	+	–	–
1: 900,000	+	+	–	+
1: 9,000,000	+	+	–	+

1) a. In Table 7.4, what is the minimal bacteriostatic concentration of each disinfectant?
   b. Which compound is bactericidal? At what concentration?

Answer:

Skill:

Table 7.5

The following results were obtained from a use–dilution test of two disinfectants. Cultures were inoculated into tubes containing varying concentrations of the disinfectants, incubated for 10 min. at 20° C, and then transferred to growth media without disinfectant. (+ = growth; – = no growth.)

Disinfectant A

Concentration	Gram–positive bacteria	Gram–negative bacteria
1:10	–	–
1:20–70	–	+
1:80	+	+
1:90	+	+
1:100	+	+
1:110–500	+	+

Disinfectant B

Concentration	Gram–positive bacteria	Gram–negative bacteria
1:10–150	–	–
1:160	+	+
1:170	+	+
1:180	+	+
1:190–500	+	+

2) a. In Table 7.5, which disinfectant is most effective?
   b. Against which group of bacteria is disinfectant A most effective?

Answer:

Skill:

3) Assume that you are responsible for decontamination of materials in a large hospital. How would you sterilize each of the following? Briefly justify your answers.
   a. A mattress used by a patient with bubonic plague.
   b. Intravenous glucose–saline solutions.
   c. Used disposable syringes.
   d. Tissues taken from patients.

Answer:

Skill:

CHAPTER 8   Microbial Genetics

OBJECTIVE QUESTIONS

1) A gene is best defined as
   A) A segment of DNA.
   B) Three nucleotides that code for an amino acid.
   C) A sequence of nucleotides in DNA that codes for a functional product.
   D) A sequence of nucleotides in RNA that codes for a functional product.
   E) A transcribed unit of DNA.

   Answer: C
   Skill: Recall

2) Which of the following pairs is mismatched?
   A) DNA polymerase — makes a molecule of DNA from a DNA template
   B) RNA polymerase — makes a molecule of RNA from an RNA template
   C) DNA ligase — joins segments of DNA
   D) Transposase — insertion of DNA segments into DNA
   E) Spliceosome — removal of introns

   Answer: B
   Skill: Recall

3) Which of the following statements is *false*?
   A) DNA polymerase joins nucleotides in one direction only.
   B) The leading strand of DNA is made continuously.
   C) The lagging strand of DNA is started by an RNA primer.
   D) DNA replication proceeds in one direction around the bacterial chromosome.
   E) Multiple replication forks are possible on a bacterial chromosome.

   Answer: D
   Skill: Analysis

4) Two E.coli strains are shown below:

Hfr: pro$^+$, arg$^+$, his$^+$, lys$^+$, met$^+$, ampicillin–sensitive

F: pro$^-$, arg$^-$, his$^-$, lys$^-$, met$^-$, ampicillin–resistant

What supplements would you add to glucose minimal salts agar to select for a recombinant cell that is lys$^+$, arg$^+$, amp–resistant?

A) Ampicillin, lysine, arginine

B) Lysine arginine

C) Ampicillin, proline, histidine, methionine

D) Proline, histidine, methionine

E) Ampicillin, prolein, histidine, lysine

Answer: C
Skill: Understanding

Figure 8.1
The following results were obtained from a replica–plating experiment.

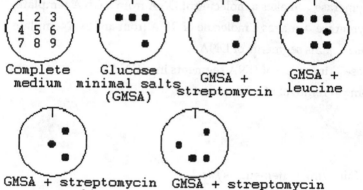

5) In Figure 8.1, which colonies are streptomycin-resistant and leucine-requiring?

A) 1, 2, 3, and 9

B) 3 and 9

C) 4, 6, and 8

D) 4 and 8

E) 5 and 6

Answer: D
Skill: Understanding

6) Which of the following is *not* a product of transcription?

A) A new strand of DNA

B) rRNA

C) tRNA

D) mRNA

E) None of the above

Answer: A
Skill: Understanding

7) All of the following are true about bacteriocins *except*

A) The genes coding for them are on plasmids.

B) They cause food poisoning symptoms.

C) Nisin is a bacteriocin used as a food preservative.

D) Bacteria that produce bacteriocins are resistant to their own bacteriocins.

E) None of the above.

Answer: B
Skill: Analysis

Table 8.1

Culture 1: F^+, leucine$^+$, histidine$^+$

Culture 2: F^-, leucine$^-$, histidine$^-$

8) In Table 8.1, what will be the result of conjugation between cultures 1 and 2?

A) 1 will remain the same;

   2 will become F^+, leucine$^-$, histidine$^-$

B) 1 will become F^-, leu$^+$, his$^+$;

   2 will become F^+, leu$^-$, his$^-$

C) 1 will become F^-, leu$^-$, his$^-$;

   2 will remain the same

D) 1 will remain the same;

   2 will become F^+, leu$^+$, his$^+$

E) 1 will remain the same;

   2 will become F^+ and recombination may occur

Answer: A
Skill: Understanding

9) In Table 8.1, if culture 1 mutates to Hfr, what will be the result of conjugation between the two cultures?

A) They will both remain the same

B) 1 will become F^+, leu$^+$, his$^+$;

   2 will become F^+, leu$^+$, his$^+$

C) 1 will remain the same;

   Recombination will occur in 2

D) 1 will become F^-, leu$^+$, his$^+$;

   2 will become Hfr, leu$^+$, his$^+$

E) Can't tell

Answer: C
Skill: Understanding

10) An enzyme produced in response to the presence of a substrate is called
    A) An inducible enzyme.
    B) A repressible enzyme.
    C) A restriction enzyme.
    D) An operator.
    E) A promoter.

Answer: A
Skill: Recall

11) Which of the following proteins are *not* coded for by genes carried on plasmids?
    A) Enzymes necessary for conjugation
    B) Enzymes that catabolize hydrocarbons
    C) Bacteriocins
    D) Enzymes that inactivate antibiotics
    E) None of the above

Answer: E
Skill: Recall

12) Transformation is the transfer of DNA from a donor to a recipient cell
    A) By a bacteriophage.
    B) As naked DNA in solution.
    C) By cell–to–cell contact.
    D) By crossing over.
    E) By sexual reproduction.

Answer: B
Skill: Recall

13) Genetic change in bacteria can be brought about by
    A) Mutation.
    B) Conjugation.
    C) Transduction.
    D) Transformation.
    E) All of the above.

Answer: E
Skill: Understanding

14) Which of the following is *not* true of a bacterium that is R+?

   A) R+ refers to the possession of a plasmid.

   B) R+ can be transferred to a recipient cell.

   C) It is resistant to certain drugs and heavy metals.

   D) It is F+.

   E) None of the above.

   Answer: D
   Skill: Analysis

15) The initial effect of ionizing radiation on a cell is that it causes

   A) DNA to break.

   B) Bonding between adjacent thymines.

   C) Base substitutions.

   D) The formation of highly reactive ions.

   E) The cells to get hot.

   Answer: D
   Skill: Recall

16) According to the operon model, for the synthesis of an inducible enzyme to occur, the

   A) End-product must not be in excess.

   B) Substrate must bind to the enzyme.

   C) Substrate must bind to the repressor.

   D) Repressor must bind to the operator.

   E) Repressor must not be synthesized.

   Answer: C
   Skill: Analysis

17) Synthesis of a repressible enzyme is stopped by

   A) The allosteric transition.

   B) The substrate binding to the repressor.

   C) The corepressor binding to the operator.

   D) The corepressor-repressor binding to the operator.

   E) The end-product binding to the promoter.

   Answer: D
   Skill: Analysis

Figure 8.2

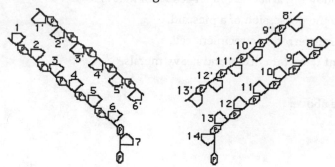

18) In Figure 8.2, if base 4 is thymine, what is base 4'?

    A) Adenine   B) Thymine   C) Cytosine   D) Guanine   E) Uracil

Answer: A
Skill: Understanding

19) In Figure 8.2, if base 4 is thymine, what is base 11'?

    A) Adenine   B) Thymine   C) Cytosine   D) Guanine   E) Uracil

Answer: B
Skill: Understanding

20) In Figure 8.2, base 2 is attached to

    A) Ribose.

    B) Phosphate.

    C) Deoxyribose.

    D) Thymine.

    E) Can't tell.

Answer: C
Skill: Recall

21) The damage caused by ultraviolet radiation is

    A) Never repaired.

    B) Repaired during transcription.

    C) Repaired during translation.

    D) Cut out and replaced.

    E) Repaired by DNA replication.

Answer: D
Skill: Recall

Table 8.2

<u>Codon on mRNA and corresponding amino acid</u>

UUA	leucine	UAA	nonsense
GCA	alanine	AAU	sparagine
AAG	lysine	UGC	cysteine
GUU	valine	UCG, UCU	serine

22) (Use Table 8.2.) If the sequence of amino acids coded for by a strand of DNA is serine–alanine–lysine–leucine, what is the order of bases in the sense strand of DNA?

A) 3′ UGUGCAAAGUUA

B) 3′ AGACGTTTCAAT

C) 3′ TCTCGTTTGTTA

D) 5′ TGTGCTTTCTTA

E) 5′ AGAGCTTTGAAT

Answer: B
Skill: Understanding

23) (Use Table 8.2.) If the sequence of amino acides coded for by a strand of DNA is serine–alanine–lysine–leucine, the coding for the antisense strand of DNA is

A) 5′ ACAGTTTCAAT

B) 5′ TCTGCAAAGTTA

C) 3′ UGUGCAAAGUUA

D) 3′ UCUCGAAAGUUA

E) 3′ TCACGUUUCAAU

Answer: B
Skill: Understanding

24) (Use Table 8.2.) The anticodon for valine is

A) GUU   B) CUU   C) CTT   D) CAA   E) GTA

Answer: D
Skill: Understanding

25) (Use Table 8.2.) What is the sequence of amino acids coded for by the following sequence of bases in a strand of DNA?

3′ ATTACGCTTTGC

A) Leucine–arginine–lysine–alanine

B) Asparagine–arginine–lysine–alanine

C) Asparagine–cysteine–valine–serine

D) Transcription would stop at the first codon

E) Can't tell

Answer: D
Skill: Understanding

26) (Use Table 8.2.) If a frameshift mutation occurred in a sequence of bases shown below, what would be the sequence of amino acids coded for?

3′ ATTACGCTTTGC

A) Leucine–arginine–lysine–alanine

B) Asparagine–arginine–lysine–alanine

C) Asparagine–cysteine–valine–serine

D) Translation would stop at the first codon

E) Can't tell

Answer: E
Skill: Understanding

Table 8.3 – Metabolic Pathway

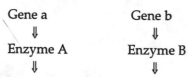

Gene a   Gene b
⇓   ⇓
Enzyme A   Enzyme B
⇓   ⇓
Compound A -----> Compound B -----> Compound C

27) In Table 8.3, if compound C reacts with the allosteric site of enzyme A, this would exemplify

A) A mutation.

B) End–product repression.

C) Feedback inhibition.

D) Competitive inhibition.

E) None of the above.

Answer: C
Skill: Understanding

28) In Table 8.3, if enzyme A is a repressible enzyme, compound C would
    A) Always be in excess.
    B) Bind to the enzyme.
    C) Bind to the corepressor.
    D) All of the above.
    E) None of the above.

Answer: C
Skill: Understanding

29) In Table 8.3, if enzyme A is an inducible enzyme,
    A) Compound C would bind to the repressor.
    B) Compound A would bind to the repressor.
    C) Compound B would bind to enzyme A.
    D) Compound A would react with enzyme B.
    E) None of the above.

Answer: B
Skill: Understanding

30) Conjugation differs from reproduction because conjugation
    A) Replicates DNA.
    B) Transfers DNA vertically, to new cells.
    C) Transfers DNA horizontally, to cells in the same generation.
    D) Transcribes DNA to RNA.
    E) None of the above.

Answer: C
Skill: Understanding

31) The necessary ingredients for DNA synthesis can be mixed together in a test tube. The DNA polymerase is from *Thermus aquaticus* and the template is from a human cell. The DNA synthesized would be most similar to
    A) Human DNA.
    B) *T. aquaticus* DNA.
    C) A mixture of human and *T. aquaticus* DNA.
    D) Human RNA.
    E) *T. aquaticus* RNA.

Answer: A
Skill: Understanding

32) An antibiotic that binds the 50S portion of the ribosome as shown above would

   A) Stop the ribosome from moving along the mRNA.

   B) Prevent tRNA attachment.

   C) Prevent peptide bond formation.

   D) Prevent transcription.

   E) None of the above.

   Answer: B
   Skill: Understanding

Figure 8.3

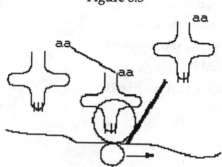

33) In Figure 8.3, the antibiotic streptomycin binds tRNA^met at the 30S ribosome. From this information you can conclude that streptomycin

   A) Prevents transcription in eucaryotes.

   B) Prevents translation in eucaryotes.

   C) Prevents transcription in procaryotes.

   D) Prevents translation in procaryotes.

   E) Prevents mRNA–ribosome binding.

   Answer: D
   Skill: Understanding

34) The mechanism by which the presence of glucose inhibits the arabinose operon.

   A) Catabolic repression

   B) Translation

   C) DNA polymerase

   D) Repression

   E) Induction

   Answer: A
   Skill: Recall

35) The mechanism by which the presence of arabinose controls the arabinose operon.

A) Catabolic repression

B) Translation

C) DNA polymerase

D) Repression

E) Induction

Answer: E
Skill: Recall

36) If you knew the sequence of nucleotides within a gene, which one of the following could you determine with the most accuracy?

A) The primary structure of the protein

B) The secondary structure of the protein

C) The tertiary structure of the protein

D) The quaternary structure of the protein

E) Can't tell

Answer: A
Skill: Analysis

37) An enzyme that covalently bonds nucleotide sequences in DNA.

A) RNA polymerase

B) DNA ligase

C) Restriction enzyme

D) Transposase

E) DNA polymerase

Answer: B
Skill: Recall

38) An enzyme that copies DNA to make a molecule of RNA.

A) RNA polymerase

B) DNA ligase

C) Restriction enzyme

D) Transposase

E) DNA polymerase

Answer: A
Skill: Recall

39) An enzyme that cuts double-stranded DNA at specific nucleotide sequences.

A) RNA polymerase

B) DNA ligase

C) Restriction enzyme

D) Transposase

E) DNA polymerase

Answer: C
Skill: Recall

40) An enzyme that cuts and seals DNA.

A) RNA polymerase

B) DNA ligase

C) Restriction enzyme

D) Transposase

E) DNA polymerase

Answer: D
Skill: Recall

Figure 8.4

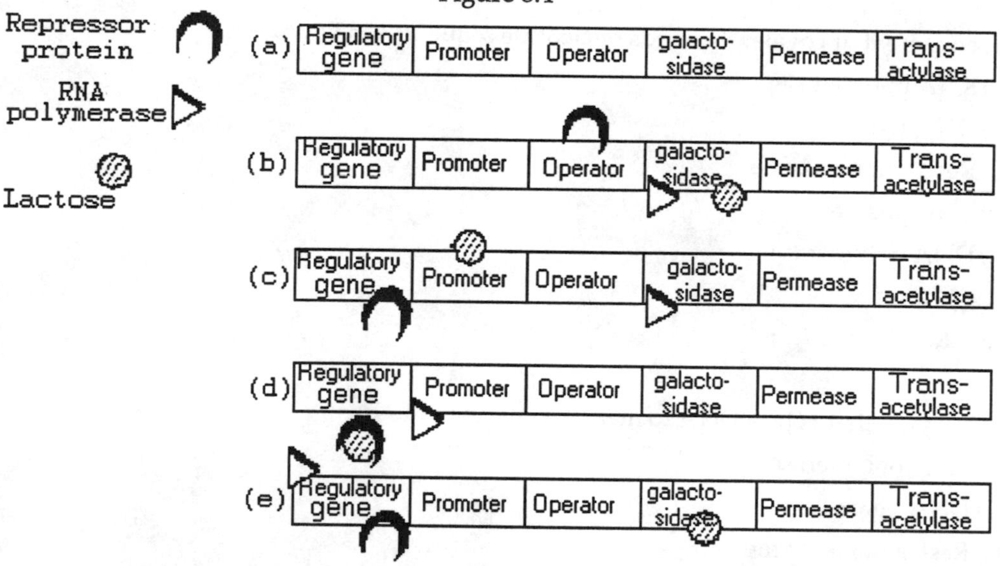

41) In Figure 8.4, which model of the *lac* operon correctly shows RNA polymerase, lactose, and repressor protein when the structural genes are being transcribed?

A) Model (a)   B) Model (b)   C) Model (c)   D) Model (d)   E) Model (e)

Answer: D
Skill: Analysis

42) In transcription,

   A) DNA is changed to RNA.

   B) DNA is copied to RNA.

   C) DNA is replicated.

   D) RNA is copied to DNA.

   E) Proteins are made.

   Answer: B
   Skill: Recall

43) DNA is constructed of

   A) A single strand of nucleotides with internal hydrogen bonding.

   B) Nucleotides bonded A—C and G—T.

   C) Two strands of nucleotides running antiparallel.

   D) Two strands of identical nucleotides with hydrogen bonds between them.

   E) None of the above.

   Answer: C
   Skill: Recall

44) The *lac* operon

   A) Hydrolyzes lactose.

   B) Produces constitutive enzymes.

   C) Produces inducible enzymes.

   D) Produces repressible enzymes.

   E) None of the above.

   Answer: C
   Skill: Recall

45) A cell that cannot make tRNA

   A) Can make proteins if amino acids are provided in the growth medium.

   B) Can make proteins if mRNA is provided in the growth medium.

   C) Can't make proteins unless aminoacyl synthetase is provided in the growth medium.

   D) Can't make proteins.

   E) None of the above.

   Answer: D
   Skill: Analysis

ESSAY QUESTIONS

1) What is the survival value of each of the following?
   a. Semiconservative replication of DNA.
   b. The degeneracy of the genetic code.

   Answer:

   Skill:

2) Scientists are concerned that bacteria will be resistant to all antibiotics within the next decade. Using your knowledge of genetics, describe how bacterial populations can develop drug resistance.

   Answer:

   Skill:

3) Explain why the following statement is false: Sexual reproduction is the only mechanism for genetic change.

   Answer:

   Skill:

CHAPTER 9   Biotechnology and Recombinant DNA

OBJECTIVE QUESTIONS

1) The following steps are used to make DNA fingerprints. What is the third step?

   A) Collect DNA.

   B) Digest with a restriction enzyme.

   C) Perform electrophoresis.

   D) Lyse cells.

   E) Add stain.

   Answer: B
   Skill: Understanding

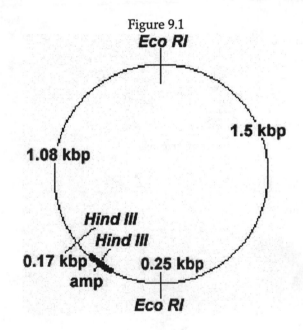

Figure 9.1

2) How many pieces will *Eco*RI produce from the plasmid shown in Figure 9.1?

   A) 1   B) 2   C) 3   D) 4   E) 5

   Answer: B
   Skill: Analysis

3) In Figure 9.1, after digestion with the appropriate restriction enzyme, what is the smallest piece containing the ampicillin–resistance (amp) gene?
   A) 0.17 kilobase pairs
   B) 0.25 kbp
   C) 1.08 kbp
   D) 1.50 kbp
   E) 3.00 kbp

   Answer: D
   Skill: Understanding

4) Which of the following can be used to make recombinant DNA?
   A) Protoplast fusion
   B) Tungsten "bullets"
   C) Microinjection
   D) Transformation
   E) All of the above.

   Answer: E
   Skill: Recall

5) The reaction catalyzed by reverse transcriptase.
   A) DNA → mRNA
   B) mRNA → cDNA
   C) mRNA → protein
   D) DNA → DNA
   E) None

   Answer: B
   Skill: Recall

6) The reaction catalyzed by DNA polymerase.
   A) DNA → mRNA
   B) mRNA → cDNA
   C) mRNA → protein
   D) DNA → DNA
   E) None

   Answer: D
   Skill: Recall

7) Which of the following is *not* a disadvantage of *E. coli* for making a human gene product?

A) Endotoxin may be in the product.

B) It doesn't secrete most proteins.

C) Its genes are well known.

D) It can't process introns.

E) None of the above.

Answer: C
Skill: Recall

8) Which of the following is *not* an agricultural product made by genetic engineering?

A) Frost retardant

B) *Bacillus thuringiensis* insecticide

C) Nitrogenase (nitrogen fixation)

D) Glyphosate resistance

E) Drought resistance

Answer: E
Skill: Recall

9) If you have inserted a gene in the Ti, the next step in genetic engineering is

A) Transformation of *E. coli* with Ti.

B) Splicing Ti into a plasmid.

C) Transformation of an animal cell.

D) Inserting Ti into *Agrobacterium*.

E) None of the above.

Answer: D
Skill: Understanding

10) Which of the following methods of genetic engineering could be described as "hit or miss"?

A) Protoplast fusion

B) Viral transduction

C) Transformation

D) Cloning

E) None of the above

Answer: A
Skill: Understanding

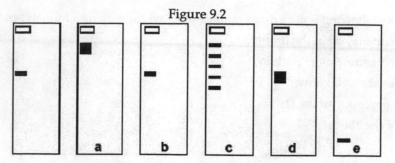
Figure 9.2

11) The figure at the left of Figure 9.2 shows a gene identified by Southern blotting. What will a Southern blot of the same gene look like after PCR?

   A) Figure a   B) Figure b   C) Figure c   D) Figure d   E) Figure e

   Answer: D
   Skill: Understanding

12) Suicide genes can be controlled by the fimbriae-gene operator. This would result in the death of

   A) All cells.

   B) Cells making flagella.

   C) Cells making fimbriae.

   D) Cells at 37°C.

   E) Conjugating cells.

   Answer: C
   Skill: Analysis

13) Subunit vaccines can be made by genetic engineering of yeast cells. A side effect of vaccination might be

   A) The disease.

   B) A yeast infection.

   C) Due to extraneous material.

   D) That the vaccine *doesn't* provide immunity.

   E) None of the above.

   Answer: E
   Skill: Analysis

14) *E. coli* makes insulin because

   A) It needs to regulate its cell-glucose level.

   B) It's an ancient gene that now has no function.

   C) The insulin gene was inserted into it.

   D) It picked up the insulin gene from another cell.

   E) No reason; it doesn't make insulin.

   Answer: C
   Skill: Recall

15) The value of cDNA in genetic engineering is that

    A) it lacks exons.   B) it lacks introns.

    C) it's really RNA.   D) None of the above.

Answer: B
Skill: Recall

16) Which enzyme does not make sticky ends?

A) Enzyme	Recognition
Bam HI	G↓GATCC
	CCTAG↑G
B) Enzyme	Recognition
*Eco*RI	G↓AATTC
	CTTAA↑G
C) Enzyme	Recognition
Hae III	GG↓CC
	CC↑GG
D) Enzyme	Recognition
*Hind*III	A↓AGCTT
	TTCGA↑A
E) Enzyme	Recognition
Pst I	CTGC↓G
	G↑ACGTC

Answer: C
Skill: Analysis

17) Which enzyme would cut this strand of DNA: GCATGGATCCCAATGC?

A) Enzyme	Recognition
Bam HI	G↓GATCC
	CCCTAG↑G
B) Enzyme	Recognition
*Eco*RI	G↓AATTC
	CTTAA↑G
C) Enzyme	Recognition
Hae III	GG↓CC
	CC↑GG
D) Enzyme	Recognition
*Hind*III	A↓AGCTT
	TTCGA↑A
E) Enzyme	Recognition
Pst I	CTGC↓G
	G↑ACGTC

Answer: A
Skill: Analysis

18) Pieces of DNA stored in yeast cells.

    A) Library

    B) Clone

    C) Vector

    D) Southern blot

    E) PCR

Answer: A
Skill: Recall

19) A population of cells carrying a desired plasmid.

    A) Library

    B) Clone

    C) Vector

    D) Southern blot

    E) PCR

Answer: B
Skill: Recall

20) Self–replicating DNA to transmit a gene from one organism to another.

    A) Library

    B) Clone

    C) Vector

    D) Southern blot

    E) PCR

Answer: D
Skill: Recall

21) A technique used to identify bacteria carrying a specific gene is

    A) Southern blot.

    B) DNA probe.

    C) Transformation.

    D) Cloning.

    E) None of the above.

Answer: B
Skill: Recall

22) A colleague has used computer modeling to design an improved enzyme. To produce this enzyme, the next step is

A) look for a bacterium that makes the improved enzyme.

B) mutate bacteria until one makes the improved enzyme.

C) determine the nucleotide sequence for the improved enzyme.

D) synthesize the gene for the improved enzyme.

E) None; the enzyme can't be produced.

Answer: C
Skill: Analysis

23) You have a small gene that you wish replicated by PCR. You add radioactively labeled nucleotides to the PCR machine. After 3 replication cycles, how many double–stranded DNA molecules do you have?

A) 2   B) 4   C) 8   D) 16   E) Thousands

Answer: C
Skill: Understanding

24) You have a small gene that you wish replicated by PCR. You add radioactively labeled nucleotides to the PCR thermocycler. After 3 replication cycles, what percentage of the DNA single–strands are radioactively labeled?

A) 0%   B) 12.5%   C) 50%   D) 87.5%   E) 100%

Answer: D
Skill: Understanding

Figure 9.3

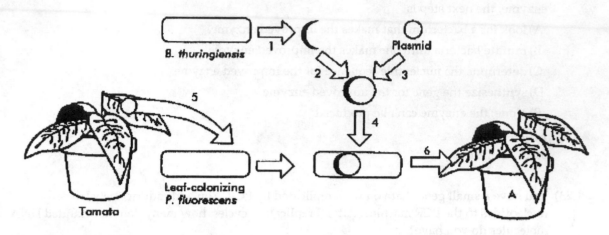

25) In Figure 9.3, the resulting organism (A) is

    A) *Bacillus thuringiensis*.

    B) *Pseudomonas fluorescens*.

    C) a tomato plant.

    D) *E. coli*.

    E) a plant × *Pseudomonas* hybrid.

Answer: C
Skill: Analysis

26) In Figure 9.3, the resulting *P. fluorescens* has

    A) a tomato gene.

    B) an *E. coli* gene.

    C) a *Bacillus* gene.

    D) a tomato and a *Bacillus* gene.

    E) no new gene.

Answer: C
Skill: Analysis

27) In Figure 9.3, the purpose of this experiment is to

    A) put a gene into a plant.

    B) put an insecticide on plant leaves.

    C) put a gene in *Bacillus*.

    D) isolate *Pseudomonas* from a plant.

    E) None of the above.

Answer: B
Skill: Analysis

28) In Figure 9.3, the vector is

A) a virus.
B) a plasmid.
C) a library.
D) None of the above.

Answer: B
Skill: Analysis

29) In Figure 9.3, the process required in step 5 is

A) Transformation.
B) Southern blotting.
C) PCR.
D) None of the above.

Answer: A
Skill: Analysis

30) Which of the following techniques is used to alter one amino acid in protein?

A) Cloning
B) PCF
C) Restriction enzymes
D) Selection
E) Site–direct mutagenesis

Answer: E
Skill: Recall

31) Source of heat-stable DNA polymerase.

A) *Agrobacterium tumefaciens*
B) *Thermus aquaticus*
C) *Saccharomyces cerevisiae*
D) *Bacillus thuringiensis*
E) *Pseudomonas*

Answer: B
Skill: Recall

32) Naturally possesses the Ti plasmid.

A) *Agrobacterium tumefaciens*
B) *Thermus aquaticus*
C) *Saccharomyces cervisiae*
D) *Bacillus thuringiensis*
E) *Pseudomonas*

Answer: A
Skill: Recall

33) Degrades PCBs; engineered to produce BT toxin.

    A) *Agrobacterium tumefaciens*

    B) *Thermus aquaticus*

    C) *Saccharomyces cerevisiae*

    D) *Bacillus thuringiensis*

    E) *Pseudomonas*

Answer: E
Skill: Recall

34) A eukaryote used in genetic engineering.

    A) *Agrobacterium tumefaciens*

    B) *Thermus aquaticus*

    C) *Saccharomyces cerevisiae*

    D) *Bacillus thuringiensis*

    E) *Pseudomonas*

Answer: C
Skill: Recall

35) An example of gene therapy is

    A) insertion of the insulin gene in *E. coli*.

    B) insertion of the insulin gene in a mammalian cell culture.

    C) insertion of the insulin gene in a diabetic person's pancreas cells

    D) injection of insulin into a diabetic person.

    E) None of the above.

Answer: C
Skill: Recall

36) The use of an antibiotic resistance gene on a plasmid used in genetic engineering makes

    A) replica plating possible.

    B) direct selection possible.

    C) the recombinant cell dangerous.

    D) the recombinant cell unable to survive.

    E) All of the above.

Answer: B
Skill: Analysis

37) The following steps must be performed to make a bacterium produce human protein X:
1–Translation; 2–Restriction enzyme; 3–Procaryotic transcription; 4–DNA ligase; 5–
Transformation; 6–Eucaryotic transcription; 7–Reverse transcription.
Put the steps in the correct sequence.

A) 5, 2, 3, 4, 7, 6, 1

B) 1, 2, 3, 5, 4, 7, 6

C) 6, 7, 2, 3, 4, 5, 1

D) 6, 7, 2, 4, 5, 3, 1

E) 6, 2, 1, 3, 4, 5, 7

Answer: D
Skill: Understanding

38) Large numbers of bacterial cells are *not* found in crown galls because

A) the plant kills the bacteria.

B) cell walls protect the plant from bacterial invasion.

C) a gene in plant cells is controlling growth.

D) the assumption is not true; many bacteria are in the galls.

E) None of the above.

Answer: C
Skill: Recall

39) A restriction fragment is

A) a gene.   B) a segment of DNA.

C) a segment of RNA.   D) None of the above.

Answer: B
Skill: Recall

40) A specific gene can be inserted into a cell by all of the following *except*

A) Protoplast fusion.

B) a gene gun.

C) Microinjection.

D) Electroporation.

E) *Agrobacterium.*

Answer: A
Skill: Analysis

41) Which of the following is *not* involved in making cDNA?

A) Reverse transcription.   B) RNA processing to remove introns.

C) Transcription.   D) Translation.

Answer: D
Skill: Understanding

42) PCR can be used to identify an unknown bacterium because

    A) the RNA primer is specific.

    B) DNA polymerase will replicate DNA.

    C) DNA can be electrophoresed.

    D) all cells have DNA.

    E) None of the above.

Answer: A
Skill: Analysis

43) PCR can be used to amplify DNA in a clinical sample. The following steps are used in PCR. What is the fourth step?

    A) Collect DNA.

    B) Incubate at 94 degrees C.

    C) Incubate at 60 degrees C.

    D) Incubate at 72 degrees C.

    E) Add DNA polymerase.

Answer: C
Skill: Understanding

44) Restriction enzymes are

    A) Bacterial enzymes that splice DNA.

    B) Bacterial enzymes that destroy phage DNA.

    C) Animal enzymes that splice RNA.

    D) Viral enzymes that destroy host DNA.

Answer: B
Skill: Recall

45) Foreign DNA can be inserted into cells by all of the following except

    A) Transformation.

    B) Electroporation.

    C) Protoplast fusion.

    D) A gene gun.

    E) None of the above.

Answer: E
Skill: Recall

ESSAY QUESTIONS

1) Some scientists are concerned that genetic engineering allows humans to tamper with evolution. Argue either for or against this position.

Answer:

Skill:

2) *Pseudomonas syringae* is found naturally in the soil. Sold as Snomax$^{(R)}$, it is used to make snow at ski resorts. The same bacterium with a gene deletion (Ice–minus$^{(R)}$) is used to prevent ice formation on plants. Should Snomax and Ice–minus be considered engineered organisms and subject to precautions of releasing genetically engineered microorganisms (GEMs)? Explain why or why not.

Answer:

Skill:

3) In the human genome project, pieces of human DNA are stored in *E. coli* or yeast. What is the purpose of this activity?

Answer:

Skill:

CHAPTER 10   Classification of Microorganisms

OBJECTIVE QUESTIONS

1) All of the following are true about archaea *except*
   A) They are prokaryotes.
   B) They lack peptidoglycan in their cell walls.
   C) Some are thermoacidophiles; others are extreme halophiles.
   D) Evidence suggests they evolved from eukaryotes.
   E) Some produce methane from carbon dioxide and hydrogen.

   Answer: D
   Skill: Recall

2) Which of the following characterizes the Domain Bacteria?
   A) Prokaryotic cells; ether linkages in phospholipids
   B) Eukaryotic cells; ester linkages in phospholipids
   C) Prokaryotic cells; ester linkages in phospholipids
   D) Complex cellular structures
   E) Multicellular

   Answer: C
   Skill: Recall

3) If two organisms have similar rRNA sequences, you can conclude all of the following *except*
   A) They are related.
   B) They evolved from a common ancestor.
   C) They will have similar G–C ratios.
   D) They will both ferment lactose.
   E) None of the above.

   Answer: D
   Skill: Understanding

4) The outstanding characteristic of the Kingdom Fungi is
   A) All members are photosynthetic.
   B) Absorption of dissolved organic matter.
   C) Absorption of dissolved inorganic matter.
   D) All members are microscopic.
   E) All members are macroscopic.

   Answer: B
   Skill: Analysis

5) All of the following are true about the members of the Kingdom Plantae *except*

A) They are multicellular.

B) They have eucaryotic cells.

C) They can photosynthesize.

D) They use organic carbon sources.

E) They use inorganic energy sources.

Answer: D
Skill: Recall

6) All of the following are true about the members of the Kingdom Animalia *except*

A) They are multicellular.

B) They have eucaryotic cells.

C) They can photosynthesize.

D) They use organic carbon sources.

E) They use organic energy sources.

Answer: C
Skill: Recall

7) A genus can best be defined as

A) A taxon composed of families.

B) A taxon comprised of one or more species and below family.

C) A taxon belonging to a species.

D) A taxon comprised of classes.

E) The most specific taxon.

Answer: B
Skill: Understanding

8) A bacterial species differs from a species of eukaryotic organisms in that a bacterial species

A) Does not breed with other species.

B) Has a limited geographical distribution.

C) Can be distinguished from other bacterial species.

D) Is a population of cells with similar characteristics.

E) All of the above are true.

Answer: D
Skill: Understanding

9) Which of the following is the best evidence for a three–domain system?
   A) There are three distinctly different cell structures.
   B) There are three distinctly different cellular chemical compositions.
   C) There are three distinctly different Gram reactions.
   D) Some bacteria live in extreme environments.
   E) None of the above.

   Answer: B
   Skill: Analysis

10) Biochemical tests are used to determine
    A) Staining characteristics.
    B) Amino acid sequences.
    C) Nucleic acid–base composition.
    D) Enzymatic activities.
    E) All of the above.

    Answer: D
    Skill: Recall

11) Amino acid sequencing provides direct information about
    A) Enzymatic activities.
    B) Nucleotide bases making up a gene.
    C) Identification of an organism.
    D) Antigenic composition.
    E) Morphology.

    Answer: B
    Skill: Analysis

12) The phylogenetic classification of bacteria is based on
    A) Cell morphology.
    B) Gram–reaction.
    C) rRNA sequences.
    D) Habitat.
    E) All of the above.

    Answer: C
    Skill: Recall

13) All of the following are reasons for classifying viruses in the five kingdoms and not in a sixth kingdom *except*

A) Some viruses can incorporate their genome into a host's genome.

B) Viruses direct anabolic pathways of host cells.

C) Viruses are obligate parasites.

D) Viruses are not composed of cells.

E) None of the above.

Answer: D
Skill: Understanding

14) Which of the following provides taxonomic information that includes the others?

A) Nucleic acid hybridization

B) Nucleic acid–base composition

C) Amino acid sequencing

D) Biochemical tests

E) Numerical taxonomy

Answer: E
Skill: Analysis

15) Which of the following is in the correct order from the most general to the most specific?

A) Kindgom—phylum—class—order—family—genus—species

B) Kingdom—class—order—family—phylum—genus—species

C) Species—genus—family—order—class—phylum—kingdom

D) Species—genus—phylum—family—order—class—kingdom

E) Phylum—kingdom—class—order—genus—family—species

Answer: A
Skill: Recall

16) Fossil evidence indicates that prokaryotic cells first existed on the Earth

A) 350 years ago.

B) 3500 years ago.

C) 3.5 million years ago.

D) 3.5 billion years ago.

E) 3.5×10^{12} years ago.

Answer: D
Skill: Recall

17) Protist is a diverse group of organisms that are similar in
    A) rRNA sequences.
    B) Metabolic type.
    C) Motility.
    D) Being unicellular.
    E) None of the above.

Answer: E
Skill: Analysis

18) Yeasts belong to the Kingdom
    A) Animalia.   B) Plantae.   C) Protista.   D) Fungi.   E) Monera.

Answer: D
Skill: Recall

19) In the scientific name *Enterobacter aerogenes*, *Enterobacter* is the
    A) Specific epithet.
    B) Genus.
    C) Family.
    D) Order.
    E) Kingdom.

Answer: B
Skill: Analysis

20) The arrangement of organisms into taxa
    A) Shows degrees of relatedness between organisms.
    B) Shows relationships to common ancestors.
    C) Was designed by Charles Darwin.
    D) Is arbitrary.
    E) Is based on evolution.

Answer: A
Skill: Analysis

21) Bacteria and archaea are similar in which of the following?
    A) Peptidoglycan cell walls.
    B) Methionine is the start signal for protein synthesis.
    C) Sensitivity to antibiotics.
    D) Possess procaryotic cells.
    E) None of the above.

Answer: D
Skill: Recall

22) Which of the following best defines a strain?

   A) A population of cells with similar characteristics.

   B) A group of organisms with a limited geographical distribution.

   C) A pure culture.

   D) A group of cells all derived from a single parent.

   E) The same as a species.

   Answer: D
   Skill: Recall

23) Serological testing is based on the fact that

   A) All bacteria have the same antigens.

   B) Antibodies react specifically with an antigen.

   C) The human body makes antibodies against bacteria.

   D) Antibodies cause the formation of antigens.

   E) Bacteria clump together when mixed with any antibodies.

   Answer: B
   Skill: Analysis

24) Phage typing is based on the fact that

   A) Bacteria are destroyed by viruses.

   B) Viruses cause disease.

   C) Bacterial viruses attack specific cells.

   D) *Staphylococcus* causes infections.

   E) Phages and bacteria are related.

   Answer: C
   Skill: Analysis

25) Organism A has 70 moles % G+C and organism B has 40 moles % G+C. Which of the following can be concluded from these data?

   A) The two organisms are related.

   B) The two organisms are unrelated.

   C) The organisms make entirely different enzymes.

   D) Their nucleic acids will not hybridize.

   E) None of the above.

   Answer: B
   Skill: Understanding

26) Nucleic acid hybridization is based on the fact that

    A) The strands of DNA can be separated.

    B) A chromosome is composed of complementary strands.

    C) Pairing between complementary bases occurs.

    D) DNA is composed of genes.

    E) None of the above.

Answer: C
Skill: Recall

27) One of the most popular taxonomic tools is DNA fingerprinting to develop profiles of organisms. These profiles provide direct information about

    A) Enzymatic activities.

    B) Protein composition.

    C) The presence of specific genes.

    D) Antigenic composition.

    E) None of the above.

Answer: E
Skill: Understanding

28) Data collected to date indicate that

    A) Humans and marine mammals cannot be infected by the same pathogens.

    B) Marine mammals do not get infectious diseases.

    C) New species of bacteria may be discovered in wild animals.

    D) Marine mammals don't have an immune system.

    E) None of the above.

Answer: E
Skill: Recall

29) Which of the following is most useful in determining whether two organisms are related?

    A) If both ferment lactose.

    B) If both are gram-positive.

    C) If both are motile.

    D) If both are aerobic.

    E) All are equally important.

Answer: B
Skill: Recall

30) A strain is

A) A taxon of cells derived from a single cell.

B) A genetically engineered cell.

C) A taxon composed of species.

D) A mound of cells on an agar medium.

E) None of the above.

Answer: A
Skill: Recall

31) Which of the following is *not* used in a standard bacteriology laboratory identification?

A) Nucleic acid–base composition

B) Differential staining

C) Serological testing

D) Morphology determination

E) Biochemical testing

Answer: A
Skill: Recall

Figure 10.1
A nucleic acid hybridization experiment produced the following results.

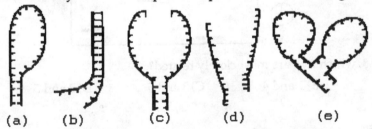

(a)　　(b)　　　(c)　　(d)　　　(e)

32) In Figure 10.1, which figure shows the most closely related organisms?

A) a　　　　　B) b　　　　　C) c　　　　　D) d　　　　　E) e

Answer: B
Skill: Analysis

Figure 10.2
1. 9+2 flagella
2. Nucleus
3. Plasma membrane
4. Peptidoglycan
5. Mitochondrion
6. Fiimbriae

33) In Figure 10.2, which is (are) found in all Eukarya?

A) 2, 3, 5　　　　B) 1, 4, 6　　　　C) 3, 5　　　　D) 2, 3　　　　E) All of them

Answer: D
Skill: Analysis

34) In Figure 10.2, which is (are) found *only* in prokaryotes?

    A) 1, 2, 3   B) 4, 6   C) 2   D) 1   E) 2, 4, 5

    Answer: B
    Skill: Analysis

Figure 10.3

This figure shows the results of a gel electrophoresis separation of restriction fragments of the DNA of different organisms.

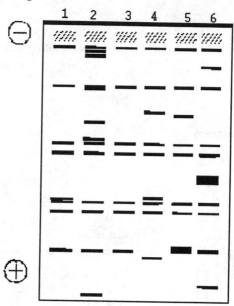

35) In Figure 10.3, which two are most closely related?

    A) 1 and 3   B) 2 and 4   C) 3 and 5   D) 2 and 5   E) 4 and 5

    Answer: A
    Skill: Analysis

36) Into which group would you place a photosynthetic cell that lacks a nucleus?

    A) Animalia   B) Bacteria   C) Fungi   D) Plantae   E) Protist

    Answer: B
    Skill: Analysis

37) Into which group would you place a multicellular amoebalike cell that produces funguslike spores?

    A) Animalia   B) Bacteria   C) Fungi   D) Plantae   E) Protist

    Answer: E
    Skill: Analysis

38) Into which group would you place a unicellular eukaryote that lives in the red blood cells of humans?

    A) Animalia   B) Bacteria   C) Fungi   D) Plantae   E) Protist

Answer: E
Skill: Analysis

39) Into which group would you place a unicellular organism that has 70S ribosomes and a peptidoglycan cell wall?

    A) Animalia   B) Bacteria   C) Fungi   D) Plantae   E) Protist

Answer: B
Skill: Analysis

Table 10.1

I. Gram–positive
   A. Catalase+
      1. Acid from glucose....................................Staphylococcus
      2. Glucose–...Micrococcus
   B. Catalase–
      1. Coccus..Streptococcus
      2. Rod...Lactobacillus
II. Gram–negative
   A. Oxidase–
      1. Acid from lactose
         a. Uses citric acid................................Citrobacter
         b. Citric acid–.....................................Escherichia
      2. Lactose–
         a. H_2S produced
            (1) Urease positive.........................Proteus
            (2) Urease negative........................Salmonella
   B. Oxidase+
      1. Rod...Pseudomonas
      2. Coccus..Neisseria

40) Use the dichotomous key in Table 10.1 to identify a gram–negative rod that ferments lactose and uses citric acid as its sole carbon source.

    A) *Citrobacter*

    B) *Escherichia*

    C) *Lactobacillus*

    D) *Pseudomonas*

    E) *Staphylococcus*

Answer: A
Skill: Understanding

41) Use the dichotomous key in Table 10.1 to identify a gram–negative coccus.

   A) *Neisseria*

   B) *Pseudomonas*

   C) *Staphylococcus*

   D) *Streptococcus*

   E) *Micrococcus*

   Answer: A
   Skill: Understanding

42) Into which group would you place a multicellular organism that has a mouth and lives inside the human liver?

   A) Animalia

   B) Fungi

   C) Plantae

   D) Firmicutes (Gram–positive bacteria)

   E) Proteobacteria (Gram–negative bacteria)

   Answer: A
   Skill: Analysis

43) Into which group would you place a photosynthetic organism that lacks a nucleus and has a thin peptidoglycan wall surrounded by an outer membrane?

   A) Animalia

   B) Fungi

   C) Plantae

   D) Firmicutes (Gram–positive bacteria)

   E) Proteobacteria (Gram–negative bacteria)

   Answer: E
   Skill: Analysis

Figure 10.4

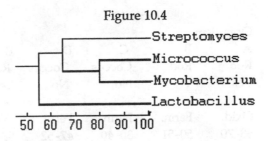

44) Using the cladogram shown in Figure 10.4, which two organisms are most closely related?

A) *Streptomyces* and *Micrococcus*

B) *Micrococcus* and *Mycobacterium*

C) *Mycobacterium* and *Lactobacillus*

D) *Streptomyces* and *Lactobacillus*

E) *Streptomyces* and *Mycobacterium*

Answer: B
Skill: Understanding

45) Which of the following indicates that two organisms are closely related?

A) Both are cocci.

B) Both ferment lactose.

C) Their DNA can hybridize.

D) Both normally live in clams.

E) All of the above.

Answer: C
Skill: Recall

ESSAY QUESTIONS

Figure 10.5

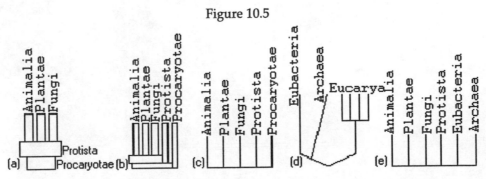

1) Choose one of the phylogenetic schemes in Figure 10.5 and explain why you feel this one is preferable to the others.

Answer:

Skill:

125

Table 10.2

Characteristic	Organism				
	A	B	C	D	E
Morphology	Rod	Rod	Coccus	Coccus	Rod
Motile	Yes	Yes	No	No	Yes
Gram Reaction	–	–	+	–	–
Glucose Utilization	Oxid.	Ferm.	Ferm.	Ferm.	None
% G+C	58–70	50–51	30–40	47–52	58–70
Cytochrome oxidase	Present	Absent	Absent	Present	Present

2) Use the information given in Table 10.2 to answer questions (a) and (b).
   a. Which organisms are most closely related? On what did you base your answer?
   b. DNA from which organisms will probably hybridize?

   Answer: To instructors: A=*Pseudomonas*; B=*Eschericia*; C=*Staphylococcus*; D=*Neisseria*;
           E=*Alcaligenes*.
   Skill:

3) One of the advantages of some newly developed rapid identification tools is that pure
   cultures aren't needed. Why is a pure culture necessary for biochemical tests such as the
   Enterotube, but not for DNA probes?

   Answer:

   Skill:

CHAPTER 11   The Prokaryotes: Domains Bacteria and Archaea

OBJECTIVE QUESTIONS

1) Which of the following are found primarily in the intestines of humans?

   A) Gram–negative aerobic rods and cocci

   B) Aerobic, helical bacteria

   C) Facultatively anaerobic gram–negative rods

   D) Gram–positive cocci

   E) None of the above

   Answer: C
   Skill: Recall

2) Which of the following is a characteistic of *Clostridium*?

   A) Endospore–forming cocci

   B) Mycobacteria

   C) Gram–negative aerobic cocci

   D) Anaerobic gram–negative rods

   E) None of the above

   Answer: A
   Skill: Recall

3) Which of the following is *not* a characteristic of the regular, nonsporing gram–positive rods?

   A) Aerotolerant

   B) Fermentative metabolism

   C) Don't produce endospores

   D) Nonpathogenic

   E) None of the above

   Answer: D
   Skill: Analysis

4) Which of the following is *not* a characteristic of *Neisseria*?

   A) Requires X and V factors

   B) Cocci

   C) Gram–negative

   D) Oxidase–positive

   E) None of the above

   Answer: A
   Skill: Analysis

5) *Staphylococcus* and *Streptococcus* can be easily differentiated in a laboratory by which one of the following?

A) Cell shape

B) Gram stain reaction

C) Growth in high salt concentrations

D) Ability to cause disease

E) None of the above

Answer: C
Skill: Analysis

6) Which of the following genera is an anaerobic gram–negative rod?

A) *Escherichia*

B) *Staphylococcus*

C) *Bacteroides*

D) *Treponema*

E) None of the above

Answer: C
Skill: Recall

7) Which of the following do you expect to be most resistant to high temperatures?

A) *Bacillus subtilis*

B) *Eschericia coli*

C) *Neisseria gonorrhoeae*

D) *Staphylococcus aureus*

E) *Streptococcocus pyogenes*

Answer: A
Skill: Analysis

8) Which of the following is *not* an enteric?

A) *Salmonella*

B) *Shigella*

C) *Escherichia*

D) *Enterobacter*

E) *Campylobacter*

Answer: E
Skill: Recall

9) Which of the following is *not* a characteristic of spirochetes?

   A) Possess an axial filament

   B) Gram–negative

   C) Helical shape

   D) Easily observed with brightfield microscopy

   E) Difficult to culture in vitro

Answer: D
Skill: Recall

10) You have isolated a bacterium that grows in a medium containing an organic substrate and nitrate in the absence of oxygen. The nitrate is reduced to nitrogen gas. This bacterium is

   A) Gram–negative.

   B) Using anaerobic respiration.

   C) A chemoautotroph.

   D) A photoauttoph.

   E) A photoheterotroph.

Answer: B
Skill: Understanding

11) Which of the following lacks a cell wall?

   A) *Borrelia*

   B) *Mycoplasma*

   C) *Mycobacterium*

   D) *Clostridium*

   E) *Nocardia*

Answer: B
Skill: Recall

12) Which of the following is *not* gram-positive?

   A) *Treponema*

   B) *Corynebacterium*

   C) *Bacillus*

   D) *Staphylococcus*

   E) *Mycobacterium*

Answer: A
Skill: Analysis

13) Which of the following form conidiospores?

    A) Endospore–forming gram–positive rods and cocci

    B) Actinomycetes and related organisms

    C) Rickettsias

    D) Anaerobic gram-negative cocci

    E) Spiral and curved bacteria

Answer: B
Skill: Recall

14) Which of the following pairs is mismatched?

    A) Dissimilatory sulfate–reducing bacteria — produce H_2S

    B) Archaea — live in extreme environments

    C) Chemoautotrophic bacteria — fix atmospheric nitrogen

    D) Actinomycetes — reproduce by fragmentation

    E) *Cytophaga* — a gliding, nonfruiting bacterium

Answer: C
Skill: Analysis

15) Rickettsias differ from chlamydias in that rickettsias

    A) Are gram-negative.

    B) Are intracellular parasites.

    C) Require an arthropod for transmission.

    D) Form elementary bodies.

    E) None of the above.

Answer: C
Skill: Analysis

16) Requirements for X and V factors are used to identify

    A) *Staphylococcus.*

    B) *Escherichia.*

    C) *Neisseria.*

    D) *Haemophilus.*

    E) None of the above.

Answer: D
Skill: Recall

17) You have isolated a bacterium that grows in a medium containing only inorganice nutrients. Ammonia is oxidized to nitrate ion. This bacterium is

A) Gram–negative.

B) Using anaerobic respiration.

C) A chemoautotroph.

D) A photoauttoph.

E) A photoheterotroph.

Answer: C
Skill: Understanding

18) Which of the following is *not* gram–negative?

A) *Pseudomonas*

B) *Salmonella*

C) *Streptococcus*

D) *Bacteroides*

E) *Rickettsia*

Answer: C
Skill: Analysis

19) *Escherichia coli* belongs to the

A) Proteobacteria.

B) Gram–positive bacteria.

C) Green sulfur bacteria.

D) Spirochetes.

E) Actinomycetes.

Answer: A, E
Skill: Recall

20) Which one of the following does *not* belong with the others?

A) *Bacillus*

B) *Escherichia*

C) *Lactobacillus*

D) *Staphylococcus*

E) *Streptococcus*

Answer: B
Skill: Analysis

21) All of the following are true about the causative agent of Rocky Mountainn spotted fever *except*

   A) It is an intracellular parasite.
   B) It is transmitted by ticks.
   C) It is in the genus *Rickettsia*.
   D) It is gram–negative.
   E) It is found in soil and water.

   Answer: E
   Skill: Recall

22) A primary difference between cyanobacteria and purple sulfur and purple nonsulfur phototrophic bacteria is

   A) Energy source.
   B) Cell wall type.
   C) Electron donor for CO_2 reduction.
   D) Cell type.
   E) Color.

   Answer: C
   Skill: Analysis

23) Which one of the following does *not* belong with the others?

   A) *Bordetella*
   B) *Burkholderia*
   C) *Campylobacter*
   D) *Pseudomonas*
   E) *Salmonella*

   Answer: E
   Skill: Analysis

24) The bacteria responsible for more infections and more different kinds of infections are

   A) *Streptococcus.*
   B) *Staphylococcus.*
   C) *Salmonella.*
   D) *Pseudomonas.*
   E) *Neisseria.*

   Answer: A
   Skill: Recall

25) Both *Beggiatoa* and the purple sulfur bacteria use H₂S. These bacteria differ because *Beggiatoa*

    A) Uses H_2S for an energy source.

    B) Uses H_2S for a carbon source.

    C) Uses light energy.

    D) Belongs to the γ–protobacteria.

    E) Is a heterotroph.

Answer: A
Skill: Understanding

26) The nonsulfur photosynthetic bacteria use organic compounds as

    A) Carbon sources.

    B) Electron donors to reduce CO_2.

    C) Energy sources.

    D) Electron acceptors.

    E) All of the above.

Answer: B
Skill: Understanding

27) Which of the following is the best reason to classify *Staphylococcus* in the Lactobacillales?

    A) Gram reaction

    B) Morphology

    C) Fermentation of lactose

    D) rRNA sequences

    E) All of the above.

Answer: D
Skill: Understanding

28) *Streptomyces* differs from *Actinomyces* because *Streptomyces*

    A) Makes antibiotics.

    B) Produces conidia.

    C) Forms filaments.

    D) Is aerobic.

    E) None of the above.

Answer: D
Skill: Recall

29) All of the following are gram–positive. Which does *not* belong with the others?

   A) *Actinomyces*

   B) *Bacillus*

   C) *Corynebacterium*

   D) *Listeria*

   E) *Mycobacterium*

   Answer: E
   Skill: Understanding

30) *Salmonella, Shigella, Yersinia,* and *Serratia* are all

   A) Pathogens.

   B) Gram–negative facultatively anaerobic rods.

   C) Gram–positive aerobic cocci.

   D) Fermentative.

   E) None of the above.

   Answer: B
   Skill: Analysis

31) You have isolated a gram–positive rod. What should you do next?

   A) Gram stain

   B) Lactose fermentation

   C) Endospore stain

   D) Flagella stain

   E) Enterotube(R)

   Answer: C
   Skill: Understanding

32) *Borrelia* is classified as a spirochete because it

   A) Is aerobic.

   B) Possesses an axial filament.

   C) Is a rod.

   D) Is a pathogen.

   E) None of the above.

   Answer: B
   Skill: Analysis

33) *Thiobacillus* oxidizes inorganic sulfur compounds and reduces CO_2. This bacterium is

    A) A chemoheterotroph

    B) A chemoautotroph

    C) A phototroph

    D) A γ–proteobacteria

    E) None of the above.

Answer: B
Skill: Understanding

34) You have isolated a prokaryotic cell. The first step in identification is

    A) Gram stain.

    B) Lactose fermentation.

    C) Endospore stain.

    D) Flagella stain.

    E) DNA fingerprint.

Answer: A
Skill: Understanding

35) Actinomycetes differ from fungi because actinomycetes

    A) Are chemoheterotrophs.

    B) Lack a membrane–bounded nucleus.

    C) Require light.

    D) Are decomposers.

    E) Cause disease.

Answer: B
Skill: Understanding

36) The bacteria presently being used to produce xanthan for food products are

    A) Genetically engineered.

    B) Natually occurring.

    C) Unable to degrade lactose.

    D) Dangerous to humans.

    E) This isn't being done.

Answer: B
Skill: Analysis

37) Which of the following pairs is mismatched?

    A) Spirochete — axial filament

    B) Aerobic, helical bacteria — gram–negative

    C) Enterics — gram–negative

    D) Mycobacteria — acid–fast

    E) None of the above

Answer: E
Skill: Analysis

38) Which one of the following does *not* belong with the others?

    A) *Coxiella*

    B) *Ehrlichia*

    C) *Rickettsia*

    D) *Staphylococcus*

    E) None of the above

Answer: D
Skill: Analysis

39) *Caulobacter* are different from most bacteria because

    A) They are gram–negative.

    B) They are gram–positive.

    C) They have stalks.

    D) They lack cell walls.

    E) None of the above.

Answer: C
Skill: Recall

40) Pathogenic bacteria are easily identified because they

    A) Are motile.

    B) Are gram–negative.

    C) Produce endospores.

    D) Grow at 37°C.

    E) None of the above.

Answer: E
Skill: Understanding

41) Which of the following does *not* belong with the others?

   A) *Halobacterium*

   B) *Halococcus*

   C) *Methanobacterium*

   D) *Staphylococcus*

   E) *Sulfolobus*

   Answer: D
   Skill: Analysis

42) What should you do if you suspect a patient has tuberculosis?

   A) A Gram stain.   B) An acid–fast stain.

   C) Check for motility.   D) Look at a wet mount.

   Answer: B
   Skill: Analysis

43) All of the following are motile; which does *not* have flagella?

   A) Enterics

   B) Helical bacteria

   C) *Pseudomonas*

   D) Spirochetes

   E) None of the above

   Answer: D
   Skill: Analysis

44) Mycoplasmas differ from other bacteria because they

   A) Grow inside host cells.

   B) Lack a cell wall.

   C) Are acid–fast.

   D) Are motile.

   E) None of the above.

   Answer: B
   Skill: Recall

45) Which of the following pairs is *not* correctly matched?

   A) Elementary body — *Escherichia*

   B) Endospore — *Bacillus*

   C) Endospore — *Clostridium*

   D) Heterocyst — cyanobacteria

   E) Myxospore — gliding bacteria

   Answer: A
   Skill: Analysis

ESSAY QUESTIONS

1) Discuss the use of *Bergey's Manual* as a tool of classification. Of identification.

   Answer:

   Skill:

2) Provide a reason to classify bacteria.

   Answer:

   Skill:

3) *Bacteroides* and *Escherichia* are both gram–negative rods found in the large intestine. Why are they in different sections of *Bergey's Manual*?

   Answer:

   Skill:

CHAPTER 12   The Eukaryotes: Fungi, Algae, Protozoa, and Helminths

OBJECTIVE QUESTIONS

1) All of the following statements about fungi are true *except*

   A) All fungi are unicellular.

   B) All fungi have eukaryotic cells.

   C) Fungi are heterotrophic.

   D) Most fungi are aerobic.

   E) Few fungi are pathogenic to humans.

   Answer: A
   Skill: Analysis

2) All of the following statements about helminths are true *except*

   A) They are heterotrophic.

   B) They are multicellular animals.

   C) They have eukaryotic cells.

   D) All are parasites.

   E) Some have male and female reproductive organs in one animal.

   Answer: D
   Skill: Analysis

3) All of the following are true about protozoa *except*

   A) They have eukaryotic cells.

   B) All make cysts.

   C) They may reproduce sexually.

   D) They may have flagella or cilia.

   E) None of the above.

   Answer: B
   Skill: Analysis

4) Fungi, more often than bacteria, are responsible for decomposition of plant material because

   A) They are aerobic.

   B) They can tolerate low–moisture conditions.

   C) They prefer a neutral environment (pH 7).

   D) They have a fermentative metabolism.

   E) They cannot tolerate high osmotic pressure.

   Answer: B
   Skill: Analysis

5) Which of the following statements is *not* true?

   A) A lichen doesn't exist if the fungal and algal partners are separated.

   B) Lichens are parasites.

   C) In a lichen, the alga produces carbohydrates.

   D) In a lichen, the fungus provides the holdfast.

   E) Lichens are important soil producers.

   Answer: B
   Skill: Recall

6) Which of the following pairs is *not* correctly matched?

   A) Plasmogamy — union of two haploid cells.

   B) Karyogamy — fusion of nucleus.

   C) Meiosis — cell division resulting in haploid cells.

   D) Anamorph — produces asexual spores.

   E) Deuteromycota — a phylum of fungi.

   Answer: E
   Skill: Recall

Table 12.1

1–Arthrospore	5–Chlamydospore
2–Ascospore	6–Conidiospore
3–Basidiospore	7–Sporangiospore
4–Blastoconidium	8–Zygospore

7) In Table 12.1, which of these spores are characteristic of *Penicillium*?

   A) 1 and 2   B) 3 and 4   C) 2 and 6   D) 1 and 4   E) 4 and 6

   Answer: C
   Skill: Recall

8) In Table 12.1, which of these spores are characteristic of *Rhizopus*?

   A) 1 and 2   B) 6 and 7   C) 2 and 8   D) 1 and 4   E) 7 and 8

   Answer: E
   Skill: Recall

9) In Table 12.1, which spore is in a sac and results from the fusion of two nuclei from different strains of the same fungi?

   A) 1   B) 2   C) 4   D) 6   E) 8

   Answer: B
   Skill: Recall

10) In Table 12.1, which spore is found externally on a pedestal?

A) 1   B) 3   C) 5   D) 7   E) None

Answer: B
Skill: Recall

11) In Table 12.1, which is a thick-walled spore formed as a segment within a hypha?

A) 1   B) 3   C) 5   D) 7   E) None

Answer: C
Skill: Recall

12) In Table 12.1, which of these spores are asexual spores?

A) 1, 4, 5, 6, 7   B) 2, 3, 6, 8   C) 1, 3, 5, 8   D) 2, 4, 6, 7, 8   E) All

Answer: A
Skill: Recall

13) Which of the following pairs is mismatched?

A) Thallus — hyphae
B) Thallus — mycelium
C) Coenocytic — lacking cross-walls
D) Septate — with cross-walls
E) Aerial mycelium — vegetative mycelium

Answer: E
Skill: Recall

14) Which of the following systems in a parasitic helminth is *not* greatly reduced compared to free-living helminths?

A) Digestive system
B) Nervous system
C) Locomotion
D) Reproductive system
E) All are reduced.

Answer: D
Skill: Recall

15) Which of the following statements is false?

A) Fungi produce sexual spores.

B) Fungi produce asexual spores.

C) Fungal spores are used in identification of fungi.

D) Fungal spores are resting spores to protect the fungus from adverse environmental conditions.

E) Fungal spores are for reproduction.

Answer: D
Skill: Analysis

16) Which of the following pairs are mismatched?
    1. Arthrospore — formed by fragmentation
    2. Sporangiospore — formed within hyphae
    3. Conidiospore — formed in a chain
    4. Blastoconidium — formed from a bud
    5. Chlamydospore — formed in a sac

A) 1 and 2   B) 2 and 3   C) 2 and 5   D) 3 and 4   E) 4 and 5

Answer: C
Skill: Recall

17) Which of the following pairs is mismatched?

A) Dinoflagellates — paralytic shellfish poisoning

B) Brown algae — algin

C) Red algae — agar

D) Diatoms — petroleum

E) Green algae — prokaryotic

Answer: E
Skill: Recall

18) Transmission of helminthic diseases to humans is usually by

A) Respiratory route.

B) Genitourinary route.

C) Gastrointestinal route.

D) Vectors.

E) All of the above.

Answer: C
Skill: Analysis

19) Which of the following is *not* a characteristic of parasitic platyhelminths?

    A) They are hermaphroditic.

    B) They are dorsoventrally flattened.

    C) They have a complete digestive system.

    D) They can be divided into flukes and tapeworms.

    E) None of the above.

Answer: C
Skill: Recall

20) Cercariae, metacercaria, miracidia, and rediae are stages in the life cycle of

    A) Cestodes.

    B) Trematodes.

    C) Nematodes.

    D) Sporozoans.

    E) Sarcodina.

Answer: C
Skill: Recall

21) Which stage immediately precedes the adult?

    A) Cercaria   B) Metacercaria   C) Miracidium   D) Redia

Answer: B
Skill: Recall

22) The encysted larva of the beef tapeworm is called a

    A) Redia.

    B) Cercaria.

    C) Cysticercus.

    D) Metacercaria.

    E) None of the above.

Answer: C
Skill: Recall

23) All of the following arthropods transmit diseases while sucking blood from a human host *except*

    A) Lice.

    B) Fleas.

    C) Houseflies.

    D) Mosquitoes.

    E) Kissing bugs.

Answer: C
Skill: Analysis

24) All of the following statements about algae are true *except*

    A) They use light as their energy source.

    B) They use CO_2 as their carbon source.

    C) They produce oxygen from hydrolysis of water.

    D) All are unicellular.

    E) Some are capable of sexual reproduction.

Answer: D
Skill: Analysis

25) Which one of the following does *not* belong with the others?

    A) *Cryptosporidium*

    B) *Entamoeba*

    C) *Giardia*

    D) *Plasmodium*

    E) *Trypanosoma*

Answer: C
Skill: Understanding

26) Which of the following statements about algae is true?

    A) All algae are unicellular.

    B) Algae are plants.

    C) All algae are green.

    D) Algae are photoautotrophs.

    E) All of the above.

Answer: D
Skill: Recall

27) A definitive host harbors which stage of a parasite?

    A) Miracidium

    B) Cyst

    C) Adult

    D) Larva

    E) All of the above

Answer: C
Skill: Recall

28) What do tapeworms eat?
   A) Intestinal bacteria
   B) Host tissues
   C) Red blood cells
   D) Intestinal contents
   E) All of the above

   Answer: D
   Skill: Recall

29) Which of the following is *not* caused by an alga?
   A) Domoic acid intoxication
   B) Possible estuary-associated syndrome
   C) Cycloporal diarrhea
   D) Paralytic shellfish poisoning
   E) Ciguatera

   Answer: C
   Skill: Recall

30) The life cycle of the fish tapeworm is similar to that of the beef tapeworm. Which of the following is the most effective preventive measure?
   A) Salting fish before eating
   B) Refrigerating stored fish
   C) Cooking fish before eating
   D) Wearing gloves while handling fish
   E) Not swimming in fish-infested waters

   Answer: C
   Skill: Understanding

31) Which of the following is the most effective control for malaria?
   A) Vaccination
   B) Treating patients
   C) Eliminate *Anopheles*
   D) Eliminate the intemediate host
   E) None of the above

   Answer: A
   Skill: Analysis

32) Multinucleated amoebalike cells that produce funguslike spores.
    A) Ascomycete
    B) Cellular slime mold
    C) Euglenozoa
    D) Tapeworm
    E) Plasmodial slime mold

    Answer: E
    Skill: Analysis

33) Amoebalike vegetative structures that produce sporangia.
    A) Ascomycete
    B) Cellular slime mold
    C) Euglenozoa
    D) Tapeworm
    E) Plasmodial slime mold

    Answer: B
    Skill: Analysis

34) A multicellular organism; the digestive tract has one opening.
    A) Ascomycete
    B) Cellular slime mold
    C) Euglenozoa
    D) Tapeworm
    E) Plasmodial slime mold

    Answer: D
    Skill: Analysis

35) A nucleated, unicellular organism; when you change the incubation temperature, it forms filaments with sporangia.
    A) Ascomycete
    B) Cellular slime mold
    C) Euglenozoa
    D) Tapeworm
    E) Plasmodial slime mold

    Answer: A
    Skill: Analysis

36) An organism that can grow photoautotrophically in the light and chemoheterotrophically in the dark.

A) Ascomycete

B) Cellular slime mold

C) Euglenozoa

D) Tapeworm

E) Plasmodial slime mold

Answer: C
Skill: Analysis

37) Which of the following pairs is mismatched?

A) Tick — Rocky Mountain Spotted Fever

B) Tick — Lyme Disease

C) Mosquito — Malaria

D) Mosquito — *Pneumocystis*

E) Mosquito — Encephalitis

Answer: D
Skill: Recall

38) All of the following groups of algae produce compounds that are toxic to humans *except*

A) Diatoms.

B) Dinoflagellates.

C) Green algae.

D) Red algae.

E) None of the above.

Answer: C
Skill: Recall

39) The cells of plasmodial slime molds can grow to several centimeters in diameter because

A) They have organelles.

B) They distribute nutrients by cytoplasmic streaming.

C) The large surface can absorb nutrients.

D) They form spores.

E) None of the above.

Answer: B
Skill: Understanding

40) Which of the following does *not* belong with the others?

   A) *Cryptosporidium*

   B) *Cyclospora*

   C) *Plasmodium*

   D) *Trypanosoma*

   E) None of the above

   Answer: D
   Skill: Analysis

41) If a larva of *Echinococcus granulosus* is found in humans, humans are the

   A) Definitive host.

   B) Infected host.

   C) Intermediate host.

   D) Reservoir.

   E) None of the above.

   Answer: C
   Skill: Analysis

42) Ringworm is caused by a(n)

   A) Ascomycete.

   B) Cestode.

   C) Nematode.

   D) Protozoan.

   E) Trematode.

   Answer: A
   Skill: Recall

43) Yeast infections are caused by

   A) *Aspergillus.*

   B) *Candida albicans.*

   C) *Histoplasma.*

   D) *Penicillium.*

   E) *Saccharomyces cerevisiae.*

   Answer: B
   Skill: Recall

44) In a food chain consisting of the following, which acts as a producer?

   A) Fungi   B) Lichens   C) Protozoa   D) Slime molds

   Answer: B
   Skill: Analysis

45) The microsporidia and archaezoa are unusual eukaryotes because they
   A) Are motile.
   B) Lack mitochondira.
   C) Lack nuclei.
   D) Don't produce cysts.
   E) Do produce cysts.

Answer: B
Skill: Analysis

ESSAY QUESTIONS

1) Provide an explanation for the complex life cycles exhibited by parasitic helminths. Cite specific examples in your discussion.
   Answer:
   Skill:

2) Explain how the presence of algae can indicate either pollution or productivity of a body of water.
   Answer:
   Skill:

CHAPTER 13   Viruses, Viroids, and Prions

OBJECTIVE QUESTIONS

1) In which of the following ways do viruses differ from bacteria?

    A) Viruses are filterable.

    B) Viruses are obligate intracellular parasites.

    C) Viruses don't have any nucleic acid.

    D) Viruses are not composed of cells.

    E) All of the above.

    Answer: D
    Skill: Recall

2) Which of the following provides the most significant support for the idea that viruses are nonliving chemicals?

    A) They are not composed of cells.

    B) They are filterable.

    C) They cannot reproduce themselves outside of a host.

    D) They cause diseases similar to those caused by chemicals.

    E) They are chemically simple.

    Answer: C
    Skill: Analysis

3) Which of the following is *not* true about spikes?

    A) They are used for penetration.

    B) They are used for absorption.

    C) They may cause hemagglutination.

    D) They are only found on enveloped viruses.

    E) None of the above.

    Answer: A
    Skill: Recall

4) Which of the following is *not* used as a criterion to classify viruses?

    A) Biochemical tests

    B) Morphology

    C) Nucleic acid

    D) Size

    E) Number of capsomeres

    Answer: A
    Skill: Analysis

5) Which of the following is *not* a method of culturing viruses?

   A) In laboratory animals

   B) In culture media

   C) In embryonated eggs

   D) In cell culture

   E) None of the above

   Answer: B
   Skill: Recall

6) Bacteriophages and animal viruses do *not* differ significantly in which one of the following steps?

   A) Adsorption

   B) Penetration

   C) Uncoating

   D) Biosynthesis

   E) Release

   Answer: D
   Skill: Understanding

7) The definition of lysogeny is

   A) Phage DNA is incorporated into host cell DNA.

   B) Lysis of the host cell due to a phage.

   C) The period during replication when virions are not present.

   D) When the burst time takes an unusually long time.

   E) None of the above.

   Answer: A
   Skill: Analysis

8) A viroid is

   A) A complete, infectious virus particle.

   B) A naked, infectious piece of RNA.

   C) A capsid without a nucleic acid.

   D) A provirus.

   E) None of the above.

   Answer: B
   Skill: Recall

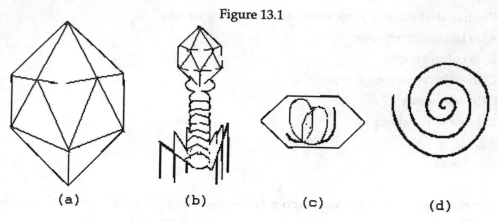

Figure 13.1

(a)   (b)   (c)   (d)

9) In Figure 13.1, which of these is a complex virus?

   A) A

   B) B

   C) C

   D) D

   E) All of the above

Answer: B
Skill: Recall

10) In Figure 13.1, the structures illustrated are composed of

   A) DNA

   B) RNA

   C) DNA or RNA

   D) Capsomeres

   E) Capsids

Answer: D
Skill: Recall

11) A clear area against a confluent "lawn" of bacteria is called a

   A) Phage.

   B) Pock.

   C) Cell lysis.

   D) Plaque.

   E) None of the above.

Answer: D
Skill: Recall

12) Continuous cell lines differ from primary cell lines in that

A) Viruses can be grown in continuous cell lines.

B) Continuous cell lines always have to be reisolated from animal tissues.

C) Continuous cell lines are derived from primary cell lines.

D) Continuous cell lines can be maintained through an indefinite number of generations.

E) Continuous cell lines are from human embryos.

Answer: D
Skill: Recall

13) Which of the following is *not* an effect of transformation?

A) Tumor–specific transplantation antigens appear on the cell.

B) Cells lose their ability to divide.

C) Cells do not exhibit contact inhibition.

D) T antigens appear in the nucleus of the cell.

E) All of the above.

Answer: B
Skill: Analysis

14) A slow viral infection is an infection in which

A) The virus remains in equilibrium with the host without causing a disease.

B) Viral replication is unusually slow.

C) The disease process occurs gradually over a long period.

D) Host cells are gradually lysed.

E) Host cells are transformed.

Answer: C
Skill: Recall

15) Which of the following statements is *not* true?

A) A prophage is phage DNA inserted into a bacterial chromosome.

B) A prophage can pop out of the chromosome.

C) Prophage genes are represented by a repressor protein coded for by the prophage.

D) A prophage may result in new properties of the host cell.

E) The prophage makes the host cell immune to infection by other phages.

Answer: E
Skill: Analysis

16) Lysogeny can result in all of the following *except*

   A) Immunity to reinfection by the same phage.

   B) Acquisition of new characteristics by the host cell.

   C) Immunity to reinfection by any phage.

   D) Transduction of specific genes.

   E) None of the above.

   Answer: C
   Skill: Analysis

17) Which of the following would be the first step in biosynthesis of a virus with a – strand of RNA?

   A) Synthesis of DNA from an RNA template

   B) Synthesis of double-stranded RNA from an RNA template

   C) Synthesis of double-stranded RNA from a DNA template

   D) Transcription of mRNA from DNA

   E) None of the above

   Answer: B
   Skill: Understanding

18) Scrapie is an example of an infection caused by

   A) A difficult–to–grow bacterium.

   B) A prion.

   C) A difficult–to–detect virus.

   D) Any of the above are possible.

   E) None of the above.

   Answer: B
   Skill: Recall

19) An envelope is acquired during which of the following steps?

   A) Penetration

   B) Adsorption

   C) Uncoating

   D) Release

   E) None of the above

   Answer: D
   Skill: Analysis

20) Which of the following contributes to the difficulty in establishing the etiology of cancer?

    A) Most viral particles can infect cells without inducing cancer.

    B) Cancer may not develop until long after infection.

    C) Cancers do not seem to be contagious.

    D) Viruses are difficult to observe.

    E) All of the above.

Answer: E
Skill: Analysis

21) An example of a latent viral infection is

    A) Subacute sclerosing panencephalitis.

    B) Cold sores.

    C) Influenza.

    D) Smallpox.

    E) None of the above.

Answer: B
Skill: Recall

22) The most common route of accidental AIDS transmission to health–care workers is

    A) Mouth to mouth.

    B) Fecal–oral.

    C) Needlestick.

    D) Aerosol.

    E) Environmental surface contact.

Answer: C
Skill: Recall

23) Assume you have isolated an unknown virus. It is a single–stranded RNA, enveloped virus. To which group does it most likely belong?

    A) Herpesvirus

    B) Picornavirus

    C) Retrovirus

    D) Togavirus

    E) None of the above

Answer: D
Skill: Recall

24) To which group does a small, nonenveloped single-stranded RNA virus most likely belong?

   A) Herpesvirus

   B) Picornavirus

   C) Retrovirus

   D) Togavirus

   E) None of the above

   Answer: B
   Skill: Recall

25) The most conclusive evidence that viruses cause cancers is provided by

   A) Finding oncogenes in viruses.

   B) The presence of antibodies against viruses in cancer patients.

   C) Cancer following injection of cell-free filtrates.

   D) Treating cancer with antibodies.

   E) Some liver cancer patients having had hepatitis.

   Answer: C
   Skill: Analysis

26) Bacteriophages derive all of the following from the host cell *except*

   A) Lysozyme.

   B) tRNA.

   C) Amino acids.

   D) Nucleotides.

   E) None of the above.

   Answer: A
   Skill: Analysis

27) Generalized transduction differs from specialized transduction in that generalized transduction

   A) Kills the host.

   B) Transfers DNA from one cell to another.

   C) Transfers specific DNA.

   D) Involves lysogeny.

   E) None of the above.

   Answer: E
   Skill: Understanding

28) Generally, in a DNA-containing virus infection, the host animal cell supplies all of the following *except*

A) RNA polymerase.

B) Nucleotides.

C) DNA polymerase.

D) tRNA.

E) None of the above.

Answer: C
Skill: Analysis

29) Put the following in the correct order for DNA–virus replication:
1–Maturation; 2–DNA synthesis; 3–Transcription; 4–Translation.

A) 1, 2, 3, 4   B) 2, 3, 4, 1   C) 3, 4, 1, 2   D) 4, 1, 2, 3   E) 4, 3, 2, 1

Answer: B
Skill: Analysis

30) A viral species is a group of viruses that

A) Have the same morphology and nucleic acid.

B) Have the same genetic information and ecological niche.

C) Infect the same cells and cause the same disease.

D) Can't be defined.

Answer: B
Skill: Recall

31) Viruses that have reverse transcriptase are in the

A) Retroviridae and Picornaviridae.

B) Herpesviridae and Retroviridae.

C) Hepadnaviridae and Retroviridae.

D) Bacteriophage families.

E) *Influenzavirus*.

Answer: C
Skill: Recall

32) DNA made from an RNA template will be incorporated into the virus capsid of

A) Retroviridae.

B) Herpesviridae.

C) Hepadnaviridae.

D) Bacteriophage families.

E) *Influenzavirus*.

Answer: C
Skill: Analysis

33) Which of the following is *not* true about viruses?

    A) Viruses contain DNA or RNA but never both.

    B) Viruses contain a protein coat.

    C) Viruses use the anabolic machinery of the cell.

    D) Viruses use their own catabolic enzymes.

    E) None of the above.

    Answer: D
    Skill: Analysis

34) Approximately how many virus particles could fit along a 1 millimeter line?

    A) 2   B) 20   C) 200   D) 20,000   E) 2,000,000

    Answer: D
    Skill: Analysis

35) Some viruses, such as Human Herpes Virus 1, infect a cell without causing symptoms; these are called

    A) Latent viruses.

    B) Lytic viruses.

    C) Phages.

    D) Slow viruses.

    E) Unconventional viruses.

    Answer: A
    Skill: Recall

36) Latent viruses are present in cells as

    A) Capsids.

    B) Enzymes.

    C) Prophages.

    D) Proviruses.

    E) None of the above.

    Answer: D
    Skill: Recall

37) Which one of the following steps does *not* occur during multiplication of a Rhabdovirus?

    A) Synthesis of + strands of RNA   B) Synthesis of – strands of RNA

    C) Synthesis of viral proteins   D) Synthesis of DNA

    Answer: D
    Skill: Analysis

38) The following steps occur during multiplication of Herpesviruses. What is the third step?

A) Attachment

B) Biosynthesis

C) Penetration

D) Release

E) Uncoating

Answer: E
Skill: Recall

39) The following steps occur during multiplication of Retroviruses. What is the fourth step?

A) Synthesis of double–stranded DNA

B) Synthesis of + RNA

C) Attachment

D) Penetration

E) Uncoating

Answer: A
Skill: Analysis

40) Nontoxic strains of *Vibrio cholerae* can become toxic when they are in the human intestine with toxic strains of bacteria. This suggests that the toxin genes are acquired by

A) Host enzymes.

B) Prions.

C) Reverse transcriptase.

D) Transduction.

E) None of the above.

Answer: D
Skill: Analysis

41) Which one of the following steps does *not* occur during multiplication of a Picornavirus?

A) Synthesis of + strands of RNA

B) Synthesis of – strands of RNA

C) Synthesis of viral proteins

D) Synthesis of DNA

E) None of the above

Answer: D
Skill: Understanding

42) An oncogenic RNA virus must have which of the following enzymes?

A) DNA–dependent DNA polymerase

B) Lysozyme

C) RNA polymerase

D) Reverse transcriptase

E) All of the above

Answer: D
Skill: Understanding

43) Which of the following is most likely a product of an early gene?

A) Capsid proteins

B) DNA polymerase

C) Envelope proteins

D) Spike proteins

E) Lysozyme

Answer: B
Skill: Understanding

44) Most RNA viruses carry which of the following enzymes?

A) DNA–dependent DNA polymerase

B) Lysozyme

C) RNA–dependent RNA polymerase

D) Reverse transcriptase

E) ATP synthase

Answer: C
Skill: Understanding

45) The following steps occur during biosynthesis of a + strand RNA virus. What is the third step?

A) Attachment

B) Penetration and uncoating

C) Synthesis of – strand RNA

D) Synthesis of + strand RNA

E) Synthesis of viral proteins

Answer: C
Skill: Analysis

ESSAY QUESTIONS

1) Bacteriophages are used as vectors in genetic engineering to insert new genes into bacteria. Describe the process that makes this genetic recombination possible.

   Answer:

   Skill:

2) Compare and contrast the lytic cycle of infection of a DNA virus and an RNA virus.

   Answer:

   Skill:

3) Why was it previously believed that only DNA viruses could cause cancer? How can RNA viruses cause cancer?

   Answer:

   Skill:

4) You are growing *Bacillus subtilis* in nine 16,000–liter fermenters to produce enzymes for industrial use. The *Bacillus* cultures had been growing for two days when the cells in one of the fermenters lysed. Explain what happened in this fermenter.

   Answer:

   Skill:

CHAPTER 14   Principles of Disease and Epidemiology

OBJECTIVE QUESTIONS

1) A commensal bacterium

   A) Does not receive any benefit from its host.

   B) Is beneficial to its host.

   C) May be an opportunistic pathogen.

   D) Does not infect its host.

   E) b and d only.

   Answer: C
   Skill: Analysis

2) Which of the following statements is true?

   A) Symbiosis refers to different organisms living together.

   B) Members of a symbiotic relationship cannot live without each other.

   C) A parasite is not in symbiosis with its host.

   D) Symbiosis refers to different organisms living together and benefiting from each other.

   E) At least one member must benefit in a symbiotic relationship.

   Answer: A
   Skill: Analysis

3) A nosocomial infection is

   A) Always present but inapparent at the time of hospitalization.

   B) Acquired during the course of hospitalization.

   C) Always caused by medical personnel.

   D) Only a result of surgery.

   E) Always caused by pathogenic bacteria.

   Answer: B
   Skill: Recall

4) The major significance of Koch's work was that

   A) Microorganisms are present in a diseased animal.

   B) Diseases can be transmitted from one animal to another.

   C) Microorganisms can be cultured.

   D) Microorganisms cause disease.

   E) Microorganisms are the result of disease.

   Answer: D
   Skill: Recall

5) Koch's postulates don't apply to all diseases because

   A) Some microorganisms can't be cultured in laboratory media.

   B) Some microorganisms don't cause the same disease in laboratory animals.

   C) Some microorganisms cause different symptoms under different conditions.

   D) Some microorganisms can't be observed.

   E) All diseases aren't caused by microorganisms.

   Answer: E
   Skill: Recall

6) Which of the following diseases is *not* spread by droplet infection?

   A) Botulism

   B) Tuberculosis

   C) Measles

   D) Common cold

   E) Diphtheria

   Answer: A
   Skill: Understanding

7) Mechanical transmission differs from biological transmission in that mechanical transmission

   A) Doesn't require an arthropod.

   B) Involves fomites.

   C) Doesn't involve specific diseases.

   D) Requires direct contact.

   E) Doesn't work with noncommunicable diseases.

   Answer: C
   Skill: Analysis

8) Which of the following definitions is incorrect?

   A) Endemic — a disease that is constantly present in a population

   B) Epidemic — fraction of the population having a disease at a specified time

   C) Pandemic — a disease that affects a large number of people in the world in a short time

   D) Sporadic — a disease that affects a population occasionally

   E) None of the above

   Answer: B
   Skill: Recall

9) Which of these infections can cause septicemia?

  A) Bacteremia

  B) Focal infection

  C) Local infection

  D) Septicemia

  E) Systemic infection

Answer: B
Skill: Understanding

10) Which type of infection can be caused by septicemia?

  A) Bacteremia

  B) Focal infection

  C) Local infection

  D) Septicemia

  E) Systemic infection

Answer: E
Skill: Understanding

11) Koch observed *Bacillus anthracis* multiplying in the blood of cattle. What is this condition called?

  A) Bacteremia

  B) Focal infection

  C) Local infection

  D) Septicemia

  E) Systemic infection

Answer: D
Skill: Understanding

12) Nosocomial infections are most often caused by

  A) *Escherichia coli.*

  B) *Staphylococcus aureus.*

  C) *Enterococcus.*

  D) *Pseudomonas.*

  E) *Klebsiella.*

Answer: A
Skill: Recall

13) Transient microbiota differ from normal microbiota because transient microbiota

A) Cause diseases.

B) Are found in a certain location on the host.

C) Are acquired by direct contact.

D) Are present for a relatively short time.

E) None of the above.

Answer: D
Skill: Recall

14) Which of the following statements about nosocomial infections is not true?

A) They occur in compromised patients.

B) They are caused by opportunists.

C) They are caused by drug-resistant bacteria.

D) They are caused by normal microbiota.

E) None of the above.

Answer: B
Skill: Recall

15) One effect of washing regularly with antibacterial agents is the removal of normal microbiota. This can result in

A) Body odor.

B) Fewer diseases.

C) Increased susceptibility to disease.

D) Normal microbiota returning immediately.

E) No bacterial growth because washing removes their food source.

Answer: C
Skill: Analysis

16) Which of the following is *not* a reservoir of infection?

A) A sick person

B) A healthy person

C) A sick animal

D) A hospital

E) None of the above

Answer: E
Skill: Analysis

17) All of the following are communicable diseases *except*

A) Malaria.

B) AIDS.

C) Tuberculosis.

D) Tetanus.

E) Typhoid fever.

Answer: D
Skill: Analysis

18) Which of the following is a fomite?

A) Water

B) Droplets from a sneeze

C) Pus

D) Insects

E) A hypodermic needle

Answer: E
Skill: Analysis

19) All of the following statements about biological transmission are true *except*

A) The pathogen reproduces in the vector.

B) The pathogen may enter the host in the vector's feces.

C) Houseflies are an important vector.

D) The pathogen may be injected by the bite of the vector.

E) The pathogen may require the vector as a host.

Answer: C
Skill: Recall

20) Which of the following definitions is incorrect?

A) Acute — a short-lasting primary infection

B) Inapparent — infection characteristic of a carrier state

C) Chronic — a disease that develops slowly and lasts for months

D) Primary infection — an initial illness

E) Secondary infection — a long-lasting illness

Answer: E
Skill: Recall

21) Symptoms of disease differ from signs of disease in that symptoms
    A) Are changes felt by the patient.
    B) Are changes observed by the physician.
    C) Are specific for a particular disease.
    D) Always occur as part of a syndrome.
    E) None of the above.

    Answer: A
    Skill: Recall

22) The science that deals with when diseases occur and how they are transmitted is called
    A) Ecology.
    B) Epidemiology.
    C) Communicable disease.
    D) Morbidity and mortality.
    E) Public health.

    Answer: B
    Skill: Recall

23) Tuberculosis may occur in individuals who have influenza. All of the following statements are true *except*
    A) Tuberculosis is a secondary infection.
    B) Tuberculosis is a predisposing factor.
    C) Tuberculosis is a chronic disease.
    D) Tuberculosis is an infectious disease.
    E) None of the above.

    Answer: B
    Skill: Analysis

24) Which of the following is *least* effective in preventing nosocomial infections?
    A) Aseptic technique
    B) Culturing fomites in a hospital
    C) Keeping insects out of hospitals
    D) Sterilizing bandages
    E) Disinfecting respirators

    Answer: C
    Skill: Analysis

25) Which of the following pairs is mismatched?

   A) Malaria — vector

   B) Salmonellosis — vehicle transmission

   C) Syphilis — direct contact

   D) Influenza — droplet infection

   E) None of the above

Answer: E
Skill: Analysis

26) All of the following can contribute to postoperative infections *except*

   A) Using syringes more than once.

   B) Normal microbiota on the operating room staff.

   C) Errors in aseptic technique.

   D) Antibiotic resistance.

   E) None of the above.

Answer: E
Skill: Analysis

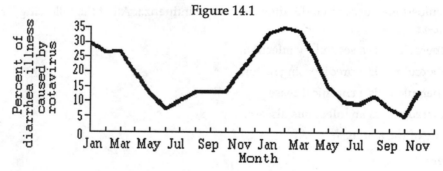

Figure 14.1

27) In Figure 14.1, what is the endemic level of rotavirus infections?

   A) 0%

   B) Approximately 10%

   C) Approximately 20%

   D) 35%

   E) The month of January

Answer: B
Skill: Analysis

28) A cold transmitted by a facial tissue.

   A) Direct contact

   B) Droplet transmission

   C) Fomite

   D) Vector

   E) Vehicle transmission

   Answer: E
   Skill: Analysis

29) Influenza transmitted by an unprotected sneeze.

   A) Direct contact

   B) Droplet transmission

   C) Fomite

   D) Vector

   E) Vehicle transmission

   Answer: B
   Skill: Analysis

30) A sexually transmitted disease.

   A) Direct contact

   B) Droplet transmission

   C) Fomite

   D) Vector

   E) Vehicle transmission

   Answer: A
   Skill: Analysis

31) Gastroenteritis acquired from roast beef.

   A) Direct contact

   B) Droplet transmission

   C) Fomite

   D) Vector

   E) Vehicle transmission

   Answer: E
   Skill: Analysis

32) A needlestick.

A) Direct contact

B) Droplet transmission

C) Fomite

D) Vector

E) Vehicle transmission

Answer: C
Skill: Analysis

33) Legionellosis transmitted by a grocery store mist machine.

A) Direct contact

B) Droplet transmission

C) Fomite

D) Vector

E) Vehicle transmission

Answer: E
Skill: Analysis

34) Plague transmitted by a flea.

A) Direct contact

B) Droplet transmission

C) Fomite

D) Vector

E) Vehicle transmission

Answer: D
Skill: Analysis

35) The most likely mode of transmission of pneumonic plague between humans.

A) Direct contact

B) Droplet transmission

C) Fomite

D) Vector

E) Vehicle transmission

Answer: B
Skill: Analysis

Situation 14.1

During a 6–month period, 239 cases of pneumonia occurred in a town of 300 people. A clinical case was defined as fever ≥39°C lasting >2 days with three or more symptoms (i.e., chills, sweats, severe headache, cough, aching muscles/joints, fatigue, or feeling ill). A laboratory–confirmed case was defined as a positive result for antibodies against *Coxiella burnetii*. Before the outbreak, 2000 sheep were kept northwest of the town. Of the 20 sheep tested from the flock, 15 were positive for *C. burnetii* antibodies. Wind blew from the northwest and rainfall was 0.5 cm compared with 7 to 10 cm during each of the previous three years.

36) Situation 14.1 is an example of
A) Human reservoirs.
B) A zoonosis.
C) A nonliving reservoir.
D) A vector.
E) A focal infection.

Answer: B
Skill: Understanding

37) In Situation 14.1, the etiologic agent of the disease is
A) Sheep.
B) Soil.
C) *C. burnetii*.
D) Pneumonia.
E) Wind.

Answer: C
Skill: Understanding

38) In Situation 14.1, the method of transmission of this disease was
A) Direct contact.
B) Droplet.
C) Indirect contact.
D) Vector–borne.
E) Vehicle.

Answer: E
Skill: Understanding

39) Which one of the following is *not* an example of microbial antagonism?
A) Acid production by bacteria
B) Bacteriocin production
C) Bacteria occupying host receptors
D) Bacteria causing disease
E) None of the above

Answer: E
Skill: Analysis

40) The yeast *Candida albicans* does not normally cause disease because of

A) Symbiotic bacteria.

B) Antagonistic bacteria.

C) Parasitic bacteria.

D) Commensal bacteria.

E) None of the above.

Answer: B
Skill: Analysis

41) *Haemophilus* bacteria require heme protein produced by *Staphylococcus* bacteria. This is an example of

A) Antagonism.

B) Commensalism.

C) Parasitism.

D) Synergism.

E) None of the above.

Answer: D
Skill: Analysis

42) Which one of the following is *not* a zoonosis?

A) Cat–scratch fever

B) *Hantavirus* pulmonary syndrome

C) Rabies

D) Tapeworm

E) None of the above

Answer: E
Skill: Recall

43) *Pseudomonas* bacteria colonized the bile duct of a patient following his liver transplant surgery. This is an example of a

A) Communicable disease.

B) Latent infection.

C) Nosocomial infection.

D) Sporadic disease.

E) None of the above.

Answer: C
Skill: Analysis

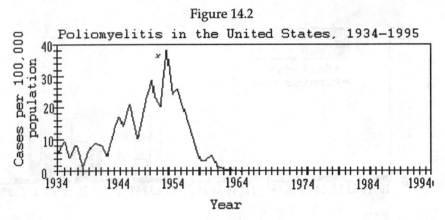

Figure 14.2

44) The graph in Figure 14.2 shows the incidence of polio in the United States. The area marked *x* indicates

   A) An endemic level.

   B) An epidemic level.

   C) A sporadic infection.

   D) A communicable disease.

   E) None of the above.

   Answer: B
   Skill: Analysis

45) Emergence of infectious diseases can be due to all of the following *except*

   A) Antibiotic resistance.

   B) Climatic changes.

   C) Digging up soil.

   D) Microbes trying to cause disease.

   E) Travel.

   Answer: D
   Skill: Understanding

ESSAY QUESTIONS

Situation 14.2

A 37–week–old infant was delivered by cesarean section and discharged from a Connecticut hospital when he was 10 days old. Two days later he was lethargic and had a fever. When he was readmitted to the hospital, he had multiple brain abscesses caused by *Citrobacter diversus*. After a prolonged illness, the baby died. A second infant with a normal pregnancy and delivery died of *C. diversus* meningitis after a short illness. Nine infants in the hospital nursery had umbilical cord colonization by *C. diversus*. Environmental cultures were negative for hospital equipment.

   1) a.   What is the normal habitat of this gram–negative, facultatively anaerobic, non–endospore–forming, lactose–positive rod?
      b.   Provide a plan for identifying the source of infection and preventing further infection.

   Answer:

   Skill:

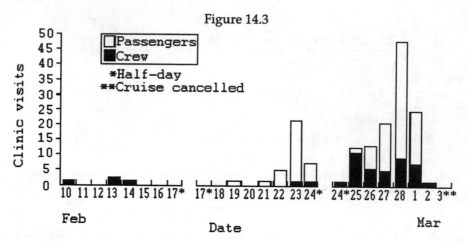

Figure 14.3

2) *Salmonella heidelberg* gastroenteritis occurred on three cruises aboard the T.S.S. *Festivale*. Figure 14.3 shows on-board clinic visits for diarrheal illness between February 10 and March 3.

a. Explain the incidence pattern shown on the graph.
b. What are probable modes of transmission?
c. What changes would you recommend before the ship books more cruises after March 3?

Answer:

Skill:

CHAPTER 15   Microbial Mechanisms of Pathogenicity

OBJECTIVE QUESTIONS

1) The most frequently used portal of entry for pathogens is the

A) Mucous membranes of the respiratory tract.

B) Mucous membranes of the gastrointestinal tract.

C) Skin.

D) Parenteral route.

E) All are used equally.

Answer: A
Skill: Recall

2) Which of the following diseases is *not* usually contracted by the respiratory route?

A) Pneumonia

B) Infectious hepatitis

C) Tuberculosis

D) Measles

E) None of the above

Answer: B
Skill: Understanding

3) Most pathogens that gain access through the skin

A) Can penetrate intact skin.

B) Just infect the skin itself.

C) Enter through hair follicles and sweat ducts.

D) Must adhere first while their invasive factors allow them to penetrate.

E) Must be injected.

Answer: C
Skill: Analysis

4) The LD$_{50}$ is a

A) Measure of pathogenicity.

B) Dose that will cause an infection in 50% of the test population.

C) Dose that will kill some of the test population.

D) Dose that will cause an infection in some of the test population.

E) Dose that will kill 50% of the test population.

Answer: E
Skill: Recall

5) Which of the following does *not* contribute to a pathogen's invasiveness?

   A) Toxins

   B) Capsule

   C) Cell wall

   D) Hyaluronidase

   E) Ligands

   Answer: A
   Skill: Understanding

6) Which of the following statements is false?

   A) Leukocidins destroy neutrophils.

   B) Hemolysins lyse red blood cells.

   C) Hyaluronidase breaks down substances between cells.

   D) Kinase destroys fibrin clots.

   E) Coagulase destroys blood clots.

   Answer: E
   Skill: Recall

7) Which of the following statements about exotoxins is generally *not* true?

   A) They are more potent than endotoxins.

   B) They are composed of proteins.

   C) They are not destroyed by heat.

   D) They have specific methods of action.

   E) They are produced by gram–positive bacteria.

   Answer: C
   Skill: Analysis

8) Endotoxins are

   A) Associated with gram–positive bacteria.

   B) Specific in their method of action.

   C) Part of the gram–negative cell wall.

   D) Excreted from the cell.

   E) None of the above.

   Answer: C
   Skill: Analysis

9) Which of the following is *not* a symptom of shigellosis?

   A) Fever   B) Weakness   C) Aches   D) Paralysis   E) Shock

   Answer: D
   Skill: Understanding

176

10) Cytopathic effects are changes in host cells due to
    A) Viral infections.
    B) Protozoan and helminthic infections.
    C) Fungal infections.
    D) Bacterial infections.
    E) All of the above.

    Answer: A
    Skill: Recall

11) Which of the following does *not* contribute to the symptoms of a fungal disease?
    A) Capsules
    B) Toxins
    C) Allergic response of the host
    D) Cell walls
    E) None of the above

    Answer: D
    Skill: Recall

12) All of the following statements about diphtheria toxin are true *except*
    A) Polypeptide A is the active component.
    B) Polypeptide A inhibits eucaryotic protein synthesis.
    C) Polypeptide B causes transport of diphtheria toxin into a target cell.
    D) The gene for diphtherotoxin is on a bacteriophage.
    E) Diphtheria is caused by a bacteriophage.

    Answer: E
    Skill: Recall

13) Which of the following is a property shared by cholera toxin, botulinum toxin, and *E. coli* enterotoxin?
    A) They are exotoxins.
    B) They require a binding polypeptide.
    C) They induce formation of cAMP in host cells.
    D) They cause diarrhea.
    E) None of the above.

    Answer: A
    Skill: Analysis

14) All of the following may be used for adherence *except*

   A) Fimbriae.

   B) Cell membrane mannose.

   C) Glycoproteins.

   D) Lipoproteins.

   E) Capsules.

   Answer: B
   Skill: Recall

15) Which of the following is *not* considered entry via the parenteral route?

   A) Injection

   B) Bite

   C) Surgery

   D) Hair follicle

   E) None of the above

   Answer: D
   Skill: Analysis

16) A cell wall can increase a bacterium's virulence because cell wall lipid A

   A) Resists phagocytosis.

   B) Helps the bacterium attach.

   C) Destroys host tissues.

   D) All bacteria have a cell wall and all are not pathogenic; therefore, cell walls do not contribute to virulence.

   E) None of the above.

   Answer: E
   Skill: Recall

17) Since botulism is caused by an exotoxin, it could easily be prevented by

   A) Boiling food prior to consumption.

   B) Administering antibiotics to patients.

   C) Not eating canned food.

   D) Preventing fecal contamination of food.

   E) None of the above.

   Answer: A
   Skill: Recall

18) Which of the following organisms *doesn't* produce an exotoxin?

    A) *Salmonella typhi*

    B) *Clostridium botulinum*

    C) *Corynebacterium diphtheriae*

    D) *Clostridium tetani*

    E) *Staphylococcus aureus*

Answer: A
Skill: Understanding

19) Which of the following cytopathic effects is cytocidal?

    A) Inclusion bodies

    B) Giant cells

    C) Antigenic changes

    D) Transformation

    E) Release of enzymes from lysosomes

Answer: E
Skill: Analysis

20) The symptoms of tetanus are due to

    A) Encapsulated *Clostridium tetani*.

    B) The growth of *Clostridium tetani* in a wound.

    C) Hemolysins produced by *Clostridium tetani*.

    D) An exotoxin produced by *Clostridium tetani*.

    E) All of the above.

Answer: D
Skill: Understanding

21) Symptoms of protozoan and helminthic diseases are due to

    A) Tissue damage due to growth of the parasite on the tissues.

    B) Waste products excreted by the parasite.

    C) Products released from damaged tissues.

    D) All of the above.

    E) None of the above.

Answer: D
Skill: Recall

22) Which of the following is *not* true of staphylococcal enterotoxin?

A) It causes vomiting.

B) It causes diarrhea.

C) It is an exotoxin.

D) It is produced by *Staphylococcus aureus* growing in the host's intestines.

E) None of the above.

Answer: D
Skill: Analysis

23) Which of the following does *not* contribute to the virulence of a pathogen?

A) Numbers of microorganisms that gain access to a host

B) Cell wall

C) Toxins

D) Enzymes

E) None of the above

Answer: E
Skill: Analysis

24) Lysogenic bacteriophages contribute to bacterial virulence because bacteriophages

A) Give new gene sequences to the host bacteria.

B) Produce toxins.

C) Carry plasmids.

D) Kill the bacteria causing release of endotoxins.

E) All of the above.

Answer: A
Skill: Analysis

25) Thirty-two people in San Francisco who ate jackfish caught at Midway Island developed malaise, nausea, blurred vision, breathing difficulty, and numbness from 3 to 6 hours after eating. The most likely cause of this food intoxication is

A) A mycotoxin.

B) Aflatoxin.

C) Staphylococcal enterotoxin.

D) Ciguatera.

E) Cholera toxin.

Answer: D
Skill: Understanding

26) All of the following are true about M protein *except*

A) It is found on *Streptococcus pyogenes*.

B) It is found on fimbriae.

C) It is heat- and acid-resistant.

D) It is readily digested by phagocytes.

E) It is a protein.

Answer: D
Skill: Recall

27) Septic shock attributed to __1__ is actually due to __2__.

A) 1 – Gram-negative bacteria; 2 – Malaria

B) 1 – Gram-negative bacteria; 2 – Tumor necrosis factor

C) 1 – Antibodies; 2 – Complement

D) 1 – Gram-positive bacteria; 2 – Complement

E) None of the above

Answer: B
Skill: Recall

28) A needlestick is an example of which portal of entry?

A) Skin

B) Parenteral route

C) Mucous membranes

D) All of the above

E) None of the above

Answer: B
Skill: Recall

29) Poliovirus is ingested and gains access to tissues by which portal of entry?

A) Skin

B) Parenteral

C) Mucous membranes

D) All of the above

E) None of the above

Answer: C
Skill: Analysis

30) *Pseudomonas aeruginosa* produces a two-part exotoxin. The most likely pathologic effect of this toxin is

A) Inhibition of protein synthesis.

B) Flaccid paralysis.

C) Tetani or lockjaw.

D) A red rash.

E) The bacteria will be able to grow in phagocytes.

Answer: A
Skill: Understanding

31) Cholera toxin polypeptide A binds to surface gangliosides on target cells. If the gangliosides were removed

A) Polypeptide A would bind to target cells.

B) Polypeptide A would enter the cells.

C) Polypeptide B would not be able to enter the cells.

D) *Vibrio* would not produce cholera toxin.

E) None of the above.

Answer: C
Skill: Analysis

32) Which is a method of avoiding phagocytosis?

A) Producing fimbriae

B) Inducing endocytosis

C) Producing toxins

D) Inducing TNF

E) All of the above

Answer: B
Skill: Understanding

33) The mechanism by which gram-negative bacteria can cross the blood-brain barrier.

A) Producing fimbriae

B) Inducing endocytosis

C) Producing toxins

D) Inducing TNF

E) All of the above

Answer: D
Skill: Understanding

34) Injectable drugs are tested for endotoxins by
    A) The *Limulus* amoebocyte lysate test.
    B) Counting the viable bacteria.
    C) Filtering out the cells.
    D) Looking for turbidity.
    E) None of the above.

    Answer: A
    Skill: Recall

35) Endotoxins in injectable drugs could cause
    A) Infection.
    B) Septic shock.
    C) Giant cell formation.
    D) Nerve damage.
    E) All of the above.

    Answer: D
    Skill: Recall

36) Septic shock results from the following events. What is the second step?
    A) Body temperature is reset in the hypothalamus.
    B) Fever occurs.
    C) IL-1 is released.
    D) LPS is released from gram-negative bacteria.
    E) Phagocytes ingest gram-negative bacteria.

    Answer: D
    Skill: Analysis

37) Antibiotics can lead to septic shock if used to treat
    A) Viral infections.
    B) Gram-negative bacterial infections.
    C) Gram-positive bacterial infections.
    D) Protozoan infections.
    E) Helminth infestations.

    Answer: B
    Skill: Recall

38) Which of the following is *not* a cytopathic effect of viruses?

A) Cell death

B) Host cells fuse to form multinucleated syncytia

C) Inclusion bodies form in the cytoplasm or nucleus

D) Increased cell growth

E) None of the above

Answer: E
Skill: Recall

39) Which of the following organisms causes the most severe disease?

A) <u>Bacteria</u>		ID_{50}
E. coli O157:H7		20
B) <u>Bacteria</u>		ID_{50}
Rhinovirus		200
C) <u>Bacteria</u>		ID_{50}
Shigella		10
D) <u>Bacteria</u>		ID_{50}
Treponema pallidum		57

E) Can't tell

Answer: E
Skill: Understanding

Table 15.1
Use these data to answer the following question:

Bacteria	Portal of entry	ID_{50}
Staphylococcus aureus	Wound	<10
S. aureus	Wound+ampicillin	300

40) The administration of ampicillin before surgery

A) Decreases the risk of staphyloccocal infection.

B) Increases the risk of staphylococcal infection.

C) Has no effect on risk of infection.

Answer: A
Skill: Understanding

41) Which organism most easily causes an infection?

   A) Bacteria ID_{50}
       E. coli O157:H7   20

   B) Bacteria ID_{50}
       Legionella pneumophila   1

   C) Bacteria ID_{50}
       Shigella   10

   D) Bacteria ID_{50}
       Treponema pallidum   57

   E) Can't tell

Answer: B
Skill: Understanding

42) Bacteria that cause periodontal disease have adhesins for receptors on streptococci that colonize on teeth. This indicates that

   A) Streptococci get bacterial infections.

   B) Streptococcal colonization is necessary for periodontal disease.

   C) Bacteria that cause periodontal disease adhere to gums and teeth.

   D) Bacteria that cause periodontal disease adhere to teeth.

   E) Streptococci cause periodontal disease.

Answer: B
Skill: Analysis

43) Nonpathogenic *V. cholerae* can acquire the cholera toxin gene by

   A) Phagocytosis.

   B) Transduction.

   C) Conjugation.

   D) Transformation.

   E) None of the above.

Answer: B
Skill: Analysis

44) In response to the presence of endotoxin, phagocytes secrete tumor necrosis factor. This causes

   A) The disease to subside.

   B) A decrease in blood pressure.

   C) A fever.

   D) A gram-negative infection.

   E) None of the above.

Answer: B
Skill: Recall

45) Which of the following is *not* used for attachment to a host?
- A) M protein
- B) Ligands
- C) Fimbriae
- D) Capsules
- E) None of the above.

Answer: E
Skill: Analysis

ESSAY QUESTIONS

1) Antibiotics can kill gram-negative bacteria, but symptoms of fever and low blood pressure can persist. Why?

Answer:

Skill:

2) Why is diagnosis of botulism difficult?

Answer:

Skill:

CHAPTER 16   Nonspecific Defenses of the Host

OBJECTIVE QUESTIONS

1) Nonspecific resistance is
   A) The body's ability to ward off diseases.
   B) The body's defenses against any kind of pathogen.
   C) The body's defense against a particular pathogen.
   D) The lack of resistance.
   E) None of the above.

   Answer: B
   Skill: Recall

2) Which of the following is *not* a mechanical factor to protect the skin and mucous membranes from infection?
   A) Layers of cells
   B) Tears
   C) Saliva
   D) Lysozyme
   E) None of the above

   Answer: D
   Skill: Analysis

3) The function of the "ciliary escalator" is to
   A) Kill microorganisms.
   B) Remove microorganisms from body cavities.
   C) Remove microorganisms from the lower respiratory tract.
   D) Remove microorganisms from the upper respiratory tract.
   E) All of the above.

   Answer: C
   Skill: Recall

4) Which of the following exhibits the highest phagocytic activity?
   A) Neutrophils
   B) Erythrocytes
   C) Lymphocytes
   D) Basophils
   E) Eosinophils

   Answer: A
   Skill: Recall

5) Which of the following is *not* a characteristic of inflammation?

A) Redness   B) Pain   C) Local heat   D) Fever   E) Swelling

Answer: D
Skill: Analysis

6) All of the following can be determined from a differential count *except*

A) The number of white blood cells.

B) The numbers of each type of white blood cell.

C) The number of red blood cells.

D) The possibility of a state of disease.

E) None of the above.

Answer: C
Skill: Recall

7) Which of these bacteria is *not* killed by phagocytes?

A) *Mycobacterium tuberculosis*

B) *Streptococcus pyogenes*

C) *Streptococcus pneumoniae*

D) *Klebsiella pneumoniae*

E) None of the above

Answer: A
Skill: Analysis

8) Which of the following does *not* cause vasodilation?

A) Kinins

B) Prostaglandins

C) Lysozymes

D) Histamine

E) None of the above

Answer: C
Skill: Recall

9) Which of the following choices shows the order in which white blood cells migrate to infected tissues?

A) Macrophages — monocytes

B) Lymphocytes — macrophages

C) Neutrophils — macrophages

D) Neutrophils — monocytes

E) Macrophages — neutrophils

Answer: D
Skill: Recall

10) "Margination" refers to

    A) The adherence of phagocytes to microorganisms.

    B) The chemotactic response of phagocytes.

    C) Adherence of phagocytes to the lining of blood vessels.

    D) Dilation of blood vessels.

    E) The movement of phagocytes through walls of blood vessels.

Answer: C
Skill: Recall

11) Which of the following statements is true?

    A) Interferon is an antiviral protein.

    B) Interferon promotes phagocytosis.

    C) Interferon causes cell lysis.

    D) Interferon acts against specific viruses.

    E) Interferon attacks invading viruses.

Answer: A
Skill: Analysis

12) Which of the following is found normally in serum?

    A) Complement

    B) Interferon

    C) Histamine

    D) Leukocytosis–promoting factor

    E) Lysozyme

Answer: A
Skill: Analysis

13) Which of the following is *not* an effect of complement activation?

    A) Interference with viral replication

    B) Bacterial cell lysis

    C) Opsonization

    D) Increased phagocytic activity

    E) Increased blood vessel permeability

Answer: A
Skill: Analysis

14) Which of the following is an effect of opsonization?

    A) Increased adherence of phagocytes to microorganisms

    B) Increased margination of phagocytes

    C) Increased diapedesis of phagocytes

    D) Inflammation

    E) None of the above

Answer: A
Skill: Recall

15) Normal microbiota provide protection from infection by all of the following *except*

    A) They provide antibacterial chemicals.

    B) They out–compete newcomers.

    C) They make the chemical environment unsuitable for nonresident bacteria.

    D) They produce lysozyme.

    E) None of the above.

Answer: D
Skill: Analysis

16) Your ability to ward off diseases is called

    A) Tolerance.

    B) Susceptibility.

    C) Resistance.

    D) Immunity.

    E) Inflammation.

Answer: C
Skill: Recall

17) Lysozyme is *not* found in

    A) Tears.

    B) Saliva.

    C) Perspiration.

    D) Nasal secretions.

    E) None of the above.

Answer: E
Skill: Analysis

18) Which of the following is *not* a chemical factor to protect the skin and mucous membranes from infection?

A) Mucus

B) Sebum

C) Gastric juices

D) pH

E) Lysozyme

Answer: A
Skill: Analysis

19) Which of the following is *not* true of fixed macrophages?

A) They are found in certain tissues and organs.

B) They develop from neutrophils.

C) They are cells of the mononuclear phagocytic system.

D) They are mature monocytes.

E) None of the above.

Answer: B
Skill: Analysis

20) Adherence of phagocytes may be accomplished by all of the following *except*

A) Trapping a bacterium against a rough surface.

B) Opsonization.

C) Chemotaxis.

D) Lysozyme.

E) Complement.

Answer: D
Skill: Analysis

21) Which of the following is *not* an effect of histamine?

A) Vasodilation

B) Fever

C) Swelling

D) Redness

E) Pain

Answer: B
Skill: Recall

22) Which of the following is *not* a function of inflammation?

A) To destroy an injurious agent

B) To remove an injurious agent

C) To wall off an injurious agent

D) To repair damaged tissue

E) None of the above

Answer: E
Skill: Recall

23) Chill is a sign that

A) Body temperature is falling.

B) Body temperature is rising.

C) Body temperature will remain the same.

D) Sweating will follow.

E) None of the above.

Answer: B
Skill: Recall

24) Which of the following statements is true?

A) There are at least twenty complement proteins.

B) All of the complement proteins are activated in serum.

C) Factors B and D are complement proteins.

D) Complement activity is antigen specific.

E) Complement increases after immunization.

Answer: A
Skill: Recall

25) Complement fixation results in

A) Activation of C3b.

B) Immune adherence.

C) Acute local inflammation.

D) Increased migration of phagocytes.

E) Cell lysis.

Answer: E
Skill: Analysis

26) The mechanism of action of interferon includes all of the following *except*

    A) It binds to the surface of uninfected cells.

    B) It inactivates viruses.

    C) It initiates manufacture of another antiviral protein.

    D) It works in cells not producing INF.

    E) None of the above.

Answer: B
Skill: Analysis

27) The alternative pathway for complement activation is initiated by

    A) Polysaccharides and C3b.

    B) C5–C9.

    C) Antigen–antibody reactions.

    D) Factors released from phagocytes.

    E) Factors released from damaged tissues.

Answer: A
Skill: Recall

28) The classical pathway for complement activation is initiated by

    A) Polysaccharides and C3b.

    B) C5–C9.

    C) Antigen–antibody reactions.

    D) Factors released from phagocytes.

    E) Factors released from damaged tissues.

Answer: C
Skill: Recall

29) Activation of C3a results in all of the following *except*

    A) Acute inflammation.

    B) Increased blood vessel permeability.

    C) Fever.

    D) Attraction of phagocytes.

    E) None of the above.

Answer: C
Skill: Analysis

30) Neutrophils with defective lysosomes are unable to

    A) Move by chemotaxis.

    B) Migrate.

    C) Produce toxic oxygen products.

    D) Live.

    E) None of the above.

Answer: C
Skill: Analysis

31) Vasodilation is caused by all of the following *except*

    A) Histamine.

    B) Complement.

    C) Prostaglandins.

    D) Leukotrienes.

    E) None of the above.

Answer: B
Skill: Recall

32) After ingesting a pathogen, lysosomal enzymes produce all of the following *except*

    A) H_2O.   B) $O_2^-\cdot$.   C) H_2O_2.   D) $OH\cdot$.   E) $HOCl$.

Answer: A
Skill: Analysis

33) Activation of C5–C9 results in

    A) Activation of C3.

    B) Fixation of complement.

    C) Leakage of cell contents.

    D) Phagocytosis.

    E) None of the above.

Answer: C
Skill: Recall

34) Involved in specific resistance (immunity).

    A) Basophil

    B) Eosinophil

    C) Lymphocyte

    D) Monocyte

    E) Neutrophil

Answer: C
Skill: Recall

35) Involved in resistance to parasitic helminths.

   A) Basophil

   B) Eosinophil

   C) Lymphocyte

   D) Monocyte

   E) Neutrophil

   Answer: B
   Skill: Recall

36) Macrophages arise from these.

   A) Basophil

   B) Eosinophil

   C) Lymphocyte

   D) Monocyte

   E) Neutrophil

   Answer: D
   Skill: Recall

37) In addition to lymphocytes, these cells do not have granules in their cytoplasm.

   A) Basophil

   B) Eosinophil

   C) Lymphocyte

   D) Monocyte

   E) Neutrophil

   Answer: D
   Skill: Recall

38) Bacteria have siderophores to capture iron; humans counter this by

   A) Producing iron.

   B) Transferrins.

   C) Toxin production.

   D) Iron–degrading enzymes.

   E) None of the above.

   Answer: B
   Skill: Recall

39) All of the following occur during inflammation. What is the first step?
   A) Diapedesis
   B) Margination
   C) Phagocyte migration
   D) Repair
   E) Vasodilation

   Answer: E
   Skill: Analysis

40) When cells are damaged, they release
   A) Antihistamines.
   B) Endotoxins.
   C) Lysosomes.
   D) Prostaglandins.
   E) All of the above.

   Answer: D
   Skill: Recall

41) The following occur during inflammation. Which one leads to the others?
   A) Abscess
   B) Edema
   C) Emigration
   D) Erythema
   E) Margination

   Answer: D
   Skill: Analysis

42) Several inherited deficiencies in the complement system occur in humans. Which of the following would be the most severe?
   A) Deficiency of C3
   B) Deficiency of C5
   C) Deficiency of C6
   D) Deficiency of C7
   E) Deficiency of C8

   Answer: A
   Skill: Understanding

43) Which of the following is *not* true about the classical pathway of complement activation?

A) C1 is the first protein activated in the classical pathway.

B) The C1 protein complex is initiated by antigen–antibody complexes.

C) C3 is not involved in the classical pathway.

D) Cleaved fragments of some of the proteins act to increase inflammation.

E) C3b causes opsonization.

Answer: C
Skill: Analysis

44) *Chlamydia trachomatis* lives inside white blood cells because it

A) Inhibits formation of phagolysosomes.

B) Inhibits phagocytosis.

C) Prevents complement activation.

D) Produces complement.

E) Produces keratinase.

Answer: A
Skill: Understanding

45) Bacterial enzymes such as catalase and superoxide dismutase can protect bacteria from

A) Complement.

B) Histamine.

C) Interferon.

D) Phagocytic digestion.

E) Phagocytosis.

Answer: D
Skill: Understanding

ESSAY QUESTIONS

1) Explain how each of the following avoids being killed by phogocytes..
   a. *Streptococcus pneumoniae*
   b. *Mycobacterium tuberculosis*
   c. *Streptococcus pyogenes*
   d. *Shigella dysenteriae*

   Answer:

   Skill:

2) A patient was seen by a physician for symptoms that included a swollen toe, a red streak along his ankle, and enlarged lymph nodes in his groin. Explain the cause of these symptoms.

   Answer:

   Skill:

CHAPTER 17   Specific Defenses of the Host: The Immune Response

OBJECTIVE QUESTIONS

1) Type of immunity resulting from vaccination.
   - A) Innate resistance
   - B) Naturally acquired active immunity
   - C) Naturally acquired passive immunity
   - D) Artificially acquired active immunity
   - E) Artificially acquired passive immunity

   Answer: D
   Skill: Understanding

2) Type of immunity resulting from transfer of antibodies from one individual to a susceptible individual by means of injection.
   - A) Innate resistance
   - B) Naturally acquired active immunity
   - C) Naturally acquired passive immunity
   - D) Artificially acquired active immunity
   - E) Artificially acquired passive immunity

   Answer: E
   Skill: Understanding

3) Immunity resulting from recovery from mumps.
   - A) Innate resistance
   - B) Naturally acquired active immunity
   - C) Naturally acquired passive immunity
   - D) Artificially acquired active immunity
   - E) Artificially acquired passive immunity

   Answer: B
   Skill: Understanding

4) A human's resistance to canine distemper.
   - A) Innate resistance
   - B) Naturally acquired active immunity
   - C) Naturally acquired passive immunity
   - D) Artificially acquired active immunity
   - E) Artificially acquired passive immunity

   Answer: A
   Skill: Understanding

5) Newborns' immunity due to the transfer of antibodies across the placenta.

   A) Innate resistance

   B) Naturally acquired active immunity

   C) Naturally acquired passive immunity

   D) Artificially acquired active immunity

   E) Artificially acquired passive immunity

Answer: C
Skill: Understanding

6) Immunity that is not due to antibodies.

   A) Innate resistance

   B) Naturally acquired active immunity

   C) Naturally acquired passive immunity

   D) Artificially acquired active immunity

   E) Artificially acquired passive immunity

Answer: A
Skill: Understanding

7) Immunity due to injection of tetanus toxoid.

   A) Innate resistance

   B) Naturally acquired active immunity

   C) Naturally acquired passive immunity

   D) Artificially acquired active immunity

   E) Artificially acquired passive immunity

Answer: D
Skill: Understanding

8) Immunity due to injection of an antigen.

   A) Innate resistance

   B) Naturally acquired active immunity

   C) Naturally acquired passive immunity

   D) Artificially acquired active immunity

   E) Artificially acquired passive immunity

Answer: E
Skill: Understanding

9) Which of the following chemicals is the least antigenic?

A) Nucleoproteins

B) Lipoproteins

C) Glycoproteins

D) Proteins

E) Sugars

Answer: E
Skill: Analysis

10) Antibodies may be called gamma globulins for all of the following reasons *except*

A) Antibodies are responsible for immunity.

B) Antibodies are globular proteins.

C) Antibodies are found in serum globulin.

D) Antibodies are found in the gamma fraction of serum.

E) Both A and C.

Answer: A
Skill: Understanding

11) The specificity of an antibody is due to

A) Its valence.

B) The H chains.

C) The L chains.

D) The constant portions of the H and L chains.

E) The variable portions of the H and L chains.

Answer: E
Skill: Recall

12) Which of the following is *not* a characteristic of B cells?

A) They originate in bone marrow.

B) They have antigen receptors on their surfaces.

C) They are responsible for the anamnestic response.

D) They are responsible for antibody formation.

E) None of the above.

Answer: E
Skill: Analysis

13) Which of the following is *not* a characteristic of cell–mediated immunity?

   A) The cells originate in bone marrow.

   B) Cells are processed in the thymus gland.

   C) It can inhibit the immune response.

   D) It includes macrophages.

   E) None of the above.

   Answer: D
   Skill: Analysis

14) Plasma cells are activated by a(n)

   A) Antigen.

   B) T cell.

   C) B cell.

   D) Memory cell.

   E) None of the above.

   Answer: A
   Skill: Recall

15) Antibodies found in mucus, saliva, and tears.

   A) IgG   B) IgM   C) IgA   D) IgD   E) IgE

   Answer: C
   Skill: Recall

16) Antibodies found on B cells.

   A) IgG   B) IgM   C) IgA   D) IgD   E) IgE

   Answer: D
   Skill: Recall

17) Antibodies that can bind to large parasites.

   A) IgG   B) IgM   C) IgA   D) IgD   E) IgE

   Answer: E
   Skill: Recall

18) In addition to IgG, these antibodies can fix complement.

   A) IgG   B) IgM   C) IgA   D) IgD   E) IgE

   Answer: B
   Skill: Recall

19) These large antibodies agglutinate antigens.

   A) IgG   B) IgM   C) IgA   D) IgD   E) IgE

   Answer: B
   Skill: Recall

20) The most abundant class of antibodies in serum.

   A) IgG   B) IgM   C) IgA   D) IgD   E) IgE

   Answer: A
   Skill: Recall

Figure 17.1

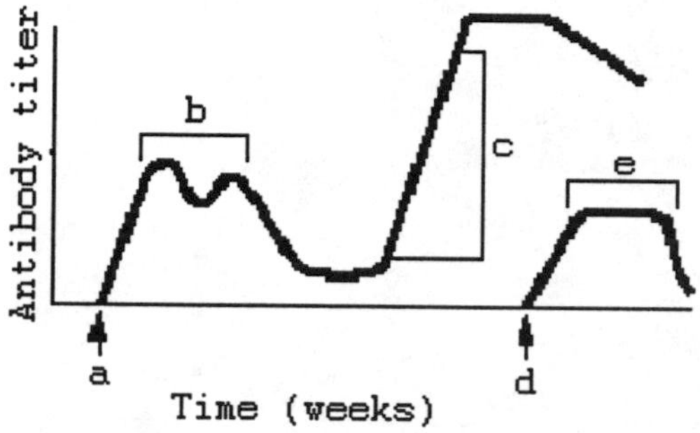

21) In Figure 17.1, the arrow *d* indicates

   A) The time of exposure to the same antigen as at *a*.

   B) The secondary response.

   C) The primary response.

   D) Exposure to a new antigen.

   E) None of the above.

   Answer: D
   Skill: Analysis

22) In Figure 17.1, which letter on the graph indicates the patient's secondary response to an antigen?

   A) a   B) b   C) c   D) d   E) e

   Answer: C
   Skill: Analysis

23) In Figure 17.1, which letter on the graph indicates the patient's response to a second antigen?

   A) a   B) b   C) c   D) d   E) e

   Answer: E
   Skill: Analysis

24) Which statement is incorrect?

   A) The variable region of a heavy chain binds with antigen.

   B) The variable region of a light chain binds with antigen.

   C) The Fc region attaches to a host cell.

   D) The constant region of a heavy chain is the same for all antibodies.

   E) None of the above.

   Answer: D
   Skill: Analysis

25) The best definition of an antigen is

   A) Something foreign in the body.

   B) A chemical that elicits an antibody response and can combine with these antibodies.

   C) A chemical that combines with antibodies.

   D) A pathogen.

   E) A protein that combines with antibodies.

   Answer: B
   Skill: Recall

26) The best definition of an antibody is

   A) A serum protein.

   B) A protein that inactivates or kills an antigen.

   C) A protein made in response to an antigen that can combine with that antigen.

   D) An immunoglobulin.

   E) None of the above.

   Answer: C
   Skill: Recall

27) The following events elicit an antibody response. What is the third step?

   A) Antigen–digest goes to surface of APC.

   B) APC phagocytizes antigen.

   C) B cell responds to antigen.

   D) T_H cell recognizes antigen–digest and MHC.

   E) T_H cell recognizes B cell.

   Answer: D
   Skill: Analysis

Figure 17.2

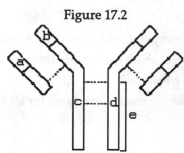

28) In Figure 17.2, which areas are similar for all IgG antibodies?

A) a and b   B) a and c   C) b and c   D) c and d   E) b and d

Answer: D
Skill: Understanding

29) In Figure 17.2, which areas are different for all IgM antibodies?

A) a and b   B) a and c   C) b and c   D) c and d

Answer: A
Skill: Analysis

30) In Figure 17.2, which areas represent antigen–binding sites?

A) a and b   B) a and c   C) b and c   D) c and d   E) b and d

Answer: A
Skill: Analysis

31) In Figure 17.2, what can attach to a host cell?

A) a and c   B) b and c   C) b   D) d   E) e

Answer: E
Skill: Analysis

32) Figure 17.2 could be all of the following *except*

A) IgM.
B) IgG.
C) IgD.
D) IgE.
E) None of the above.

Answer: A
Skill: Analysis

33) Which of the following statements about septic shock is *not* true?

A) Symptoms are caused by the host's immune system.

B) It is treated with antibiotics.

C) Symptoms include fever and low blood pressure.

D) Live bacteria are not necessary for the symptoms to occur.

E) None of the above.

Answer: B
Skill: Recall

34) Destroys virus–infected cells.

A) T_C B) T_D C) T_H D) T_S

Answer: A
Skill: Recall

35) Responsible for hypersensitivity (allergy).

A) T_C B) T_D C) T_H D) T_S

Answer: B
Skill: Recall

36) Involved in antibody production.

A) T_C B) T_D C) T_H D) T_S

Answer: C
Skill: Recall

37) Which one of the following causes transmembrane channels in target cells?

A) Antigen   B) Hapten   C) IL–1   D) IL–2   E) Perforin

Answer: E
Skill: Recall

38) Patients with an inherited type of colon cancer called familial adenomatous polyposis have a mutation in the gene that codes for

A) Apoptosis.

B) IgE antibodies.

C) Helper T cells.

D) ADCC.

E) Phagocytosis.

Answer: A
Skill: Understanding

39) Chemical signals are sent between leukocytes by
    A) Cytokines.
    B) Interferons.
    C) Interleukins.
    D) Tumor necrosis factor.
    E) Colony-stimulating factor.

    Answer: B
    Skill: Recall

40) All of the following are true about natural killer cells *except*
    A) They destroy virus-infected cells.
    B) They destroy tumor cells.
    C) They lyse their target cells.
    D) They are stimulated by an antigen.
    E) None of the above.

    Answer: D
    Skill: Analysis

41) An antibody's Fc region can be bound by
    A) Antibodies.
    B) Macrophages.
    C) Helper T cells.
    D) B cells.
    E) None of the above.

    Answer: B
    Skill: Analysis

42) Antigens coated with antibodies are susceptible to
    A) Further antibody attack.
    B) Phagocytosis.
    C) Helper T cells.
    D) B cells.
    E) None of the above.

    Answer: B
    Skill: Analysis

43) Monoclonal antibodies against CD4 antigens might be used to treat AIDS because
   A) The HIV has CD4 surface antigens.
   B) Susceptible host cells would be killed.
   C) Receptor sites would not be available for the virus.
   D) The virus would attach to the antibodies.
   E) None of the above.

   Answer: C
   Skill: Understanding

44) IL–2 is produced by T_H cells, to
   A) Activate macrophages.
   B) Stimulate T_H cell maturation.
   C) Cause phagocytosis.
   D) Activate antigen–presenting cells.
   E) None of the above.

   Answer: B
   Skill: Recall

45) Antigen–antibody binding may result in all of the following *except*
   A) Agglutination of the antigens.
   B) Complement activation.
   C) IL–2 production.
   D) Neutralization of the antigen.
   E) Opsonization of the antigen.

   Answer: C
   Skill: Recall

ESSAY QUESTIONS

1) A person has antibodies against the measles virus. Identify three ways in which these antibodies could be acquired.
   Answer:
   Skill:

2) Describe the production of antibodies using the clonal selection theory.
   Answer:
   Skill:

3) Positive diagnosis of AIDS is made when a patient has antibodies against the human immunodeficiency virus (HIV). Why does a patient have an immune deficiency if he or she is making antibodies?

Answer:

Skill:

CHAPTER 18   Practical Applications of Immunology

OBJECTIVE QUESTIONS

1) Which of the following is *not* normally used in a vaccine?

   A) Toxoid

   B) Parts of bacterial cells

   C) Live, attenuated bacteria

   D) Inactivated viruses

   E) Antibodies

   Answer: E
   Skill: Recall

2) Patient's serum, influenza virus, and red blood cells are mixed in a tube. What happens if the patient has antibodies against influenza virus?

   A) Agglutination

   B) Hemagglutination

   C) Complement fixation

   D) Hemolysis

   E) None of the above

   Answer: E
   Skill: Analysis

3) If a patient shows the presence of antibodies against Herpes simplex, this indicates all of the following *except*

   A) The patient may have the disease.

   B) The patient may have had the disease and has recovered.

   C) The patient may have been vaccinated.

   D) A recent transfusion may have passively introduced the antibodies.

   E) The patient was near someone who had the disease.

   Answer: E
   Skill: Recall

4) In an agglutination test, eight serial dilutions to determine antibody titer were set up: tube #1 contained a 1:2 dilution; tube #2, a 1:4, etc. If tube #6 is the last tube showing agglutination, what is the antibody titer?

   A) 6   B) 1:6   C) 64   D) 1:32   E) 32

   Answer: C
   Skill: Analysis

5) All of the following are disadvantages of a live virus vaccine *except*

   A) The live vaccine may revert to a more virulent form.

   B) Exogenous protein contaminants may be present.

   C) Antibody response is not as good as with inactivated viruses.

   D) Live viruses generally require refrigeration.

   E) None of the above.

   Answer: C
   Skill: Recall

6) Antibodies for serological testing can be obtained from all of the following *except*

   A) Vaccinated humans.

   B) Vaccinated animals.

   C) Monoclonal antibodies.

   D) Viral cultures.

   E) None of the above.

   Answer: D
   Skill: Analysis

7) Reaction between an antibody and soluble antigen–forming lattices.

   A) Agglutination reaction

   B) Complement fixation

   C) Immunofluorescence

   D) Neutralization reaction

   E) Precipitation reaction

   Answer: E
   Skill: Recall

8) Reaction between antibody and particulate antigen.

   A) Agglutination reaction

   B) Complement fixation

   C) Immunofluorescence

   D) Neutralization reaction

   E) Precipitation reaction

   Answer: A
   Skill: Recall

9) Reaction using red blood cells as the indicator and hemolysis indicates an antigen–antibody reaction.

    A) Agglutination reaction

    B) Complement fixation

    C) Immunofluorescence

    D) Neutralization reaction

    E) Precipitation reaction

Answer: B
Skill: Recall

10) An indirect version using antihuman globulin may be used to detect patient's antibodies against *Treponema pallidum*.

    A) Agglutination reaction

    B) Complement fixation

    C) Immunofluorescence

    D) Neutralization reaction

    E) Precipitation reaction

Answer: C
Skill: Recall

11) The Schick test in which diphtheria exotoxin is injected subcutaneously to determine whether a person has antibodies against diphtheria toxin.

    A) Agglutination reaction

    B) Complement fixation

    C) Immunofluorescence

    D) Neutralization reaction

    E) Precipitation reaction

Answer: D
Skill: Understanding

12) *Streptococcus pyogenes* capsule.

    A) Conjugated vaccine

    B) Subunit vaccine

    C) Nucleic acid vaccine

    D) Attenuated whole–agent vaccine

    E) Toxoid vaccine

Answer: B
Skill: Recall

13) Host synthesizes viral antigens.

  A) Conjugated vaccine

  B) Subunit vaccine

  C) Nucleic acid vaccine

  D) Attenuated whole-agent vaccine

  E) Toxoid vaccine

Answer: C
Skill: Recall

14) Purified protein from *B. pertussis*.

  A) Conjugated vaccine

  B) Subunit vaccine

  C) Nucleic acid vaccine

  D) Attenuated whole-agent vaccine

  E) Toxoid vaccine

Answer: B
Skill: Recall

15) Live measles virus.

  A) Conjugated vaccine

  B) Subunit vaccine

  C) Nucleic acid vaccine

  D) Attenuated whole-agent vaccine

  E) Toxoid vaccine

Answer: D
Skill: Recall

16) Test used to identify antibodies against *Treponema pallidum* in a patient.

  A) Direct fluorescent antibody

  B) Indirect fluorescent antibody

  C) None of the above

Answer: B
Skill: Analysis

17) Test used to identify *Streptococcus pyogenes* in a patient.

  A) Direct fluorescent antibody

  B) Indirect fluorescent antibody

  C) None of the above

Answer: A
Skill: Analysis

18) Test used to detect anti-*Rickettsia* antibodies in a patient.

    A) Direct fluorescent antibody

    B) Indirect fluorescent antibody

    C) None of the above

Answer: B
Skill: Analysis

19) A pregnancy test used to find the fetal hormone HCG in a woman's urine uses anti–HCG and latex spheres.

    A) Direct agglutination reaction

    B) Passive agglutination reaction

    C) Immunofluorescence

    D) Neutralization reaction

    E) Precipitation reaction

Answer: B
Skill: Analysis

20) A test to determine patient's blood type by mixing patient's red blood cells with antisera.

    A) Direct agglutination reaction

    B) Passive agglutination reaction

    C) Immunofluorescence

    D) Neutralization reaction

    E) Precipitation reaction

Answer: A
Skill: Analysis

21) A test to determine the presence of soluble AB antigens in patient's saliva.

    A) Direct agglutination reaction

    B) Passive agglutination reaction

    C) Immunofluorescence

    D) Neutralization reaction

    E) Precipitation reaction

Answer: E
Skill: Analysis

22) Patient's serum, *Rickettsia*, guinea pig complement, sheep red blood cells, and anti–sheep red blood cells are mixed in a tube. What happens if the patient has epidemic typhus?
   A) Bacteria fluoresce
   B) Hemagglutination
   C) Hemagglutination–inhibition
   D) Hemolysis
   E) None of the above

   Answer: E
   Skill: Understanding

23) A vaccine against HIV proteins made by vaccinia virus.
   A) Conjugated vaccine
   B) Subunit vaccine
   C) Nucleic acid vaccine
   D) Inactivated whole–agent vaccine
   E) Toxoid vaccine

   Answer: B
   Skill: Analysis

24) Inactivated tetanus toxin.
   A) Conjugated vaccine
   B) Subunit vaccine
   C) Nucleic acid vaccine
   D) Inactivated whole–agent vaccine
   E) Toxoid vaccine

   Answer: E
   Skill: Analysis

25) *Haemophilus influenzae* b capsular polysaccharide with a protein.
   A) Conjugated vaccine
   B) Subunit vaccine
   C) Nucleic acid vaccine
   D) Inactivated whole–agent vaccine
   E) Toxoid vaccine

   Answer: A
   Skill: Analysis

	Table 18.1			
	Antibody Titer			
	Day 1	Day 7	Day 14	Day 21
Patient A	0	0	256	512
Patient B	128	256	512	1024
Patient C	0	0	0	0
Patient D	128	128	128	128

26) In Table 18.1, who probably has the disease?

A) A and B   B) B and C   C) A and C   D) C and D   E) A and D

Answer: A
Skill: Analysis

27) In Table 18.1, who is most likely protected from the disease?

A) A

B) B

C) C

D) D

E) All of the above

Answer: D
Skill: Analysis

28) In Table 18.1, who showed seroconversion during these observations?

A) A

B) B

C) C

D) D

E) All of the above

Answer: A
Skill: Analysis

29) In a direct ELISA test, what are you looking for in the patient?

A) Antibodies   B) Antigen

C) Either antigen or antibodies   D) None of the above

Answer: B
Skill: Recall

30) Which of the following tests is *not* correctly matched to its positive reaction?

   A) Hemagglutination — clumping of red blood cells

   B) Complement fixation — no hemolysis

   C) Neutralization — no tissue/animal death

   D) ELISA — enzyme–substrate reaction

   E) None of the above

   Answer: E
   Skill: Recall

31) The circumsporozoite antigen of *Plasmodium* can be used for all of the following *except* to

   A) Vaccinate healthy people.

   B) Cure infected people.

   C) Produce monoclonal antibodies.

   D) Decrease recurring infections.

   E) All of the above.

   Answer: B
   Skill: Recall

32) For years, scientists believed that the immune system didn't respond to parasitic infestations. It is now known that

   A) This is correct.

   B) Parasites change their surface antigens.

   C) The response is nonspecific phagocytosis.

   D) The response is only cell–mediated, not antibodies.

   E) None of the above.

   Answer: B
   Skill: Recall

33) To detect botulinum toxin in food, suspect food is injected into two guinea pigs. The guinea pig that was vaccinated against botulism survives, while the one that was not vaccinated dies. This is an example of

   A) Agglutination.

   B) Neutralization.

   C) Hemagglutination.

   D) Fluorescent antibodies.

   E) ELISA.

   Answer: B
   Skill: Understanding

34) Live polio virus.

   A) Inactivated whole–agent vaccine

   B) Attenuated whole–agent vaccine

   C) Conjugated vaccine

   D) Subunit vaccine

   E) Toxoid vaccine

   Answer: B
   Skill: Analysis

35) *Haemophilus* capsule polysaccharide plus diphtheria toxoid.

   A) Inactivated whole–agent vaccine

   B) Attenuated whole–agent vaccine

   C) Conjugated vaccine

   D) Subunit vaccine

   E) Toxoid vaccine

   Answer: C
   Skill: Analysis

36) Dead *Bordetella pertussis*.

   A) Inactivated whole–agent vaccine

   B) Attenuated whole–agent vaccine

   C) Conjugated vaccine

   D) Subunit vaccine

   E) Toxoid vaccine

   Answer: A
   Skill: Analysis

37) Hepatitis B virus surface antigen.

   A) Inactivated whole–agent vaccine

   B) Attenuated whole–agent vaccine

   C) Conjugated vaccine

   D) Subunit vaccine

   E) Toxoid vaccine

   Answer: D
   Skill: Analysis

38) Which is the third step in a direct ELISA test?

   A) Substrate for the enzyme   B) Antigen

   C) Antihuman immune serum   D) Antibodies against the antigen

   Answer: C
   Skill: Understanding

217

39) Which item is from the patient in a direct ELISA test?

    A) Substrate for the enzyme

    B) Antigen

    C) Antihuman immune serum

    D) Antibodies against the antigen

Answer: B
Skill: Understanding

40) Which of the following is most useful in determining the presence of AIDS antibodies?

    A) Agglutination

    B) Complement fixation

    C) Neutralization

    D) Indirect ELISA

    E) Direct fluorescent antibody

Answer: D
Skill: Understanding

41) Uses radioactively labeled viruses.

    A) Agglutination

    B) Complement fixation

    C) Precipitation

    D) Flow cytometry

    E) Radioimmunoassay

Answer: E
Skill: Recall

42) Uses fluorescent–labeled antibodies.

    A) Agglutination

    B) Complement fixation

    C) Precipitation

    D) Flow cytometry

    E) Radioimmunoassay

Answer: D
Skill: Recall

43) Uses red blood cells as the indicator.

    A) Agglutination

    B) Complement fixation

    C) Precipitation

    D) Flow cytometry

    E) Radioimmunoassay

Answer: B
Skill: Recall

In an immunodiffusion test to diagnose histoplasmosis, patient's serum is placed in a well in an agar plate. In a positive test, a precipitate forms as the serum diffuses from the well and meets material diffusing from a second well.

44) In Situation 18.1, what is in the second well?

   A) Antibodies

   B) A fungal antigen

   C) Fungal cells

   D) Mycelia

   E) Red blood cells

   Answer: B
   Skill: Understanding

45) The immunodiffusion test described in Situation 18.1 is

   A) An agglutination reaction.

   B) A precipitation reaction.

   C) A complement–fixation test.

   D) An ELISA test.

   E) A direct test.

   Answer: B
   Skill: Understanding

ESSAY QUESTIONS

1) Explain the ELISA test to detect the presence of HIV antibodies in a patient.
   Answer:
   Skill:

2) Design a serological test to detect botulinum toxin in food.
   Answer:
   Skill:

3) A person has an antibody titer of 28. What do you know about this person?
   Answer:
   Skill:

CHAPTER 19 Disorders Associated with the Immune System

OBJECTIVE QUESTIONS

1) Hypersensitivity is due to
    A) The presence of an antigen.
    B) Immunity.
    C) The presence of antibodies.
    D) An altered immune response.
    E) Allergies.

    Answer: D
    Skill: Recall

2) The chemical mediators of anaphylaxis are
    A) Found in basophils and mast cells.
    B) Antibodies.
    C) Antigens.
    D) Antigen–antibody complexes.
    E) The proteins of the complement system.

    Answer: A
    Skill: Analysis

3) Which of the following may result from systemic anaphylaxis?
    A) Hay fever
    B) Asthma
    C) Shock
    D) Hives
    E) None of the above

    Answer: C
    Skill: Analysis

4) Which antibodies will be in the serum of a person with blood type B, Rh⁻?
    A) Anti A, anti B, anti Rh
    B) Anti A, anti Rh
    C) Anti A
    D) Anti B, anti Rh
    E) Anti B

    Answer: C
    Skill: Understanding

5) Which type of graft is least compatible?

   A) Autograft

   B) Allograft

   C) Isograft

   D) Xenograft

   E) None of the above

   Answer: D
   Skill: Analysis

6) Which of the following is *not* used to determine relatedness between a donor and a recipient for transplants?

   A) ABO antigens

   B) ABO antibodies

   C) MHC antigens

   D) MHC antibodies

   E) None of the above

   Answer: E
   Skill: Analysis

7) Graft–versus–host disease will most likely be a complication of

   A) A skin graft.

   B) A bone marrow transplant.

   C) A blood transfusion.

   D) An Rh incompatibility between mother and fetus.

   E) All of the above.

   Answer: B
   Skill: Analysis

8) Which of the following is *not* an immune complex disease?

   A) Rheumatic fever

   B) Systemic lupus erythematosus

   C) Hemolytic disease of the newborn

   D) Glomerulonephritis

   E) None of the above

   Answer: C
   Skill: Analysis

9) Cancer cells may escape the immune system because

   A) They are recognized as "self."

   B) Antibodies are not formed against cancer cells.

   C) Killer T cells react with tumor–specific antigens.

   D) Tumor cells shed their specific antigens.

   E) None of the above.

   Answer: D
   Skill: Recall

10) The symptoms of an immune complex reaction are due to

   A) Destruction of the antigen.

   B) Complement fixation.

   C) Phagocytosis.

   D) Antibodies against self.

   E) None of the above.

   Answer: B
   Skill: Recall

11) Autoimmunity is due to

   A) IgG and IgM antibodies.

   B) IgA antibodies.

   C) IgD antibodies.

   D) IgE antibodies.

   E) All of the above.

   Answer: A
   Skill: Recall

12) Allergic contact dermatitis is due to

   A) Sensitized T cells.

   B) IgG antibodies.

   C) IgE antibodies.

   D) IgM antibodies.

   E) All of the above.

   Answer: A
   Skill: Analysis

13) Immunotoxins can be used to treat cancer because they

   A) Phagocytize foreign cells.

   B) Fix complement.

   C) Poison cells.

   D) Agglutinate cells.

   E) None of the above.

   Answer: C
   Skill: Analysis

14) The T-cell response can be suppressed by all of the following *except*

   A) HIV infection.

   B) Certain genes.

   C) Anti-transplant-rejection drugs.

   D) Immune complex formation.

   E) Antibodies against CD3.

   Answer: D
   Skill: Understanding

15) A hypersensitivity reaction occurs

   A) During the first exposure to an antigen.

   B) On a second or subsequent exposure to an antigen.

   C) In immunologically tolerant individuals.

   D) During autoimmune diseases.

   E) In individuals with diseases of the immune system.

   Answer: B
   Skill: Recall

16) All of the following statements about type I hypersensitivities are true *except*

   A) They are cell mediated.

   B) They involve IgE antibodies.

   C) The symptoms are due to histamine.

   D) Antibodies are bound to host cells.

   E) The symptoms occur soon after exposure to an antigen.

   Answer: A
   Skill: Analysis

17) All of the following statements about type IV hypersensitivities are true *except*

A) They are cell mediated.

B) The symptoms occur within a few days after exposure to an antigen.

C) They can be passively transferred with serum.

D) The symptoms are due to lymphokines.

E) They contribute to the symptoms of certain diseases.

Answer: C
Skill: Analysis

18) Which of the following blood transfusions are incompatible?

	Donor	Recipient
1.	AB, Rh⁻	AB, Rh⁺
2.	A, Rh⁺	A, Rh⁻
3.	A, Rh⁺	O, Rh⁺
4.	B, Rh⁻	B, Rh⁺
5.	B, Rh⁺	A, Rh⁺

A) 2 and 5

B) 1, 2, and 3

C) 2, 3, and 5

D) 3 and 4

E) All of the above

Answer: C
Skill: Understanding

19) Hemolytic disease of the newborn can result from

A) An Rh⁺ mother with an Rh⁻ fetus.

B) An Rh⁻ mother with an Rh⁺ fetus.

C) An AB mother with a B fetus.

D) An AB mother with an O fetus.

E) None of the above.

Answer: B
Skill: Understanding

20) Reaction of antigen with IgE antibodies attached to mast cells causes

A) Lysis of the cells.

B) Release of chemical mediators.

C) Complement fixation.

D) Agglutination.

E) None of the above.

Answer: B
Skill: Analysis

21) Uses a monoclonal anti–tumor antibody and a toxin.

    A) Immunologic enhancement   B) Immunologic surveillance

    C) Immunotherapy   D) Immunosuppression

Answer: C
Skill: Analysis

22) May be inherited or result from HIV infection.

    A) Immunologic enhancement   B) Immunologic surveillance

    C) Immunotherapy   D) Immunosuppression

Answer: D
Skill: Analysis

23) Due to treatment with certain drugs to reduce transplant rejection.

    A) Immunologic enhancement   B) Immunologic surveillance

    C) Immunotherapy   D) Immunosuppression

Answer: D
Skill: Analysis

24) How cancer cells avoid the immune system.

    A) Immunologic enhancement   B) Immunologic surveillance

    C) Immunotherapy   D) Immunosuppression

Answer: D
Skill: Analysis

25) Body's response to tumor-specific antigen.

    A) Immunologic enhancement   B) Immunologic surveillance

    C) Immunotherapy   D) Immunosuppression

Answer: B
Skill: Analysis

26) Results in increased susceptibility to infection.

    A) Immunologic enhancement   B) Immunologic surveillance

    C) Immunotherapy   D) Immunosuppression

Answer: D
Skill: Analysis

27) Hay fever.

    A) Type I hypersensitivity

    B) Type II hypersensitivity

    C) Type III hypersensitivity

    D) Type IV hypersensitivity

    E) All of the above

Answer: A
Skill: Analysis

28) Transfusion reactions.

    A) Type I hypersensitivity

    B) Type II hypersensitivity

    C) Type III hypersensitivity

    D) Type IV hypersensitivity

    E) All of the above

Answer: B
Skill: Analysis

29) Transplant rejection.

    A) Type I hypersensitivity

    B) Type II hypersensitivity

    C) Type III hypersensitivity

    D) Type IV hypersensitivity

    E) All of the above

Answer: D
Skill: Analysis

30) Which one of the following statements about HIV is *not* true?

    A) The T-cell response triggers viral multiplication.

    B) HIV can be transmitted by cell–to–cell contact.

    C) Bone marrow can be a reservoir for future infection.

    D) Viral infection of T_H cells results in signs elsewhere in the patient.

    E) HIV infection directly causes death.

Answer: E
Skill: Understanding

31) Someone with AIDS will probably

   A) Not make any antibodies.

   B) Make T–dependent antibodies.

   C) Make T–independent antibodies.

   D) Make T_C– and T_D–dependent antibodies.

   E) None of the above.

   Answer: C
   Skill: Understanding

32) Which of the following is the least likely vaccine against HIV?

   A) Attenuated virus

   B) Glycoprotein

   C) Protein core

   D) Subunit

   E) None of the above

   Answer: A
   Skill: Analysis

33) Antibodies against HIV are ineffective for all of the following reasons *except*

   A) Antibodies aren't made against HIV.

   B) Transmission by cell–to–cell fusion.

   C) Antigenic change.

   D) Latency.

   E) Virus particles staying in vesicles.

   Answer: A
   Skill: Analysis

34) The outcome of an HIV infection could be all of the following *except*

   A) Latency.

   B) Slow production of new viruses.

   C) T_C–killing of infected cells.

   D) Viral–killing of infected cells.

   E) None of the above.

   Answer: E
   Skill: Recall

35) Which of these causes of glomerulonephritis leads to all the others?
  A) Antibodies against *Streptococcus*
  B) Circulating immune complexes
  C) Complement fixation
  D) Formation of immune complexes
  E) Production of IgG

Answer: A
Skill: Understanding

36) Which of these causes the damage to kidney cells?
  A) Antibodies against *Streptococcus*
  B) Circulating immune complexes
  C) Complement fixation
  D) Formation of immune complexes
  E) Production of IgG

Answer: C
Skill: Understanding

37) HIV is transmitted by all of the following *except*
  A) Homosexual activity.
  B) Heterosexual activity.
  C) Hypodermic needles.
  D) Mosquitoes.
  E) None of the above.

Answer: D
Skill: Recall

38) Drugs, such as AZT and ddC, currently used to treat AIDS act by
  A) Stimulatory T_H cells.
  B) Stopping DNA synthesis.
  C) Promoting antibody formation.
  D) Neutralizing the virus.
  E) All of the above.

Answer: B
Skill: Analysis

39) Tissues for transplantation are typed for
   A) Rh.
   B) ABO and Class I and Class II HLA antigens.
   C) Rh and ABO.
   D) DR antigen.

   Answer: B
   Skill: Analysis

40) Which of the following is *not* considered a type I hypersensitivity?
   A) Asthma
   B) Dust allergies
   C) Penicillin allergic reactions
   D) Pollen allergies
   E) Transplant rejections

   Answer: E
   Skill: Recall

41) Immune deficiencies are caused by all of the following. Which one does *not* cause an acquired immune deficiency?
   A) Chromosomal–linked B–cell deficiency
   B) Cyclosporine to inhibit IL–2 secretion
   C) HIV infection
   D) Rapamycin to inhibit IL–2 action
   E) None of the above

   Answer: A
   Skill: Analysis

42) Which of the following describes a Type III autoimmune reaction?
   A) Antibodies react to cell–surface antigens.
   B) Antibodies are not made.
   C) Antibodies to a pathogen cross–react with self.
   D) Immune complexes form.
   E) Mediate by T cells.

   Answer: C
   Skill: Recall

43) Clinical AIDS is diagnosed when

   A) A patient has lymphadenopathy.

   B) HIV is found in a patient by Western blotting.

   C) The CD4 T–cell count is <200/mm^3.

   D) The patient has persistent diarrhea.

   E) The patient has antibodies against HIV.

   Answer: C
   Skill: Recall

44) A person with Graves' disease makes antibodies against thyroid hormone receptors on cells. This is an example of

   A) Type I autoimmunity.

   B) Type II autoimmunity.

   C) Type III autoimmunity.

   D) Type IV autoimmunity.

   E) Acquired immunodeficiency.

   Answer: B
   Skill: Analysis

45) MMR vaccine contains hydrolyzed gelatin. A person receiving this vaccine could develop an anaphylactic reaction if the person has

   A) An immunodeficiency.

   B) Antibodies against eggs.

   C) Antibodies against gelatin.

   D) Been vaccinated previously.

   E) None of the above.

   Answer: C
   Skill: Analysis

ESSAY QUESTIONS

1) Differentiate between type II and type III hypersensitivity reactions.
   Answer:
   Skill:

2) Simian immune deficiency disease is caused by a virus that is closely related to HIV. This disease occurs naturally in African monkeys only. Recently a primate center that raises Rhesus monkeys found an immune deficiency disease in its Indian Rhesus monkeys. Provide a hypothesis to explain how the Rhesus monkeys acquired this disease.
   Answer:
   Skill:

3) What is desensitization? Explain how this treatment can induce systemic anaphylaxis. How is systemic anaphylaxis treated?

Answer:

Skill:

4) The ratio of CD4:CD8 is 2.0 in normal individuals. What can you conclude if a patient has a CD4:CD8 ratio of 0.5?

Answer:

Skill: Recall

CHAPTER 20   Antimicrobial Drugs

OBJECTIVE QUESTIONS

1) Penicillin was considered a "miracle drug" for all of the following reasons *except*

   A) It was the first antibiotic.

   B) It doesn't affect eucaryotic cells.

   C) It inhibits gram–positive cell wall synthesis.

   D) It has selective toxicity.

   E) None of the above.

   Answer: A
   Skill: Analysis

2) The first antibiotic discovered was

   A) Quinine.

   B) Salvarsan.

   C) Streptomycin.

   D) Sulfa drugs.

   E) Penicillin.

   Answer: E
   Skill: Recall

3) Most of the available antimicrobial agents are effective against

   A) Viruses.

   B) Bacteria.

   C) Fungi.

   D) Protozoa.

   E) All of the above.

   Answer: B
   Skill: Recall

4) Antimicrobial peptides work by

   A) Inhibiting protein synthesis.

   B) Disrupting the plasma membrane.

   C) Complementary base–pairing with DNA.

   D) Inhibiting cell–wall synthesis.

   E) Hydrolyzing peptidoglycan.

   Answer: B
   Skill: Recall

5) Semisynthetic penicillins differ from natural penicillins in all of the following respects *except* that both are

   A) Broad spectrum.

   B) Resistant to penicillinase.

   C) Resistant to stomach acids.

   D) Bactericidal.

   E) None of the above.

   Answer: D
   Skill: Recall

6) Which of the following antibiotics is *not* bactericidal?

   A) Aminoglycosides

   B) Cephalosporins

   C) Polyenes

   D) Rifampins

   E) Penicillin

   Answer: C
   Skill: Analysis

7) Which one of the following does not belong with the others?

   A) Bacitracin

   B) Cephalosporin

   C) Monobactam

   D) Penicillin

   E) Streptomycin

   Answer: E
   Skill: Analysis

Figure 20.1

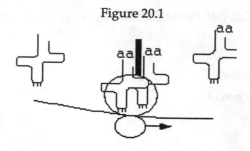

8) The antibiotic chloramphenicol binds to the 50S subunit of the ribosome as shown in Figure 20.1. The effect is to

A) Prevent attachment of tRNA.

B) Prevent peptide bond formation.

C) Prevent transcription.

D) Stop the ribosome from moving along the mRNA.

E) None of the above.

Answer: B
Skill: Understanding

9) The antibiotic cycloheximide binds to the 60S subunit of the ribosome as shown in Figure 20.1. The effect is to

A) Prevent mRNA–ribosome binding in eucaryotes.

B) Prevent peptide bond formation in procaryotes.

C) Prevent polypeptide elongation in eucaryotes.

D) Prevent transcription in procaryotes.

E) None of the above.

Answer: C
Skill: Understanding

10) Which of these antimicrobial agents has the fewest side effects?

A) Streptomycin

B) Tetracycline

C) Penicillin

D) Erythromycin

E) Chloramphenicol

Answer: C
Skill: Recall

11) All of the following act by competitive inhibition *except*

   A) Ethambutol.

   B) Isoniazid.

   C) Streptomycin.

   D) Sulfonamide.

   E) None of the above.

   Answer: C
   Skill: Analysis

12) Which of the following methods of action would be bacteriostatic?

   A) Competitive inhibition with folic acid synthesis

   B) Inhibition of RNA synthesis

   C) Injury to plasma membrane

   D) Inhibition of cell wall synthesis

   E) None of the above

   Answer: B
   Skill: Understanding

13) Which of the following antibiotics is recommended for use against gram–negative bacteria?

   A) Polyenes

   B) Bacitracin

   C) Cephalosporin

   D) Penicillin

   E) Polymyxin

   Answer: E
   Skill: Recall

14) Which of the following antimicrobial agents is recommended for use against fungal infections?

   A) Amphotericin B

   B) Bacitracin

   C) Cephalosporin

   D) Penicillin

   E) Polymyxin

   Answer: A
   Skill: Recall

Table 20.1
The following data were obtained from a broth dilution test.

Concentration of Antibiotic X	Growth	Growth in Subculture
2 µg/ml	+	+
10 µg/ml	–	+
15 µg/ml	–	–
25 µg/ml	–	–

15) In Table 20.1, the minimal bactericidal concentration of antibiotic X is

A) 2 µg/ml.   B) 10 µg/ml.   C) 15 µg/ml.   D) 25 µg/ml.   E) Can't tell.

Answer: C
Skill: Understanding

16) In Table 20.1, the minimal inhibitory concentration of antibiotic X is

A) 2 µg/ml.   B) 10 µg/ml.   C) 15 µg/ml.   D) 25 µg/ml.   E) Can't tell.

Answer: B
Skill: Understanding

17) More than half of our antibiotics are

A) Produced by fungi.

B) Produced by bacteria.

C) Synthesized in laboratories.

D) Produced by Fleming.

E) None of the above.

Answer: B
Skill: Understanding

Figure 20.2

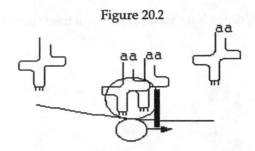

18) The antibiotic erythromycin binds to the 50S subunit of the ribosome as shown in Figure 20.2. The effect is to

A) Prevent attachment of tRNA.

B) Prevent peptide bond formation.

C) Prevent transcription.

D) Stop the ribosome from moving along the mRNA.

E) None of the above.

Answer: D
Skill: Understanding

Figure 20.3

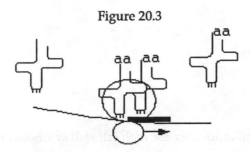

19) The antibiotic tetracycline binds to the 30S subunit of ribosome as shown in Figure 20.3. The effect is to

A) Prevent mRNA–ribosome binding in eucaryotes.

B) Prevent binding of tRNA in procaryotes.

C) Prevent polypeptide elongation in eucaryotes.

D) Prevent peptide bond formation in procaryotes.

E) None of the above.

Answer: B
Skill: Understanding

20) Which of the following drugs is *not* used primarily to treat tuberculosis?

A) Ethambutol

B) Isoniazid

C) Rifampin

D) Sulfonamide

E) None of the above

Answer: D
Skill: Recall

21) Which of the following antibiotics are used to treat fungal infections?
1. Aminoglycosides
2. Cephalosporins
3. Griseofulvin
4. Polyenes
5. Bacitracin

A) 1, 2, and 3

B) 3 and 4

C) 3, 4, and 5

D) 4 and 5

E) All of the antibiotics

Answer: B
Skill: Recall

22) All of the following antibiotics interfere with cell wall synthesis *except*

A) Cephalosporins.

B) Macrolides.

C) Natural penicillins.

D) Semisynthetic penicillins.

E) Vancomycin.

Answer: B
Skill: Recall

23) The antimicrobial drugs with the broadest spectrum of activity are

A) Aminoglycosides.

B) Chloramphenicol.

C) Lincomycin.

D) Macrolides.

E) Tetracyclines.

Answer: E
Skill: Recall

24) Which of the following statements is false?

   A) Fluoroquinolone inhibits DNA synthesis.

   B) Acyclovir inhibits DNA synthesis.

   C) Amantadine inhibits release of viral nucleic acid.

   D) Interferon inhibits glycolysis.

   E) None of the above.

   Answer: D
   Skill: Recall

25) Protozoan and helminthic diseases are difficult to treat because

   A) Their cells are structurally and functionally similar to human cells.

   B) They replicate inside human cells.

   C) They don't have ribosomes.

   D) They don't reproduce.

   E) None of the above.

   Answer: A
   Skill: Understanding

26) Which of the following organisms would most likely be sensitive to natural penicillin?

   A) L forms

   B) *Streptococcus pyogenes*

   C) Penicillinase–producing *Neisseria gonorrhoeae*

   D) *Penicillium*

   E) *Mycoplasma*

   Answer: B
   Skill: Understanding

27) All of the following statements about drug resistance are true *except*

   A) It may be carried on a plasmid.

   B) It may be transferred from one bacterium to another during conjugation.

   C) It may be due to enzymes that degrade some antibiotics.

   D) It is found only in gram–negative bacteria.

   E) It may be due to increased uptake of a drug.

   Answer: D
   Skill: Recall

28) All of the following are advantages of using two antibiotics together *except*

A) Prevention of drug resistance.

B) Lessening the toxicity of individual drugs.

C) Two are always twice as effective as one.

D) Providing treatment prior to diagnosis.

E) None of the above.

Answer: C
Skill: Analysis

29) Drug resistance occurs

A) Because bacteria are normal microbiota.

B) When antibiotics are used indiscriminately.

C) Against antibiotics and not against synthetic chemotherapeutic agents.

D) When antibiotics are taken after the symptoms disappear.

E) All of the above.

Answer: B
Skill: Analysis

Table 20.2

The following results were obtained from a disk–diffusion test for microbial susceptibility to antibiotics. *Staphylococcus aureus* was the test organism.

Antibiotic	Zone of Inhibition
A	3 mm
B	7 mm
C	0 mm
D	10 mm

30) In Table 20.2, the most effective antibiotic tested was

A) A.   B) B.   C) C.   D) D.   E) Can't tell.

Answer: D
Skill: Understanding

31) In Table 20.2, the antibiotic that exhibited bactericidal action was

A) A.   B) B.   C) C.   D) D.   E) Can't tell.

Answer: E
Skill: Understanding

32) In Table 20.2, which antibiotic would be most useful for treating a *Salmonella* infection?

A) A   B) B   C) C   D) D   E) Can't tell

Answer: E
Skill: Understanding

33) Use of antibiotics in animal feed leads to antibiotic-resistant bacteria because

   A) Bacteria from other animals replace those killed by the antibiotics.
   B) The few surviving bacteria that are affected by the antibiotics develop immunity to the antibiotics, which they pass on to their progeny.
   C) The antibiotics cause new mutations to occur in the surviving bacteria, which results in resistance to antibiotics.
   D) The antibiotics kill susceptible bacteria, but the few that are naturally resistant live and reproduce, and their progeny repopulate the host animal.
   E) The antibiotics persist in soil and water.

   Answer: D
   Skill: Understanding

34) In the presence of penicillin a cell dies because

   A) It lacks a cell wall.
   B) It plasmolyzes.
   C) It undergoes osmotic lysis.
   D) It lacks a cell membrane.
   E) None of the above.

   Answer: C
   Skill: Understanding

35) Most of the antiviral drugs

   A) Damage the cell wall.
   B) Are nucleoside analogs.
   C) Are enzyme inhibitors.
   D) Prevent viruses from entering cells.
   E) Damage the plasma membrane.

   Answer: B
   Skill: Recall

36) Niclosamide prevents ATP generation in mitochondria. You would expect this drug to be effective against

   A) Gram–negative bacteria.
   B) Gram–positive bacteria.
   C) Helminths.
   D) *Mycobacterium tuberculosis*.
   E) Viruses.

   Answer: C
   Skill: Understanding

Table 20.3
The following data were obtained from a broth dilution test:

Concentration of Antibiotic X	Growth
2.0 μg/ml	−
1.0 μg/ml	−
0.5 μg/ml	−
0.25 μg/ml	+
0.125 μg/ml	+
0	+

Bacteria from the 0.25 μg/ml tube were transferred to new growth media containing antibiotic X with the following results:

Concentration of Antibiotic X	Growth
2.0 μg/ml	−
1.0 μg/ml	+
0.5 μg/ml	+
0.25 μg/ml	+

37) The experiment in Table 20.3 shows that these bacteria

   A) Can be subcultured.

   B) Developed resistance to antibiotics.

   C) Were killed by 0.125 μg/ml of antibiotic X.

   D) Were killed by 0.5 μg/ml of antibiotic X.

   E) Were resistant to 1.0 μg/ml at the start of the experiment.

Answer: B
Skill: Analysis

38) Which of the following statements about drugs that competitively inhibit DNA polymerase or RNA polymerase is *not* true?

   A) They cause mutations.

   B) They are used against viral infections.

   C) They affect host cell DNA.

   D) They are too dangerous to be used.

   E) None of the above.

Answer: D
Skill: Understanding

Figure 20.4

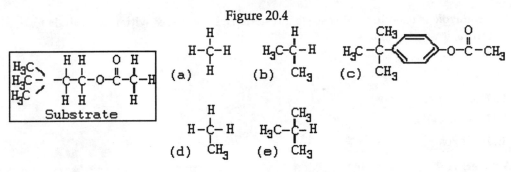

39) If the compound shown in Figure 20.4 is the substrate for a particular enzyme, which of the drugs would be the most effective competitive inhibitor?

A) Graph (a)

B) Graph (b)

C) Graph (c)

D) Graph (d)

E) None of the above.

Answer: C
Skill: Understanding

40) Which of the following is *not* bactericidal?

A) Tetracycline

B) Natural penicillin

C) Streptomycin

D) Bacitracin

E) Semisynthetic penicillin

Answer: A
Skill: Recall

41) Which of the following does *not* affect eukaryotic cells?

A) Antiprotozoan drugs

B) Antihelminthic drugs

C) Antifungal drugs

D) Nucleotide analogs

E) None of the above

Answer: E
Skill: Analysis

42) Mebendazole is used to treat cestode infestations. It interferes with microtubule formation; therefore, it would *not* affect

A) Bacteria.

B) Fungi.

C) Helminths.

D) Human cells.

E) Protozoa.

Answer: A
Skill: Understanding

43) Which one of the following does *not* belong with the others?

A) AZT

B) ddC

C) Protease inhibitor

D) Ribavirin

E) Trifluridine

Answer: C
Skill: Analysis

44) The antibiotic actinomycin D binds between adjacent G–C pairs, thus interfering with

A) Transcription.

B) Translation.

C) Cellular respiration.

D) Plasma membrane function.

E) Peptide bond formation.

Answer: A
Skill: Understanding

45) Which of the following would be selective against the tubecle bacillus?

A) Bacitracin — inhibits peptidoglycan synthesis

B) Ethambutol — inhibits mycolic acid synthesis

C) Streptogramin — inhibits protein synthesis

D) Streptomycin — inhibits protein synthesis

E) Vancomycin — inhibits peptidoglycan synthesis

Answer: B
Skill: Understanding

ESSAY QUESTIONS

1) Discuss why penicillin was called the "miracle drug" when it was first used in the 1940s.
   Answer:
   Skill:

2) Penicillin does not directly kill bacteria. Why do cells usually die in the presence of penicillin?

Answer:

Skill:

OBJECTIVE QUESTIONS

1) Which of the following is *not* normal microbiota of the skin?

   A) *Streptococcus*

   B) *Pityrosporum*

   C) *Staphylococcus*

   D) *Propionibacterium*

   E) *Corynebacterium*

   Answer: A
   Skill: Recall

2) An 8–year–old female has scabs and pus–filled vesicles on her face and throat. Three weeks earlier she had visited her grandmother who had shingles. What infection does the 8–year–old have?

   A) Chickenpox

   B) Measles

   C) Fever blisters

   D) Shingles

   E) German measles

   Answer: A
   Skill: Analysis

3) Which of the following pairs is mismatched?

   A) *S. aureus* — impetigo

   B) *S. pyogenes* — erysipelas

   C) *P. acnes* — acne

   D) *P. aeruginosa* — otitis externa

   E) None of the above

   Answer: E
   Skill: Analysis

4) The etiologic agent of warts is

   A) Papovavirus.

   B) Poxvirus.

   C) Herpesvirus.

   D) Parvovirus

   E) None of the above.

   Answer: A
   Skill: Recall

5) Which of the following is *not* a characteristic of *P. aeruginosa*?
   A) Gram–positive rods
   B) Oxidative metabolism
   C) Oxidase positive
   D) Produce pyocyanin
   E) None of the above

   Answer: A
   Skill: Recall

6) Which of the following pairs is mismatched?
   A) Pustular rash — smallpox
   B) Koplik spots — rubella
   C) Papular rash — measles
   D) Vesicular rash — chickenpox
   E) None of the above

   Answer: B
   Skill: Recall

7) Which of the following is *not* transmitted by the respiratory route?
   A) Smallpox
   B) Chickenpox
   C) German measles
   D) Cold sore
   E) None of the above

   Answer: D
   Skill: Analysis

8) Which of these is *not* caused by herpesvirus?
   A) Chickenpox
   B) Shingles
   C) Keratoconjunctivitis
   D) Smallpox
   E) None of the above

   Answer: D
   Skill: Recall

9) Thrush and vaginitis are caused by

   A) Herpesvirus.

   B) *Chlamydia trachomatis.*

   C) *Candida albicans.*

   D) *Staphylococcus aureus.*

   E) *Streptococcus pyogenes.*

   Answer: C
   Skill: Recall

10) The greatest single cause of blindness in the world is

   A) Neonatal gonorrheal ophthalmia.

   B) Keratoconjunctivitis.

   C) Trachoma.

   D) Inclusion conjunctivitis.

   E) Pinkeye.

   Answer: C
   Skill: Recall

11) Which of the following can be treated with topical chemotherapeutic agents?

   A) Herpes gladiatorium

   B) Sporotrichosis

   C) Dermatomycosis

   D) Rubella

   E) None of the above

   Answer: C
   Skill: Analysis

12) Which of the following is *not* a cause of ringworm?

   A) *Microsporum*

   B) *Trichophyton*

   C) Tinea capitis

   D) *Epidermophyton*

   E) None of the above

   Answer: C
   Skill: Analysis

13) Newborns' eyes are treated with an antibiotic when
    A) *N. gonorrhoeae* is isolated from the eyes.
    B) The mother is blind.
    C) The mother has genital herpes.
    D) The mother has gonorrhea.
    E) Always.

    Answer: E
    Skill: Recall

14) A possible complication of herpetic keratitis is
    A) Encephalitis.
    B) Fever blisters.
    C) Subacute sclerosing panencephalitis.
    D) Congenital rubella syndrome.
    E) None of the above.

    Answer: A
    Skill: Recall

15) Which of the following is sensitive to penicillin?
    A) *Chlamydia*
    B) Herpesvirus
    C) *Candida*
    D) *Streptococcus*
    E) *Pseudomonas*

    Answer: D
    Skill: Understanding

16) Which region of the skin supports the largest bacterial population?
    A) Axilla
    B) Scalp
    C) Forearms
    D) Legs
    E) All are equal

    Answer: A
    Skill: Recall

17) Which infection is caused by *S. aureus*?

    A) Pimples

    B) Sty

    C) Furuncle

    D) Carbuncle

    E) All of the above

Answer: E
Skill: Recall

18) Which of the following is *not* a characteristic used to identify *S. pyogenes*?

    A) Coagulase–positive

    B) Group A cell wall antigen

    C) Group M proteins

    D) Beta–hemolytic

    E) None of the above

Answer: A
Skill: Recall

19) Which of the following is *not* a causative agent of conjunctivitis?

    A) *Chlamydia trachomatis*

    B) Herpes simplex

    C) Adenovirus

    D) *Neisseria gonorrhoeae*

    E) *Hemophilus aegyptii*

Answer: D
Skill: Recall

20) In which of the following respects is measles similar to German measles?

    A) Rash

    B) Etiologic agent

    C) Encephalitis as a complication

    D) Congenital complications

    E) In name only

Answer: E
Skill: Analysis

21) Vaccination for rubella is

    A) Not necessary because the disease is mild.

    B) Not necessary if a person has had an infection.

    C) Recommended only for pregnant women.

    D) Recommended for newborns to prevent congenital disease.

    E) None of the above.

Answer: E
Skill: Analysis

22) All of the following statements about congenital rubella syndrome are true *except*

    A) It is contracted in utero.

    B) It may be fatal.

    C) It may result in deafness, blindness, and mental retardation.

    D) It doesn't occur with subclinical infections.

    E) All of the above are true.

Answer: D
Skill: Recall

23) The etiologic agent of chickenpox.

    A) Herpes simplex

    B) Herpes zoster

    C) HHV–6

    D) Parvovirus

    E) Poxvirus

Answer: B
Skill: Recall

24) The etiologic agent of fifth disease.

    A) Herpes simplex

    B) Herpes zoster

    C) HHV–6

    D) Parvovirus

    E) Poxvirus

Answer: D
Skill: Recall

25) The etiologic agent of roseola.

A) Herpes simplex

B) Herpes zoster

C) HHV-6

D) Parvovirus

E) Poxvirus

Answer: C
Skill: Recall

26) The etiologic agent of fever blisters.

A) Herpes simplex

B) Herpes zoster

C) HHV-6

D) Parvovirus

E) Poxvirus

Answer: A
Skill: Recall

27) Used to treat epidemic herpetic keratitis.

A) Penicillin

B) Sulfonamide

C) Trifluridine

D) Fungicide

E) None of the above

Answer: C
Skill: Understanding

28) Used to treat smallpox.

A) Penicillin

B) Sulfonamide

C) Trifluridine

D) Fungicide

E) None of the above

Answer: E
Skill: Understanding

29) Used to treat sporotrichosis.

    A) Penicillin

    B) Sulfonamide

    C) Trifluridine

    D) Fungicide

    E) None of the above

Answer: D
Skill: Understanding

30) Used to treat candidiasis.

    A) Penicillin

    B) Sulfonamide

    C) Trifluridine

    D) Fungicide

    E) None of the above

Answer: D
Skill: Understanding

31) Scabies is a skin disease caused by

    A) A slow virus.

    B) A protozoan.

    C) A mite.

    D) A bacterium.

    E) A prion.

Answer: C
Skill: Recall

32) Scabies is transmitted by

    A) Fomites.

    B) Food.

    C) Water.

    D) Soil.

    E) None of the above.

Answer: A
Skill: Recall

33) A patient has pus–filled vesicles and scabs on her face, throat, and lower back. She most likely has

A) Measles.

B) Mumps.

C) Chickenpox.

D) Rubella.

E) Smallpox.

Answer: C
Skill: Recall

34) Which of the following leads to all the others?

A) Toxemia

B) Scalded skin syndrome

C) Staphylococcal infection

D) TSST–1

E) Sudden drop in blood pressure

Answer: C
Skill: Understanding

35) Herpes gladiatorium is transmitted by

A) Direct contact.

B) The respiratory route.

C) The fecal–oral route.

D) Insect bites.

E) All of the above.

Answer: A
Skill: Recall

36) The patient has a papular rash. Microscopic examination of skin scrapings reveals small 8–legged animals.

A) *Candida*

B) *Microsporum*

C) *P. aeruginosa*

D) *S. aureus*

E) Scabies

Answer: E
Skill: Analysis

37) The patient has vesicles and scabs over her forehead. Microscopic examination of skin scrapings shows gram–positive cocci in clusters.

   A) *Candida*

   B) *Microsporum*

   C) *P. aeruginosa*

   D) *S. aureus*

   E) Scabies

   Answer: D
   Skill: Analysis

38) The patient has scaling skin on his fingers. Conidiospores are seen in microscopic examination of skin scrapings.

   A) *Candida*

   B) *Microsporum*

   C) *P. aaeruginosa*

   D) *S. aureus*

   E) Scabies

   Answer: B
   Skill: Analysis

39) A 45–year–old male has pus–filled vesicles distributed over his back in the upper right quadrant, over his right shoulder, and upper right quadrant of his chest. His symptoms are most likely due to

   A) *Candida albicans.*

   B) Herpes simplex virus.

   C) *Staphylococcus aureus.*

   D) *Streptococcus pyogenes.*

   E) Varicella–zoster virus.

   Answer: E
   Skill: Analysis

40) A 35–year–old female has a red, raised rash on the inside of her thighs. Gram–stained skin scrapings show large budding cells with pseudohyphae. The infection is caused by

   A) *Candida albicans.*

   B) Herpes simplex virus.

   C) *Staphylococcus aureus.*

   D) *Streptococcus pyogenes.*

   E) Varicella–zoster virus.

   Answer: A
   Skill: Analysis

41) Cytoplasmic inclusions were observed in a fetus that died in utero after 6 months' gestation. The probable cause of the fetus's death was
   A) Fifth disease.
   B) Herpes simplex.
   C) Measles.
   D) *Staphylococcus aureus*.
   E) *Streptococcus pyogenes*.

   Answer: A
   Skill: Analysis

42) Assume that your lab partner swabs the side of his face and used the swab to inoculate a nutrient agar plate. The next day, he performs a Gram stain on the colonies. They are gram-positive cocci. You advise him that he should next look for
   A) An acid–fast reaction.
   B) A coagulase reaction.
   C) Conidiospores.
   D) Pseudohyphae.
   E) Pseudopods.

   Answer: B
   Skill: Understanding

43) Which of the following is incorrectly matched?
   A) Chickenpox — Poxvirus
   B) Conjunctivitis — *Chlamydia trachomatis*
   C) Conjunctivitis — *Pseudomonas*
   D) Infected tissue fluoresces — Dermatomycosis
   E) Opportunistic infection in AIDS patients — Candidiasis

   Answer: A
   Skill: Recall

44) A 17–year–old male has pus–filled cysts on his face and upper back. Microscopic examination reveals gram–positive rods. This infection is
   A) Acne.
   B) Boils.
   C) Carbuncles.
   D) Impetigo.
   E) Pimples.

   Answer: A
   Skill: Analysis

45) A 17–year–old male has pus–filled cysts on his face and upper back. Microscopic examination reveals gram–positive rods. This infection is caused by

A) *Acanthamoeba.*

B) Herpes simplex virus.

C) *Propionibacterium acnes.*

D) *Staphylococcus aureus.*

E) *Streptococcus pyogenes.*

Answer: C
Skill: Analysis

ESSAY QUESTIONS

1) A teenaged boy knew an over–the–counter cortisone preparation would decrease the swelling and redness of insect bites and poison ivy, so he used it to decrease the swelling and redness of pimples. Why, in 24 hours, were his pimples more inflamed?

Answer:

Skill:

2) A 56-year-old Army officer received a smallpox vaccination at a military vaccination clinic. Within 2 weeks, a painful ulcer was noted at the vaccination site. Because of the appearance of an increasing number of peripheral lesions and because of continued enlargement of the initial ulcer, he was treated. Eventual recovery was complicated by *Pseudomonas* sepsis and the need for a skin graft at the vaccination site. What was the cause of the ulcer and lesions, and what were the treatments? What caused the *Pseudomonas* infection?

Answer:

Skill:

CHAPTER 22   Microbial Diseases of the Nervous System

OBJECTIVE QUESTIONS

1) Which of the following is true about the normal microbiota of the nervous system?

   A) Only transient microbiota are present.

   B) There are no normal microbiota.

   C) Normal microbiota are present in the central nervous system only.

   D) Normal microbiota are present in the peripheral nervous system only.

   E) None of the above.

   Answer: B
   Skill: Recall

2) Encephalitis and meningitis are difficult to treat because

   A) They are not caused by bacteria.

   B) Antibiotics damage tissues.

   C) Antibiotics cannot penetrate the blood–brain barrier.

   D) The infections move along peripheral nerves.

   E) All of the above.

   Answer: C
   Skill: Recall

3) Which of the following organisms is *not* capable of causing meningitis?

   A) *Neisseria meningitidis*

   B) *Haemophilus influenzae*

   C) *Cryptococcus neoformans*

   D) *Streptococcus pneumoniae*

   E) None of the above

   Answer: E
   Skill: Recall

4) All of the following are true about *H. influenzae, except*

   A) A healthy carrier state can exist.

   B) is encapsulated.

   C) It requires a blood supplement in media.

   D) It usually infects children.

   E) It is used in a whole bacterial vaccine.

   Answer: E
   Skill: Analysis

5) Which of the following pairs is mismatched?

    A) *Neisseria meningitidis* — cultured in a candle jar

    B) *Haemophilus influenzae* — virulence due to capsule

    C) *Mycobacterium leprae* — cultured in armadillos

    D) *Cryptococcus neoformans* — acid–fast rod

    E) *Naegleria fowleri* — causes amoebic encephalitis

Answer: D
Skill: Recall

6) Which of the following organisms is *not* correctly matched to the recommended treatment?

    A) *Neisseria meningitidis* — cephalosporins

    B) *Haemophilus influenzae* — cephalosporins

    C) *Cryptococcus neoformans* — amphotericin B

    D) *Mycobacterium leprae* — dapsone

    E) Poliovirus — Salk vaccine

Answer: E
Skill: Understanding

7) All of the following are true about leprosy *except*

    A) It is rarely fatal.

    B) Patients with leprosy must be isolated.

    C) It is transmitted by direct contact.

    D) Diagnosis may be based on the lepromin test.

    E) None of the above.

Answer: B
Skill: Recall

8) Which of the following is *not* transmitted by the respiratory route?

    A) *N. meningitidis*

    B) *H. influenzae*

    C) *L. monocytogenes*

    D) *C. neoformans*

    E) None of the above

Answer: C
Skill: Recall

9) All of the following are true about rabies *except*

    A) It is caused by Rhabdovirus.

    B) Hydrophobia is an early symptom.

    C) The reservoir is mainly rodents.

    D) Diagnosis is based on immunofluorescent techniques.

    E) It is not fatal in bats.

Answer: C
Skill: Recall

10) The symptoms of tetanus are due to

    A) Deep puncture wounds.

    B) Hemolysins.

    C) Lack of oxygen.

    D) Clostridial neurotoxin.

    E) All of the above.

Answer: D
Skill: Recall

11) The treatment for tetanus is

    A) Penicillin.

    B) Antibodies.

    C) Toxoid.

    D) Cleansing the wound.

    E) There is no treatment.

Answer: B
Skill: Analysis

12) A 30–year–old female was hospitalized after she experienced convulsions. On examination, she was alert and oriented and complained of a fever, headache, and stiff neck. Her symptoms could be due to all of the following *except*

    A) *Clostridium botulinum.*

    B) *Listeria monocytogenes.*

    C) *Naegleria fowleri.*

    D) *Streptococcus pneumoniae.*

    E) None of the above.

Answer: A
Skill: Analysis

13) The most effective control of a vectorborne disease is

    A) Treatment of infected humans.

    B) Treatment of infected wild animals.

    C) Elimination of the vector.

    D) Avoidance of endemic areas.

    E) None of the above.

Answer: C
Skill: Understanding

14) Treatment for tetanus in an unimmunized person is

    A) Tetanus toxoid.

    B) Tetanus immune globulin.

    C) Penicillin.

    D) DTP.

    E) None of the above.

Answer: B
Skill: Analysis

15) Treatment for tetanus in an immunized person is

    A) Tetanus toxoid.

    B) Tetanus immune globulin.

    C) Penicillin.

    D) DTP.

    E) None of the above.

Answer: A
Skill: Analysis

16) The most common route of central nervous system invasion by pathogens is through

    A) The skin.

    B) The circulatory system.

    C) The gastrointestinal system.

    D) The parenteral route.

    E) Direct penetration into nerves.

Answer: B
Skill: Recall

17) The prodromal symptom(s) of meningitis is (are)

A) Like a mild cold.

B) Fever and headache.

C) Stiff neck and back pains.

D) Convulsions.

E) Paralysis.

Answer: A
Skill: Understanding

18) All of the following are requirements for an outbreak of botulism *except*

A) Killing bacteria that compete with *Clostridium*.

B) An anaerobic environment.

C) An incubation period.

D) A nutrient medium with a pH below 4.5.

E) None of the above.

Answer: D
Skill: Understanding

19) The most common cause of meningitis in children is

A) *Mycobacterium tuberculosis*.

B) *Streptococcus pneumoniae*.

C) *Cryptococcus neoformans*.

D) *Haemophilus influenzae*.

E) *Neisseria meningitidis*.

Answer: D
Skill: Understanding

20) Meningitis that begins as an infection of the lungs is caused by

A) *Flavobacterium meningosepticum*.

B) *Streptococcus pneumoniae*.

C) *Cryptococcus neoformans*.

D) *Haemophilus influenzae*.

E) *Neisseria meningitidis*.

Answer: C
Skill: Recall

21) Which of the following pairs is mismatched?

    A) Leprosy — direct contact

    B) Poliomyelitis — respiratory route

    C) Meningococcal meningitis — respiratory route

    D) Rabies — direct contact

    E) None of the above

Answer: B
Skill: Analysis

22) A 30–year–old female was hospitalized after she experienced convulsions. On examination, she was alert and oriented and complained of a fever, headache, and stiff neck. Which of the following is most likely to provide rapid identification of the cause of her symptoms?

    A) Gram stain of cerebrospinal fluid

    B) Gram stain of throat culture

    C) Biopsy of brain tissue

    D) Check serum antibodies

    E) None of the above; it can't be diagnosed

Answer: A
Skill: Analysis

23) All of the following are caused by prions *except*

    A) Sheep scrapie.

    B) Kuru.

    C) Creutzfeldt–Jakob disease.

    D) Transmissible mink encephalopathy.

    E) Rabies.

Answer: E
Skill: Analysis

24) Which of the following is (are) *not* correctly matched?

Salk Vaccine		Sabin Vaccine	
1.	Consists of a formalin–inactivated virus	4.	Consists of a live, attenuated polio virus
2.	Administered orally	5.	Administered orally
3.	Requires booster doses	6.	May cause polio

A) 1, 2, and 3

B) 4, 5, and 6

C) 1 and 3

D) 2

E) None of the above

Answer: C
Skill: Analysis

25) All of the following are true about the lepromin test *except*

A) It consists of human tissue extract.

B) It detects the presence of anti–*M. leprae* antibodies.

C) It is negative in the lepromatous form.

D) It consists of *M. leprae*.

E) None of the above.

Answer: D
Skill: Analysis

26) Purplish spots on the skin are characteristic of an infection by

A) *C. neoformans*.

B) *H. influenzae*.

C) *N. meningitidis*.

D) *S. pneumoniae*.

E) *M. leprae*.

Answer: C
Skill: Recall

27) A 1–year–old female was hospitalized with fever, lethargy, and rash. Gram– negative, oxidase-positive cocci were cultured from her cerebrospinal fluid. Her symptoms were caused by

A) A prion.

B) *Clostridium tetani*.

C) *Mycobacterium leprae*.

D) *Neisseria meningitidis*.

E) Rabies.

Answer: D
Skill: Analysis

28) A 1–year–old female was hospitalized with fever, lethargy, and rash. Gram- negative, oxidase-positive cocci were cultured from her cerebrospinal fluid. All of the following are true about the microbe responsible for her symptoms *except* it may

  A) Be normal in the cerebrospinal fluid.

  B) Be normal in the throat.

  C) Be treated with antibiotics.

  D) Cause epidemics.

  E) None of the above.

Answer: A
Skill: Analysis

29) On June 30, a 47–year–old man was hospitalized with dizziness, blurred vision, slurred speech, difficulty swallowing, and nausea. Examination revealed facial paralysis. The patient had partially healed superficial knee wounds incurred while laying cement. Cultures taken from the knee wounds should be incubated

  A) Aerobically.

  B) Anaerobically.

  C) In 5–10% CO_2.

  D) In animal cell culture.

  E) Any of the above will work.

Answer: B
Skill: Analysis

30) A diagnosis of rabies is confirmed by

  A) Gram stain.

  B) Direct fluorescent–antibody test.

  C) Patient's symptoms.

  D) Passive agglutination.

  E) Patient's death.

Answer: B
Skill: Recall

31) Which of the following is treated with antibiotics?

  A) Botulism

  B) Tetanus

  C) Streptococcal pneumonia

  D) Polio

  E) All of the above

Answer: C
Skill: Analysis

32) Microscopic examination of cerebrospinal fluid reveals gram–negative rods:

   A) *Hemophilus*

   B) *Listeria*

   C) *Naegleria*

   D) *Neisseria*

   E) *Streptococcus*

   Answer: A
   Skill: Analysis

33) Microscopic examination of cerbrospinal fluid reveals amoebae:

   A) *Hemophilus*

   B) *Listeria*

   C) *Naegleria*

   D) *Neisseria*

   E) *Streptococcus*

   Answer: C
   Skill: Analysis

34) Microscopic examination of cerebrospinal fluid reveals gram–positive rods:

   A) *Hemophilus*

   B) *Listeria*

   C) *Naegleria*

   D) *Neisseria*

   E) *Streptococcus*

   Answer: B
   Skill: Analysis

35) On June 30, a 47–year–old man was hospitalized with dizziness, blurred vision, slurred speech, difficulty swallowing, and nausea. Examination revealed facial paralysis. The patient had partially healed superficial knee wounds incurred while laying cement. He reported eating home–canned green beans and stew containing roast beef and potatoes 24 hours before onset of symptoms. The patient should be treated with

   A) Antibiotics.   B) Antitoxin.   C) Surgery.   D) Vaccination.

   Answer: C
   Skill: Understanding

36) On October 5, a pet store sold a kitten that subsequently died. On October 22, rabies was diagnosed in the kitten. Between September 19 and October 23, the pet store had sold 34 kittens. Approximately 1000 people responded to health–care providers following local media alerts. These people were given

 A) Antibiotics.

 B) Antirabies immunoglobulin.

 C) Rabies vaccination.

 D) Serological tests for rabies.

 E) Treatment if they tested positive.

Answer: B
Skill: Understanding

37) A vaccine is available for all of the following *except*

 A) *Hemophilus* meningitis.

 B) *Neisseria* meningitis.

 C) Tetanus.

 D) Rabies.

 E) Botulism.

Answer: E
Skill: Recall

38) Patients with leprosy usually die from

 A) Brain damage.

 B) Loss of nerve function.

 C) Tuberculosis.

 D) Influenza.

 E) Leprosy.

Answer: C
Skill: Recall

39) Which of the following is *not* acquired by ingestion?

 A) Botulism

 B) Cryptococcosis

 C) Listeriosis

 D) Poliomyelitis

 E) None of the above

Answer: B
Skill: Recall

40) Which of the following statements is *not* true?

   A) The lepromin test is positive during tuberculoid leprosy.

   B) Leprosy is highly contagious.

   C) Loss of nerve sensation occurs in tuberculoid leprosy.

   D) Disfiguring nodules and deformation occur in lepromatous leprosy.

   E) Spontaneous recovery occurs in tuberculoid and borderline leprosy.

   Answer: B
   Skill: Recall

41) Which of the following pairs is *not* correctly matched?

   A) Tetanus — blocks relaxation nerve impluse

   B) Botulism — stimulates transmission of nerve impulse

   C) Poliomyelitis — kills CNS cells

   D) Rabies virus — grows in brain cells

   E) None of the above

   Answer: B
   Skill: Recall

42) Which of the following is *not* transmitted by ingestion?

   A) Poliomyelitis

   B) Listeriosis

   C) Botulism

   D) Meningococcal meningitis

   E) Creutzfeldt–Jakob disease

   Answer: A
   Skill: Recall

43) Which of the following vaccines is a cause of the disease it is designed to prevent?

   A) Tetanus toxoid vaccine   B) Oral polio vaccine

   C) Inactivated polio vaccine   D) *Hemophilus* influenza capsule vaccine

   Answer: B
   Skill: Recall

Situation 22.1

On July 5, an 11-year-old girl complained of pain in the knuckles of her left hand. During July 6–7, she had increasing pain that extended up to the left shoulder. A throat culture was obtained and amoxacillin was prescribed. On July 9, she had difficulty walking and hallucinations. The throat culture was positive for *Streptococcus pyogenes*. She was treated with ceftiaxone. On July 11, she was hospitalized with a temperature of 40.7°C and she could not drink. She developed respiratory distress and tachycardia; she died from cardiac arrest. Fluorescent antibody testing of brain tissue revealed inclusions in the brain stem.

44) In Situation 22.1, the antibiotics did not cure her disease because the pathogen was

   A) A virus.

   B) Already growing in her brain.

   C) Part of her normal microbiota.

   D) Protected by the blood-brain barrier.

   E) Resistant to antibiotics.

   Answer: A
   Skill: Analysis

45) The disease described in Situation 22.1 is

   A) Botulism.

   B) Meningitis.

   C) Rabies

   D) Streptococcal sore throat.

   E) Tetanus.

   Answer: C
   Skill: Analysis

ESSAY QUESTIONS

1) There is an antitoxin for botulism. Why, then, is the outcome of botulism often fatal?

   Answer:

   Skill:

2) On August 20, a man died of presumed Guillain-Barre syndrome. Within 90 minutes of his death, his eyes were removed and refrigerated. The following day, a cornea from one eye was transplanted into the right eye of a woman. The woman's postoperative course was uneventful until 30 days after the transplant, when she developed right retroorbital headache. Over the next few days her headache worsened, and she developed neurologic symptoms on the right side of her face and difficulty walking. She was hospitalized on September 27. Thereafter she developed flaccid paralysis, had loss of mental acuity, and died on October 10. Serum collected on October 2 was negative for rabies antibody, but serum collected on October 5 had a 23 titer.

   What is the etiology of this disease? Identify the periods of incubation, prodromal, illness, and decline. What is the mode of transmission?

   Answer:

   Skill:

CHAPTER 23   Microbial Diseases of the Cardiovascular and Lymphatic Systems

OBJECTIVE QUESTIONS

1) All of the following statements about septicemia are true *except*

   A) Symptoms include fever and decreased blood pressure.

   B) Lymphangitis may occur.

   C) Symptoms are due to bacterial endotoxin.

   D) It usually is caused by gram–positive bacteria.

   E) It may be aggravated by antibiotics.

   Answer: D
   Skill: Recall

2) Which of the following pairs is mismatched?

   A) Subacute bacterial endocarditis — alpha–hemolytic streptococci

   B) Acute bacterial endocarditis — *Staphylococcus aureus*

   C) Pericarditis — *Streptococcus pneumoniae*

   D) Puerperal sepsis — *Staphylococcus aureus*

   E) None of the above

   Answer: D
   Skill: Recall

3) All of the following grow inside host cells *except*

   A) *Babesia.*

   B) *Brucella.*

   C) Dengue fever virus.

   D) *Leishmania.*

   E) None of the above.

   Answer: E
   Skill: Understanding

4) Which of the following is *not* treated with penicillin?

   A) Pericarditis

   B) Tularemia

   C) Anthrax

   D) Listeriosis

   E) None of the above

   Answer: B
   Skill: Analysis

5) All of the following statements about tularemia are true *except*

   A) It is caused by *Francisella tularensis*.

   B) The reservoir is rabbits.

   C) It may be transmitted by arthropods.

   D) It may be transmitted by direct contact.

   E) It occurs only in California.

   Answer: E
   Skill: Recall

6) Which of the following is a Undulant symptom of brucellosis?

   A) A local infection

   B) Septicemia

   C) Undulant fever

   D) Pneumonia

   E) Jaundice

   Answer: C
   Skill: Recall

7) Which of the following is *not* transmitted in raw milk?

   A) Toxoplasmosis

   B) Anthrax

   C) Brucellosis

   D) Listeriosis

   E) None of the above

   Answer: A
   Skill: Recall

8) Which of the following is *not* a characteristic of *Bacillus anthracis*?

   A) Aerobic

   B) Gram–positive

   C) Forms endospores

   D) Found in soil

   E) All of the above are characteristics of *B. anthracis*

   Answer: E
   Skill: Recall

9) The symptoms of gas gangrene are due to all of the following *except*

   A) Microbial fermentation.

   B) Necrotizing exotoxins.

   C) Proteolytic enzymes.

   D) Hyaluronidase.

   E) Anaerobic environment.

   Answer: E
   Skill: Analysis

10) Infections by all of the following bacteria may result from dog or cat bites *except*

   A) *Pasteurella multocida.*

   B) *Streptobacillus.*

   C) *Bacteroides.*

   D) *Fusobacterium.*

   E) None of the above.

   Answer: B
   Skill: Recall

11) Which of the following pairs is mismatched?

   A) Malaria — *Anopheles* (mosquito)

   B) Dengue — *Aedes* (mosquito)

   C) Epidemic typhus — *Pediculus* (louse)

   D) Rocky Mountain spotted fever — *Dermacentor* (tick)

   E) Encephalitis — *Ixodes* (tick)

   Answer: E
   Skill: Recall

12) The incidence of all of the following diseases is increased by unsanitary and crowded conditions *except*

   A) Plague.

   B) Epidemic typhus.

   C) Endemic murine typhus.

   D) Rocky Mountain spotted fever.

   E) Relapsing fever.

   Answer: D
   Skill: Analysis

13) Which of the following is *not* true about toxoplasmosis?

    A) It is caused by a protozoan.

    B) The reservoir is cats.

    C) It is transmitted by the gastrointestinal route.

    D) It is a severe illness in adults.

    E) It can be congenital.

Answer: D
Skill: Recall

14) All of the following facts about American trypanosomiasis are true *except*

    A) Causative agent — *T. cruzi*.

    B) Vector — kissing bug.

    C) Reservoir — rodents.

    D) Diagnosis — serological tests for antibodies.

    E) Treatment — Nifurtimox.

Answer: D
Skill: Recall

15) Which of the following is *not* caused by a bacterium?

    A) Epidemic typhus

    B) Tickborne typhus

    C) Malaria

    D) Plague

    E) Relapsing fever

Answer: C
Skill: Recall

16) Septicemia may result from all of the following *except*

    A) A focal infection.

    B) Pneumonia.

    C) A nosocomial infection.

    D) Contamination through the parenteral route.

    E) None of the above.

Answer: E
Skill: Recall

17) All of the following statements about puerperal sepsis are true *except*

   A) It is transmitted from mother to fetus.

   B) It is caused by health–care personnel.

   C) It begins as a focal infection.

   D) It is a complication of abortion or childbirth.

   E) It doesn't occur anymore because of antibiotics and aseptic techniques.

   Answer: A
   Skill: Recall

18) All of the following statements about schistosomiasis are true *except*

   A) The cercariae penetrate human skin.

   B) A parasite of birds causes swimmer's itch in humans.

   C) The intermediate host is an aquatic snail.

   D) It is caused by a flatworm.

   E) None of the above.

   Answer: E
   Skill: Recall

19) All of the following statements about rheumatic fever are true *except*

   A) It is a complication of a Group A β–hemolytic streptococcal infection.

   B) It is an inflammation of the heart.

   C) It is an inflammation of the joints.

   D) It is cured with penicillin.

   E) The incidence has declined in the last 10 years.

   Answer: D
   Skill: Recall

20) Which of the following pairs is mismatched?

   A) *Rickettsia* — intracellular parasite

   B) *Brucella* — gram–negative aerobic rods

   C) *Francisella* — gram–positive facultatively anaerobic pleomorphic rods

   D) *Bacillus* — gram–positive endospore–forming rods

   E) None of the above

   Answer: C
   Skill: Recall

21) Tetracycline is used to treat all of the following *except*

A) Yellow fever.

B) Plague.

C) Relapsing fever.

D) Tickborne typhus.

E) Cat–scratch fever.

Answer: A
Skill: Analysis

22) Which of the following pairs is mismatched?

A) *Borrelia* — spirochete

B) *Yersinia* — gram-negative rod

C) *Rickettsia* — intracellular parasite

D) *Pasteurella* — gram-negative rod

E) *Spirillum* — spirochete

Answer: E
Skill: Analysis

23) Plague is

A) Endemic in wild rodents.

B) Endemic in rats.

C) Epidemic in humans.

D) The presence of *Y. pestis* in the lungs.

E) Characterized by buboes.

Answer: A
Skill: Analysis

24) Human–to–human transmission of plague is usually by

A) Rat flea.

B) Dog flea.

C) The respiratory route.

D) Wounds.

E) None of the above.

Answer: C
Skill: Analysis

25) A characteristic symptom of plague is
    A) Small red spots on the skin.
    B) Bruises on the skin.
    C) Rose-colored spots.
    D) Recurrent fever.
    E) None of the above.

    Answer: B
    Skill: Recall

26) Which of the following pairs regarding the epidemiology of malaria is mismatched?
    A) Vector — *Anopheles*
    B) Etiology — *Plasmodium*
    C) Found in liver — sporozoites
    D) Diagnosis — presence of merozoites
    E) Treatment — antibiotics

    Answer: E
    Skill: Recall

27) A predisposing factor for infection by *Clostridium perfringens* is
    A) Gangrene.
    B) Burns.
    C) Debridement.
    D) Hyperbaric treatment.
    E) An infected finger.

    Answer: A
    Skill: Analysis

28) Which of the following is not a *zoonosis*?
    A) Puerperal sepsis
    B) *Hantavirus* infection
    C) Anthrax
    D) Brucellosis
    E) Tularemia

    Answer: A
    Skill: Recall

29) Arthropods can serve as a reservoir for which of the following diseases?

   A) Plague

   B) Brucellosis

   C) Epidemic typhus

   D) Yellow fever

   E) Malaria

   Answer: B
   Skill: Recall

30) Which of the following pairs is mismatched?

   A) Cat–scratch fever — malignant pustule developing into septicemia

   B) Brucellosis — a temperature of 40°C each evening

   C) Tularemia — a localized infection appearing as a small ulcer

   D) *Borrelia* — rash and flulike

   E) Toxoplasmosis — congenital brain damage

   Answer: A
   Skill: Recall

31) All of the following can be transmitted to humans from domestic cats *except*

   A) Toxoplasmosis.

   B) Plague.

   C) American trypanosomiasis.

   D) Cat–scratch fever.

   E) None of these diseases is transmitted by cats.

   Answer: C
   Skill: Analysis

32) Relapsing fever and undulant fever differ in all of the following ways *except*

   A) Mode of transmission.

   B) Presence of rash.

   C) Reservoir.

   D) Etiology.

   E) None of the above.

   Answer: E
   Skill: Analysis

33) All of the following are treated with antibiotics *except*

   A) Plague.

   B) Tularemia.

   C) Lyme disease.

   D) Yellow fever.

   E) None of the above.

Answer: D
Skill: Understanding

34) All of the following enter across mucous membranes of the gastrointestinal tract *except*

   A) *F. tularensis.*

   B) *B. anthracis.*

   C) *B. melitensis.*

   D) EB virus.

   E) None of the above.

Answer: E
Skill: Analysis

35) EB virus causes all of the following *except*

   A) Endocarditis.

   B) Infectious mononucleosis.

   C) Burkitt's lymphoma.

   D) Nasopharyngeal carcinoma.

   E) None of the above.

Answer: A
Skill: Recall

36) Which of the following leads to all the others?

   A) Subcutaneous hemorrhaging

   B) Presence of antirickettsial antibodies

   C) Blockage of capillaries

   D) Bacterial growth in endothelial cells

   E) Breakage of capillaries

Answer: D
Skill: Analysis

37) A patient complains of fever, severe muscle and joint pain, and a rash. The patient reports returning from a Caribbean vacation one week ago. Which one of the following do you suspect?

A) Bolivian hemorrhagic fever

B) Dengue

C) *Hantavirus* hemorrhagic fever

D) Typhus

E) Yellow fever

Answer: B
Skill: Analysis

38) Which of the following pairs is *not* correctly matched for Gram reaction?

A) Lyme disease — gram–negative

B) Tularemia — gram–negative

C) Anthrax — gram–positive

D) Rocky Mountain spotted fever — gram–negative

E) All of the above are correctly matched

Answer: E
Skill: Recall

39) Scrapings from a patient's rash reveal cercariae. The disease is most likely

A) Lyme disease.

B) Rocky Mountain spotted fever.

C) Relapsing fever.

D) Swimmer's itch.

E) Chagas' disease.

Answer: D
Skill: Analysis

40) You advise your pregnant friend to give her cat away because

A) She could contract plague.

B) She could give the cat tularemia.

C) She could get toxoplasmosis.

D) She could get listeriosis.

E) You don't like cats and want to see your friend without one.

Answer: C
Skill: Analysis

41) Which of the following is evidence that the arthritis afflicting children in Lyme, Connecticut, was due to bacterial infection?

A) Treatable with penicillin

B) Not contagious

C) Accompanied by a rash

D) Affected mostly children

E) All of the above

Answer: A
Skill: Analysis

42) Which of the following is the usual cause of septic shock?

A) Endotoxin

B) Exotoxin

C) Lymphangitis

D) Septicemia

E) None of the above

Answer: A
Skill: Analysis

43) A 62–year–old man was hospitalized with an 8–day history of fever, chills, sweats, and vomiting. His temperature on admission was 40°C. A routine peripheral blood smear revealed ring–shaped bodies in the RBCs. What treatment would you prescribe?

A) Hyperbaric oxygen

B) Mefloquine

C) No treatment

D) Penicillin

E) Streptomycin

Answer: B
Skill: Understanding

44) A patient presents with fever and enlarged lymph nodes; the recommended treatment is

A) Anti–inflammatory drugs.

B) Streptomycin.

C) Chloroquine.

D) Hyperbaric chamber.

E) Praziquantel.

Answer: B
Skill: Understanding

45) A patient presents with inflammation of the heart valves, fever, malaise, and subcutaneous nodules at joints; the recommended treatment is

A) Anti–inflammatory drugs.

B) Streptomycin.

C) Chloroquine.

D) Hyperbaric chamber.

E) Praziquantel.

Answer: A
Skill: Understanding

ESSAY QUESTIONS

1) Humans are not the normal hosts for *Ixodes* and *Xenopsylla*. How then do humans contract Lyme disease and plague?

Answer:

Skill:

2) On June 1, a 32–year–old hiker was bitten by a tick. After one week, he noticed an erythrematous ring at the location of the bite. Four weeks later, a physician found a large, macular, centrifugally spreading ring. During the next month, it expanded to 35 cm and faded. Over the next two–and–a–half years, the man experienced recurrent inflammation of a knee.

Identify the periods of incubation, prodromal, illness, and decline. What is the etiology? What caused the symptoms? What treatment should have been administered in July? What treatment should be administered two years later?

Answer:

Skill:

CHAPTER 24   Microbial Diseases of the Respiratory System

OBJECTIVE QUESTIONS

1) All of the following statements about otitis media are true *except*
   A) It is caused by *Streptococcus pyogenes*.
   B) It is a complication of tonsillitis.
   C) It is transmitted by swimming pool water.
   D) It is caused by rhinovirus.
   E) It is caused by *Staphylococcus aureus*.

   Answer: D
   Skill: Recall

2) A diagnosis of strep throat is confirmed by all of the following *except*
   A) Hemolytic reaction.
   B) Bacitracin inhibition.
   C) Symptoms.
   D) Serological tests.
   E) Gram stain.

   Answer: C
   Skill: Recall

3) Penicillin is used to treat all of the following *except*
   A) Streptococcal sore throat.
   B) Diphtheria.
   C) Pneumococcal pneumonia.
   D) Mycoplasmal pneumonia.
   E) Scarlet fever.

   Answer: D
   Skill: Analysis

4) Mycoplasmal pneumonia differs from viral pneumonia in that
   A) It doesn't have any known etiologic agent.
   B) It is treated with tetracyclines.
   C) Viral pneumonia is treated with tetracyclines.
   D) The symptoms are distinctly different.
   E) None of the above.

   Answer: B
   Skill: Analysis

5) Which of the following diseases is *not* correctly matched to a virulence factor?

    A) Diphtheria — exotoxin

    B) Scarlet fever — exotoxin

    C) Pneumococcal pneumonia — exotoxin

    D) *Hemophilus* pneumonia — endotoxin

    E) Whooping cough — endotoxin

Answer: C
Skill: Understanding

6) Which of the following pairs is mismatched?

    A) *Corynebacterium* — gram-positive rod

    B) *Mycobacterium* — acid-fast rod

    C) *Mycoplasma* — gram-positive pleomorphic rod

    D) *Bordetella* — gram-negative pleomorphic rod

    E) *Hemophilus* — gram-negative rod

Answer: C
Skill: Recall

7) Which of the following microorganisms causes symptoms most like tuberculosis?

    A) *Histoplasma*

    B) *Coccidioides*

    C) *Legionella*

    D) *Mycoplasma*

    E) Influenza virus

Answer: A
Skill: Analysis

8) A person can have a positive tuberculin skin test because

    A) She has been vaccinated.

    B) She has tuberculosis.

    C) She had tuberculosis.

    D) She is immune to tuberculosis.

    E) All of the above.

Answer: E
Skill: Understanding

9) Which of the following diseases is *not* correctly matched to its vaccine?

   A) Tuberculosis — toxoid

   B) Whooping cough — heat–killed bacteria

   C) Diphtheria — toxoid

   D) Influenza — viruses grown in embryonated eggs

   E) None of the above

   Answer: A
   Skill: Recall

10) Which of the following diseases has a cutaneous form, especially in individuals over 30 years of age?

   A) Coccidioidomycosis

   B) Diphtheria

   C) Legionellosis

   D) Scarlet fever

   E) None of the above

   Answer: B
   Skill: Recall

11) Which of the following causes an infection of the respiratory system that is transmitted by the gastrointestinal route?

   A) *Streptococcus pyogenes*

   B) *Mycobacterium tuberculosis*

   C) *Mycoplasma pneumoniae*

   D) *Haemophilus influenzae*

   E) *Streptococcus pneumoniae*

   Answer: B
   Skill: Understanding

12) Which of the following pairs is mismatched?

   A) Epiglottitis — *Hemophilus*

   B) Q fever — *Rickettsia*

   C) Psittacosis — *Chlamydia*

   D) Whooping cough — *Bordetella*

   E) None of the above

   Answer: B
   Skill: Recall

13) Pneumonia can be caused by all of the following *except*

A) *Legionella.*

B) *Hemophilus.*

C) *Mycoplasma.*

D) *Streptococcus.*

E) None of the above.

Answer: E
Skill: Recall

14) Which of the following causes opportunistic infections in AIDS patients?

A) *Pneumocystis*

B) *Aspergillus*

C) *Rhizopus*

D) *Mucor*

E) All of the above

Answer: E
Skill: Understanding

15) Which of the following diseases is *not* correctly matched to its reservoir?

A) Tuberculosis — cattle

B) Histoplasmosis — soil

C) Psittacosis — parakeets

D) Coccidioidomycosis — air

E) *Pneumocystis* — humans

Answer: D
Skill: Analysis

16) Does not produce any exotoxin.

A) *Bordetella pertussis*

B) *Corynebacterium diptheriae*

C) *Mycobacterium tuberculosis*

D) *Streptococcus pygones*

E) None of the above

Answer: C
Skill: Recall

17) Causes a disease characterized by the catarrhal, paraxysmal, and convalescent stages.

   A) *Bordetella pertussis*

   B) *Corynebacterium diphtheriae*

   C) *Mycobacterium tuberculosis*

   D) *Streptococcus pyogenes*

   E) None of the above

   Answer: A
   Skill: Recall

18) Classified as an irregular, gram–positive rod.

   A) *Bordetella pertussis*

   B) *Corynebacterium diptheriae*

   C) *Myobacterium tuberculosis*

   D) *Streptococcus pyogenes*

   E) None of the above

   Answer: B
   Skill: Recall

19) Infection results in the formation of Ghon complexes.

   A) *Bordetella pertussis*

   B) *Corynebacterium diptheriae*

   C) *Mycobacterium tuberculosis*

   D) *Streptococcus pyogenes*

   E) None of the above

   Answer: C
   Skill: Recall

20) Produces the most potent exotoxin of these bacteria.

   A) *Bordetella pertussis*

   B) *Corynebacterium diphtheriae*

   C) *Mycobacterium tuberculosis*

   D) *Streptococcus pyogenes*

   E) None of the above

   Answer: B
   Skill: Recall

21) The recurrence of influenza epidemics is due to

   A) Lack of antiviral drugs.

   B) The Guillain–Barré syndrome.

   C) Antigenic shift.

   D) Lack of naturally acquired active immunity.

   E) All of the above.

   Answer: C
   Skill: Recall

22) Which of the following is an opportunistic pathogen?

   A) *Pneumocystis*

   B) *Legionella*

   C) *Histoplasma*

   D) *Mycoplasma*

   E) Rhinovirus

   Answer: A
   Skill: Analysis

23) Which of the following etiologic agents results in the formation of abscesses?

   A) *Staphylococcus*

   B) *Mycoplasma*

   C) *Streptococcus*

   D) *Blastomyces*

   E) None of the above

   Answer: D
   Skill: Recall

24) Which of the following is most susceptible to destruction by phagocytes?

   A) *Chlamydia psittaci*

   B) *Streptococcus pneumoniae*

   C) *Streptococcus pyogenes*

   D) Influenza virus

   E) All of the above

   Answer: C
   Skill: Understanding

25) A healthy carrier state exists for
    A) *Corynebacterium diphtheriae.*
    B) *Streptococcus pneumoniae.*
    C) β–hemolytic streptococci.
    D) *Hemophilus influenzae.*
    E) All of the above.

    Answer: E
    Skill: Understanding

26) Often confused with viral pneumonia.
    A) *Blastomyces*
    B) *Coccidioides*
    C) *Mycoplasma*
    D) *Streptococcus*
    E) None of the above

    Answer: C
    Skill: Analysis

27) Causes a disease characterized by a red rash.
    A) *Blastomyces*
    B) *Coccidioides*
    C) *Mycoplasma*
    D) *Streptococcus*
    E) None of the above

    Answer: D
    Skill: Analysis

28) Inhalation of arthrospores is responsible for infection by this organism.
    A) *Blastomyces*
    B) *Coccidioides*
    C) *Mycoplasma*
    D) *Streptococcus*
    E) None of the above

    Answer: B
    Skill: Analysis

29) Which of the following pairs is mismatched?

    A) Q fever — ticks

    B) Psittacosis — parrots

    C) *Pneumocystis* pneumonia — nosocomial

    D) *Coccidioides* — soil

    E) None of the above

Answer: E
Skill: Recall

30) *Legionella* is transmitted by

    A) Airborne transmission.

    B) Foodborne transmission.

    C) Person–to–person contact.

    D) Fomites.

    E) Vectors.

Answer: A
Skill: Recall

31) The patient is suffocating due to an inflamed epiglottis.

    A) *Corynebacterium*

    B) *Hemophilus*

    C) *Bordetella*

    D) *Mycobacterium*

    E) Can't tell

Answer: B
Skill: Analysis

32) The patient has a sore throat:

    A) *Corynebacterium*

    B) *Hemophilus*

    C) *Bordetella*

    D) *Mycobacterium*

    E) Can't tell

Answer: E
Skill: Analysis

33) The patient is suffocating due to the accumulation of dead tissue and fibrin in her throat.

A) *Corynebacterium*

B) *Hemophilus*

C) *Bordetella*

D) *Mycobacterium*

E) Can't tell

Answer: A
Skill: Analysis

34) Infection begins in lungs and spreads to skin.

A) *Blastomyces*

B) *Coccidioides*

C) *Histoplasma*

D) *Mycobacterium*

E) *Pneumocystis*

Answer: A
Skill: Analysis

35) This organism does *not* belong with the others.

A) *Blastomyces*

B) *Coccidioides*

C) *Histoplasma*

D) *Mycobacterium*

E) *Pneumocystis*

Answer: D
Skill: Analysis

36) Microscopic examination of a lung biospsy shows thick–walled cysts.

A) *Blastomyces*

B) *Coccidioides*

C) *Histoplasma*

D) *Mycobacterium*

E) *Pneumocystis*

Answer: E
Skill: Analysis

37) Microscopic examination of a lung biopsy shows spherules.

   A) *Blastomyces*

   B) *Coccidioides*

   C) *Histoplasma*

   D) *Mycobacterium*

   E) *Pneumocystis*

   Answer: B
   Skill: Analysis

38) You are trying to identify the cause of a patient's middle ear infection. After 24 hours, there is no growth on blood agar incubated aerobically at 37°C. Your next step is to try again,

   A) Using nutrient agar.

   B) Incubating at 25°C.

   C) Incubating anaerobically.

   D) Incubating at 45°C.

   E) Then give up.

   Answer: C
   Skill: Understanding

39) A patient has a paroxysmal cough and mucus accumulation. What is the etiology of the symptoms?

   A) *Bordetella*

   B) *Corynebacterium*

   C) *Klebsiella*

   D) *Mycobacterium*

   E) *Mycoplasma*

   Answer: A
   Skill: Analysis

40) A patient who presents with red throat and tonsils can be diagnosed as having

   A) Streptococcal pharyngitis.

   B) Scarlet fever.

   C) Diphtheria.

   D) Common cold.

   E) There is insufficient information.

   Answer: E
   Skill: Analysis

41) A patient has fever, difficulty breathing, chest pains, fluid in the alveoli, and a positive tuberculin skin test. Gram–positive cocci are isolated from the sputum. The patient most likely has

A) Tuberculosis.

B) Influenza.

C) Pneumococcal pneumonia.

D) Mycoplasmal pneumonia.

E) Common cold.

Answer: C
Skill: Understanding

42) Which of the following respiratory infections can be contracted by ingestion?

A) Streptococcal pharyngitis

B) Diphtheria

C) Tuberculosis

D) Mycoplasmal pneumonia

E) *Hemophilus* pneumonia

Answer: C
Skill: Analysis

43) Which of the following is *not* an intracellular parasite?

A) *Chlamydia*

B) *Coccidioides*

C) *Coxiella*

D) Influenza virus

E) RSV

Answer: B
Skill: Recall

44) Which one of the following produces small "fried–egg" colonies on medium containing horse serum–yeast extract?

A) *Chlamydia*

B) *Legionella*

C) *Mycobacterium*

D) *Mycoplasma*

E) *Streptococcus*

Answer: D
Skill: Recall

45) Which organism does *not* belong with the others?

   A) *Bordetella pertussis*   B) *Corynebacterium diphtheriae*

   C) *Mycobacterium tuberculosis*   D) *Streptococcus pyogenes* (scarlet fever)

Answer: C
Skill: Understanding

ESSAY QUESTIONS

1) Provide reasons why influenza vaccination is not recommended for everyone.
Answer:
Skill:

2) Pneumonia is diagnosed by the presence of fluid (dark shadows in an X ray) in the alveoli. Since pneumonia usually is caused by a microorganism, what causes the fluid accumulation? Name a bacterium, a virus, a fungus, a protozoan, and a helminth that can cause pneumonia.
Answer:
Skill:

CHAPTER 25   Microbial Diseases of the Digestive System

OBJECTIVE QUESTIONS

1) All of the following are required for tooth decay *except*

   A) Sucrose.

   B) Glucose.

   C) Capsule–forming bacteria.

   D) Acid–producing bacteria.

   E) None of the above.

   Answer: E
   Skill: Analysis

2) Most cases of posttransfusion hepatitis are caused by

   A) Hepatitis A virus.

   B) Hepatitis B virus.

   C) Hepatitis C virus.

   D) All of the above.

   E) None of the above; modern screening techniques have eliminated posttransfusion hepatitis.

   Answer: C
   Skill: Recall

3) All of the following statements about salmonellosis are true *except*

   A) It is a bacterial infection.

   B) It requires a large infective dose.

   C) A healthy carrier state exists.

   D) The mortality rate is high.

   E) It is often associated with poultry products.

   Answer: D
   Skill: Recall

4) Which of the following does *not* produce a gastrointestinal disease due to an exotoxin?

   A) *Clostridium perfringens*

   B) *Vibrio cholerae*

   C) *Shigella dysenteriae*

   D) *Staphylococcus aureus*

   E) *Clostridium botulinum*

   Answer: C
   Skill: Analysis

5) Which of the following diseases of the gastrointestinal system is transmitted by the respiratory route?

   A) Cytomegalovirus inclusion disease
   B) Mumps
   C) Vibrio gastroenteritis
   D) Bacillary dysentery
   E) Traveler's diarrhea

Answer: B
Skill: Recall

6) Amoebic dysentery and bacillary dysentery differ in the

   A) Mode of transmission.
   B) Appearance of the patient's stools.
   C) Etiologic agent.
   D) All of the above.
   E) None of the above.

Answer: C
Skill: Analysis

7) The symptoms of trichinosis are due to the

   A) Growth of larval *Trichinella* in the large intestine.
   B) Growth of adult *Trichinella* in the large intestine.
   C) Formation of cysticerci.
   D) Encystment of adult *Trichinella* in muscles.
   E) Encystment of larval *Trichinella* in muscles.

Answer: E
Skill: Analysis

8) Poultry products are a likely source of infection by

   A) *Staphylococcus aureus*.
   B) *Salmonella* spp.
   C) *Vibrio cholera*.
   D) *Shigella* spp.
   E) *Clostridium perfringens*.

Answer: B
Skill: Recall

9) Which of the following is diagnosed by the presence of flagellates in the patient's feces?

   A) *Cyclospora* infection

   B) Giardiasis

   C) Trichinosis

   D) Cholera

   E) Cryptosporidiosis

   Answer: B
   Skill: Recall

10) Which of the following feeds on red blood cells?

   A) *Giardia lamblia*

   B) *Escherichia coli*

   C) *Taenia* spp.

   D) *Vibrio parahaemolyticus*

   E) *Entamoeba histolytica*

   Answer: E
   Skill: Recall

11) In humans, beef tapeworm infestations are acquired by

   A) Ingesting the eggs of *Taenia saginata*.

   B) Ingesting segments of adult tapeworms.

   C) Ingesting contaminated water.

   D) Ingesting cysticerci in the intermediate host.

   E) Ingesting contaminated definitive hosts.

   Answer: D
   Skill: Recall

12) All of the following statements about staphylococcal food poisoning are true *except*

   A) Suspect foods are those not cooked before eating.

   B) It can be prevented by refrigeration.

   C) It can be prevented by boiling foods for 5 minutes before eating.

   D) It is treated by replacing water and electrolytes.

   E) It is characterized by rapid onset and short duration of symptoms.

   Answer: C
   Skill: Recall

13) The most common cause of traveler's diarrhea is probably

   A) Shigella spp.

   B) *Salmonella typhi.*

   C) *Giardia lamblia.*

   D) *Escherichia coli.*

   E) *Entamoeba coli.*

   Answer: D
   Skill: Recall

14) Which of the following can be transferred from an infected mother to her fetus across the placenta?

   A) Mumps

   B) Cytomegalovirus inclusion disease

   C) Infectious hepatitis

   D) Typhoid fever

   E) Enteroinvasive *E. coli*

   Answer: B
   Skill: Analysis

15) Which of the following pairs is *not* correctly matched?

   A) Hydatid disease — humans are the definitive host

   B) *Taenia* infestation — humans are the definitive host

   C) Trichinosis — humans eat larva of parasite

   D) Pinworm infestation — humans ingest parasite's eggs

   E) Hookworm infestation — parasite bores through skin

   Answer: A
   Skill: Understanding

16) Thorough cooking of food will prevent all of the following *except*

   A) Trichinosis.

   B) Beef tapeworm.

   C) Staphylococcal food poisoning.

   D) Salmonellosis.

   E) All of the above.

   Answer: C
   Skill: Understanding

17) Most of the normal microbiota of the digestive system are found in the
   A) Mouth.
   B) Stomach.
   C) Small intestine.
   D) Large intestine.
   E) C and D.

   Answer: D
   Skill: Analysis

18) Which of the following organisms is most likely to be responsible for periodontal disease?
   A) Gram–positive cocci
   B) Gram–positive rods
   C) Gram–negative cocci
   D) Gram–negative rods
   E) Gingivavirus

   Answer: B
   Skill: Recall

19) Typhoid fever differs from salmonellosis in that in typhoid fever
   A) The microorganisms become invasive.
   B) The symptoms are due to an exotoxin.
   C) The symptoms are due to infection of the gallbladder.
   D) The classic symptom is diarrhea.
   E) Chemotherapy is highly effective.

   Answer: A
   Skill: Analysis

20) Which of the following organisms is likely to be transmitted via contaminated shrimp?
   A) *Trichinella*
   B) *Vibrio*
   C) *Giardia*
   D) *Clostridium perfringens*
   E) *Staphylococcus aureus*

   Answer: B
   Skill: Analysis

21) Which of the following organisms is likely to be transmitted via contaminated pork?

A) *Salmonella*

B) *Staphylococcus*

C) *Trichinella*

D) *Entamoeba*

E) *Shigella*

Answer: C
Skill: Recall

22) A vaccine to provide active immunity to serum hepatitis is being prepared from

A) Viruses grown in tissue culture.

B) Genetically engineered yeast.

C) Pooled gamma globulin.

D) Viruses grown in embryonated eggs.

E) None of the above.

Answer: B
Skill: Analysis

23) Which of the following causes a congenital disease?

A) *Salmonella*

B) *Balantidium*

C) *Trichinella*

D) Cytomegalovirus

E) Hepatitis A virus

Answer: D
Skill: Recall

24) Which of the following causes an infection of the liver?

A) *Salmonella*

B) *Shigella*

C) Hepatitis A virus

D) *Vibrio*

E) *Escherichia*

Answer: C
Skill: Analysis

25) "Rice water stools" are characteristic of

A) Salmonellosis.

B) Cholera.

C) Bacillary dysentery.

D) Amoebic dysentery.

E) Tapeworm infestation.

Answer: B
Skill: Analysis

26) Epidemics of bacterial infections of the digestive system are transmitted by

A) Food.

B) Water.

C) Milk.

D) The respiratory route.

E) All of the above.

Answer: B
Skill: Understanding

27) Most gastrointestinal infections are treated with

A) Antitoxin.

B) Penicillin.

C) Water and electrolytes.

D) Quinacrine.

E) None of the above.

Answer: C
Skill: Understanding

28) Which of the following is treated with tetracycline?

A) Staphylococcal food poisoning

B) *Vibrio parahaemolyticus* gastroenteritis

C) Infectious hepatitis

D) *Escherichia coli* gastroenteritis

E) Trichinosis

Answer: B
Skill: Understanding

29) Which of the following pairs is *not* correctly matched?

    A) Ergot — gangrene

    B) *Salmonella endotoxin* — coagulates blood

    C) *Vibrio* enterotoxin — secretion of Cl–, CO_3^{2-}, and H_2O

    D) Aflatoxin — liver cancer

    E) All of the above

    Answer: B
    Skill: Understanding

30) Bacterial intoxications differ from bacterial infections of the digestive system in that intoxications

    A) Are transmitted via water.

    B) Are more severe.

    C) Have shorter incubation times.

    D) Are treated with antibiotics.

    E) Are accompanied by fever.

    Answer: C
    Skill: Analysis

31) The most common mode of HAV transmission is

    A) Contamination of food during preparation.

    B) Contamination of food before it reaches a food service establishment.

    C) Blood transfusion.

    D) Contaminated hypodermic needles.

    E) Airborne.

    Answer: A
    Skill: Recall

32) With which of the following substrates can *Streptococcus mutans* make a capsule?

    A) Fructose

    B) Glucose

    C) Mannitol

    D) Sucrose

    E) All of the above

    Answer: D
    Skill: Analysis

33) Most bacteria associated with the teeth and gums are

A) Aerobes.

B) Anaerobes.

C) Facultative anaerobes.

D) None of the above.

Answer: B
Skill: Recall

34) A 38-year-old man had onset of fever, chills, nausea, and myalgia. On April 29, he had eaten raw oysters. On May 2, he was admitted to a hospital because of a fever of 39°C and two circular necrotic lesions on the left leg. He had alcoholic liver disease. He was transferred to the ICU; therapy with ciprofloxacin was initiated. On May 4, he died. Which of the following is the most likely etiology?

A) *Bacillus cereus*

B) *Cyclospora*

C) *Salmonella*

D) *Vibrio vulnificus*

E) *Yersinia enterocolitica*

Answer: D
Skill: Analysis

35) The easiest way to prevent outbreaks of gram-negative gastroenteritis is to

A) Cook foods thoroughly.

B) Salt foods.

C) Add vinegar and spices to foods.

D) Refrigerate foods.

E) All of the above.

Answer: A
Skill: Analysis

36) Microscopic examination of a patient's fecal culture shows spiral bacteria. The bacteria probably belong to the genus

A) *Campylobacter*.

B) *Escherichia*.

C) *Salmonella*.

D) *Shigella*.

E) *Vibrio*.

Answer: A
Skill: Recall

37) Feces from a patient with diarrhea lasting for weeks with frequent, watery stools should be examined for

A) Bacillus *cereus*.

B) *Cyclospora*.

C) *Salmonella*.

D) *Vibrio vulnificus*.

E) *Yersinia enterocolitica*.

Answer: B
Skill: Analysis

38) All of the following are gram-negative rods that cause gastroenteritis *except*

A) *Clostridium*.   B) *Escherichia*.   C) *Salmonella*.   D) *Shigella*.   E) *Yersinia*.

Answer: A
Skill: Analysis

39) *Helicobacter* can grow in the stomach because it

A) Hides in macrophages.

B) Makes a capsule.

C) Makes NH_3.

D) Makes HCl.

E) None of the above.

Answer: C
Skill: Analysis

40) All of the following are eucaryotes that cause gastroenteritis *except*

A) *Cryptosporidium*.

B) *Entamoeba*.

C) *Giardia*.

D) *Pneumocystis*.

E) None of the above.

Answer: D
Skill: Analysis

41) Which of the following produces a permanent carrier state following infection?

A) Cytomegalovirus

B) *E. coli* O157:H7

C) *Salmonella*

D) *Shigella dysenteriae*

E) *Vibrio cholerae*

Answer: A
Skill: Recall

42) Most cases of posttransfusion hepatitis are caused by

    A) Hepatitis A virus.

    B) Hepatitis B virus.

    C) Hepatitis C virus.

    D) Hepatitis D virus.

    E) Hepatitis E virus.

Answer: C
Skill: Recall

43) Which of the following pairs is incorrectly matched?

    A) Beef — *E. coli* O157:H7

    B) Delicatessen meats — *Listeria*

    C) Eggs — *Trichinella*

    D) Milk — *Campylobacter*

    E) Oysters — *Vibrio*

Answer: C
Skill: Analysis

Situation 25.1

Following a county fair, 160 persons complained of gastrointestinal symptoms. Symptoms included diarrhea (84%), abdominal cramps (96%), nausea (84%), vomiting (82%), body aches (50%), fever (60%; median body temperature = 38.3°C); median duration of illness 6 days (range 10 hr to 13 days).

44) In Situation 25.1, fecal samples should be cultured for all of the following *except*

    A) *Salmonella*.

    B) *Shigella*.

    C) *Campylobacter*.

    D) Enteropatheogenic *Escherichia coli*.

    E) *Giardia*.

Answer: E
Skill: Understanding

45) In Situation 25.1, assume the samples were culture–negative. The next step is

    A) Begin antibiotic therapy.

    B) Blood cultures.

    C) Microscopic examination for oocysts.

    D) Microscopic examination for viruses.

    E) Muscle biopsy.

Answer: C
Skill: Understanding

ESSAY QUESTIONS

1) *Escherichia coli* is normally in the large intestines of humans. How can this bacterium be the etiologic agent of most cases of traveler's diarrhea?

Answer:

Skill:

2) Discuss why cholera epidemics are often associated with floods.

Answer:

Skill:

3) How can you avoid contracting the following diseases?
   a. Trichinosis
   b. Hydatidosis
   c. Tapeworm infestations
   d. Staphylococcal food poisoning
   e. Traveler's diarrhea
   f. Giardiasis

Answer:

Skill:

CHAPTER 26   Microbial Diseases of the Urinary and Reproductive Systems

OBJECTIVE QUESTIONS

1) Normal microbiota of the adult vagina consist primarily of
   A) Lactobacillus.
   B) Streptococcus.
   C) Mycobacterium.
   D) Neisseria.
   E) All of the above.

   Answer: A
   Skill: Recall

2) All of the following are predisposing factors to urinary tract infection except
   A) Toxemia.
   B) Tumors.
   C) Diabetes mellitus.
   D) Kidney stones.
   E) None of the above.

   Answer: E
   Skill: Recall

3) Pyelonephritis may result from
   A) Urethritis.
   B) Cystitis.
   C) Ureteritis.
   D) Systemic infections.
   E) All of the above.

   Answer: E
   Skill: Recall

4) Cystitis is most often caused by
   A) Gram-negative cocci.
   B) Gram-negative rods.
   C) Gram-positive cocci.
   D) Gram-positive rods.
   E) All of the above.

   Answer: B
   Skill: Recall

5) Pyelonephritis usually is caused by
    A) Pseudomonas aeruginosa.
    B) Proteus spp.
    C) Escherichia coli.
    D) Enterobacter aerogenes.
    E) Streptococcus pyogenes.

    Answer: C
    Skill: Recall

6) The reservoir for leptospirosis is
    A) Humans.
    B) Water.
    C) Domestic dogs.
    D) Domestic cats.
    E) None of the above.

    Answer: C
    Skill: Recall

7) Which of the following is not primarily a sexually transmitted disease (STD)?
    A) Lymphogranuloma venereum
    B) Genital herpes
    C) Gonorrhea
    D) Chancroid
    E) Trichomoniasis

    Answer: E
    Skill: Analysis

8) Which of the following is treated with penicillin?
    A) Lymphogranuloma venereum
    B) Genital warts
    C) Candidiasis
    D) Syphilis
    E) Trichomoniasis

    Answer: D
    Skill: Analysis

9) Which of the following pairs is mismatched?

A) Trichomoniasis — fungus
B) Gonorrhea — gram–negative cocci
C) Chancroid — gram–negative rod
D) Gardnerella — clue cells
E) Syphilis — gram–negative spirochete

Answer: A
Skill: Recall

10) One form of NGU is lymphogranuloma venereum caused by

A) Leptospira interrogans.
B) Chlamydia trachomatis.
C) Neisseria gonorrhoeae.
D) Treponema pallidum.
E) Candida albicans.

Answer: B
Skill: Recall

11) All of the following can cause congenital infections or infections of the newborn except

A) Syphilis.
B) Gonorrhea.
C) Nongonococcal urethritis.
D) Genital herpes.
E) Lymphogranuloma venereum.

Answer: E
Skill: Recall

12) Which of the following recurs at the initial site of infection?

A) Gonorrhea
B) Syphilis
C) Genital herpes
D) Chancroid
E) All of the above

Answer: C
Skill: Analysis

13) Which of the following is diagnosed by detection of antibodies against the causative agent?

A) Nongonococcal urethritis

B) Gonorrhea

C) Syphilis

D) Lymphogranuloma venereum

E) Candidiasis

Answer: C
Skill: Recall

14) Nongonococcal urethritis can be caused by all of the following *except*

A) Mycoplasma homini.

B) Candida albicans.

C) Trichomonas vaginalis.

D) Streptococci.

E) Neisseria gonorrhoeae.

Answer: E
Skill: Analysis

15) Which of the following is caused by an opportunistic pathogen?

A) Trichomoniasis

B) Genital herpes

C) Candidiasis

D) Gonorrhea

E) None of the above

Answer: C
Skill: Analysis

16) The pH of the adult vagina is acidic due to the conversion of _1_ to _2_ by bacteria.

A) 1–Glucose; 2–Ethyl alcohol

B) 1–Protein; 2–Acetic acid

C) 1–Glycogen; 2–Lactic acid

D) 1–Mucosal cells; 2–Lactic acid

E) 1–Urine; 2–Lactic acid

Answer: C
Skill: Analysis

17) A normal urine sample collected by urinating into a sterile collection cup
   A) Is sterile.
   B) Contains less than 100 bacteria/ml.
   C) Contains less than 10,000 bacteria/ml.
   D) Contains more than 100,000 bacteria/ml.
   E) None of the above.

   Answer: C
   Skill: Recall

18) Most nosocomial infections of the urinary tract are caused by
   A) E. coli.
   B) Enterococcus.
   C) Proteus.
   D) Klebsiella.
   E) Pseudomonas.

   Answer: A
   Skill: Recall

19) Glomerulonephritis is
   A) Caused by Streptococcus pyogenes.
   B) An immune complex disease.
   C) Treated with penicillin.
   D) Transmitted by contaminated water.
   E) A and C.

   Answer: B
   Skill: Recall

20) All of the following are predisposing factors to cystitis in females except
   A) The proximity of the anus to the urethra.
   B) The length of the urethra.
   C) Sexual intercourse.
   D) Poor personal hygiene.
   E) None of the above.

   Answer: E
   Skill: Analysis

21) The most common reportable disease in the United States is

   A) Cystitis.

   B) Lymphogranuloma venereum.

   C) Gonorrhea.

   D) Syphilis.

   E) Candidiasis.

Answer: C
Skill: Recall

22) Which of the following is not a complication of gonorrhea?

   A) Sterility

   B) Pelvic inflammatory disease

   C) Endocarditis

   D) Meningitis

   E) None of the above

Answer: E
Skill: Analysis

23) Which of the following is the most difficult to treat with chemotherapeutic agents?

   A) Genital herpes

   B) Gonorrhea

   C) Syphilis

   D) Trichomoniasis

   E) All of the above

Answer: A
Skill: Understanding

24) Itching and cheesy discharge are symptoms of:

   A) *Gardnerella* vaginosis

   B) Genital herpes

   C) Candidiasis

   D) Trichomoniasis

   E) Lymphogranuloma venereum

Answer: C
Skill: Recall

25) Recurring vesicles are symptoms of:

    A) *Gardnerella* vaginosis

    B) Genital herpes

    C) Candidiasis

    D) Trichomoniasis

    E) Lymphogranuloma venereum

    Answer: B
    Skill: Recall

26) Leukocytes at the infected site is a symptom of:

    A) *Gardnerella* vaginosis

    B) Genital herpes

    C) Candidiasis

    D) Trichomoniasis

    E) Lymphogranuloma venereum

    Answer: D
    Skill: Recall

27) This is caused by *Chlamydia*:

    A) *Gardnerella* vaginosis

    B) Genital herpes

    C) Candidiasis

    D) Trichomoniasis

    E) Lymphogranuloma venereum

    Answer: E
    Skill: Recall

28) Which of the following forms lesions similar to those of tuberculosis?

    A) Gonorrhea

    B) Lymphogranuloma venereum

    C) Chancroid

    D) Syphilis

    E) Genital herpes

    Answer: D
    Skill: Analysis

29) Which of the following is treated with spectinomycin because the organism produces penicillinase?

   A) Mycoplasma hominis
   B) Hemophilus ducreyi
   C) Neisseria gonorrhoeae
   D) Treponema pallidum
   E) None of the above

   Answer: C
   Skill: Recall

30) Which of the following diseases causes a skin rash, hair loss, malaise, and fever?

   A) Gonorrhea
   B) Syphilis
   C) NGU
   D) Trichomoniasis
   E) None of the above

   Answer: B
   Skill: Recall

31) Staphylococcus saprophyticus causes

   A) Cystitis.
   B) Pyelonephritis.
   C) Vaginitis.
   D) Gonorrhea.
   E) Syphilis.

   Answer: A
   Skill: Recall

32) Which one of the following statements about genital warts is not true?

   A) It is transmitted by direct contact.
   B) It is caused by papillomaviruses.
   C) It is always precancerous.
   D) It is treated by removing them.
   E) All of the above.

   Answer: C
   Skill: Analysis

33) The most common STD in the United States is treated with
    A) Penicillin.
    B) Tetracycline.
    C) Acyclovir.
    D) AZT.
    E) None of the above; it's untreatable.

    Answer: B
    Skill: Analysis

34) Pelvic inflammatory disease
    A) Affects one million women annually.
    B) Can be caused by N. gonorrhoeae.
    C) Can be transmitted sexually.
    D) Can be caused by C. trachomatis.
    E) All of the above.

    Answer: E
    Skill: Recall

35) Nongonococcal urethritis can be caused by all of the following *except*
    A) Chlamydia.
    B) Mycoplasma.
    C) Neisseria.
    D) Ureaplasma.
    E) None of the above.

    Answer: C
    Skill: Analysis

36) Infants born to mothers with recurrent genital herpes do not usually acquire herpes at birth if the mother is asymptomatic because
    A) Maternal antibodies offer protection.
    B) The disease cannot be transmitted to newborns.
    C) The disease is not communicable.
    D) Prophylactic antibiotics are administered to the newborn.
    E) None of the above.

    Answer: A
    Skill: Understanding

37) Which of the following is greater?
    A) The number of reported cases of gonorrhea last year.
    B) The number of reported cases of AIDS last year.

    Answer: A
    Skill: Understanding

38) A patient presents with fever and extensive lesions of the labia minora. Her VDRL test was negative. What is the most likely treatment?

A) Metronidazole

B) Cephalosporins

C) Acyclovir

D) Miconazole

E) No treatment is available

Answer: C
Skill: Understanding

39) A patient is experiencing profuse greenish-yellow, foul-smelling discharge from her vagina; she is complaining of itching and irritation. What is the most likely treatment?

A) Metronidazole

B) Cephalosporins

C) Acyclovir

D) Miconazole

E) No treatment is available

Answer: A
Skill: Understanding

40) A 25-year-old male presented with fever, malaise, and a rash on his chest, arms, and feet. The etiology could be any of the following *except*

A) Borrelia.

B) Mumps virus.

C) Rickettsia.

D) Streptococcus.

E) Treponema.

Answer: B
Skill: Analysis

41) A 25-year-old male presented with fever, malaise, and a rash on his chest, arms, and feet. Which of the following will be most useful for a rapid diagnosis?

A) Bacterial culture

B) Microscopic examination of blood

C) Serological test for antibodies

D) Serological test for antigen

E) Viral culture

Answer: C
Skill: Understanding

42) A 25–year–old male presented with fever, malaise, and a rash on his chest, arms, and feet. Diagnosis was made based on serological testing. The patient then reported that he had an ulcer on his penis two months earlier. What stage of disease is the patient in?

A) NGU   B) Primary   C) Secondary   D) Tertiary

Answer: C
Skill: Analysis

43) A 25–year–old male presented with fever, malaise, and a rash on his chest, arms, and feet. Diagnosis was made based on serological testing. The patient then reported that he had an ulcer on his penis two months earlier. This disease can be treated with

A) Acyclovir.

B) Metronidazole.

C) Miconazole.

D) Penicillin.

E) Surgery.

Answer: D
Skill: Analysis

44) A pelvic examination of a 23–year–old female showed vesicles and ulcerated lesions on her labia. Cultures were negative for Neisseria and Chlamydia; the VDRL test was negative. Which treatment is appropriate?

A) Acyclovir

B) Metronidazole

C) Miconazole

D) Penicillin

E) Surgery

Answer: A
Skill: Analysis

45) A pelvic examination of a 23–year–old female showed vesicles and ulcerated lesions on her labia. Cultures were negative for Neisseria and Chlamydia; the VDRL test was negative. Which of the following is probable?

A) Candidiasis

B) Genital herpes

C) Gonorrhea

D) NGU

E) Syphilis

Answer: B
Skill: Analysis

ESSAY QUESTIONS

1) Why are the unreported cases of STDs an important public health concern?

   Answer:

   Skill:

2) Nearly 70% of the patients seen in STD clinics are male.

   a. Offer a reason men are more likely to seek treatment than women.

   b. Why is it important that women seek treatment for STDs?

   Answer:

   Skill:

CHAPTER 27   Environmental Microbiology

OBJECTIVE QUESTIONS

1) Which of the following is *not* a habitat for an extremophile?
   A) Acid mine wash
   B) The Atlantic Ocean
   C) Inside rock
   D) Salt–evaporating pond
   E) 100°C water

   Answer: B
   Skill: Recall

2) Which of the following organisms uses sulfur as a source of energy?
   A) Thiobacillus — $H_2S \rightarrow S^0$

   B) Desulfovibrio — $SO_4^{2-} \rightarrow H_2S$

   C) Proteus — Amino acids $\rightarrow H_2S$

   D) Redwood tree — $SO_4^{2-} \rightarrow$ Amino acids

   E) None of the above

   Answer: A
   Skill: Understanding

3) Which of the following is *not* a symbiotic pair of organisms?
   A) Elk and rumen bacteria
   B) Orchid and mycorrhizae
   C) Onions and arbuscules
   D) Bean plant and Rhizobium
   E) Sulfur and Thiobacillus

   Answer: E
   Skill: Understanding

4) In which of the following animals would you expect to find a specialized organ that holds cellulose–degrading bacteria and fungi?
   A) Cat   B) Dog   C) Termite   D) Human   E) Wolf

   Answer: C
   Skill: Understanding

5) Which of the following pairs is mismatched?

   A) Rhizobium — legumes

   B) Bacillus thuringiensis — insect control

   C) Thiobacillus ferrooxidans — uranium mining

   D) Frankia — alders

   E) Lichen — an alga and a bacterium

Answer: E
Skill: Analysis

6) The addition of untreated sewage to a freshwater lake would cause the biochemical oxygen demand to

   A) Increase.   B) Decrease.   C) Stay the same.   D) Can't tell.

Answer: A
Skill: Analysis

7) What conditions are necessary for the following: coal formation

   A) If the process takes place under aerobic conditions.

   B) If the process takes place under anaerobic conditions.

   C) If the amount of oxygen doesn't make any difference.

Answer: B
Skill: Understanding

8) What conditions are necessary for the following: Trickling filter

   A) If the process takes place under aerobic conditions.

   B) If the process takes place under anaerobic conditions.

   C) If the amount of oxygen doens't make any difference.

Answer: A
Skill: Understanding

9) What conditions are necessary for the following:

   $$6H_2S + 6CO_2 \xrightarrow{\text{light}} 6S^0 + \text{glucose}$$

   A) If the process takes place under aerobic conditions.

   B) If the process takes place under anaerobic conditions.

   C) If the amount of oxygen doesn't make any difference.

Answer: B
Skill: Understanding

10) What conditions are necessary for the following: Nitrogen fixation
   A) If the process takes place under aerobic conditions.
   B) If the process takes place under anaerobic conditions.
   C) If the amount of oxygen doesn't make any difference.

Answer: B
Skill: Understanding

11) What conditions are necessary for the following: Sludge digestion
   A) If the process takes place under aerobic conditions.
   B) If the process takes place under anaerobic conditions.
   C) If the amount of oxygen doesn't make any difference.

Answer: B
Skill: Understanding

12) What conditions are necessary for the following: Primary sewage treatment
   A) If the process takes place under aerobic conditions.
   B) If the process takes place under anaerobic conditions.
   C) If the amount of oxygen doens't make any difference

Answer: C
Skill: Understanding

13) What conditions are necessary for the following: $NH_3 \rightarrow NO_2^-$

   A) If the process takes place under aerobic conditions.
   B) If the process takes place under anaerobic conditions.
   C) If the amount of oxygen doesn't make any difference.

Answer: A
Skill: Understanding

14) What conditions are necessary for the following: Bioremediation of petroleum
   A) If the process takes place under aerobic conditions.
   B) If the process takes place under anaerobic conditions.
   C) If the amount of oxygen doesn't make any difference.

Answer: A
Skill: Understanding

15) Which of the following equations represents anaerobic respiration?

A) $H_2S \rightarrow S^0$

B) $NO_3^- \rightarrow NO_2^-$

C) Amino acids $\rightarrow H_2S$

D) $SO_4^{2-} \rightarrow$ Amino acids

E) None of the above

Answer: B
Skill: Understanding

16) Biochemical oxygen demand is a measure of

A) The number of bacteria present in a water sample.

B) The amount of oxygen present in a water sample.

C) The amount of organic matter present in a water sample.

D) The amount of undissolved solid matter present in a water sample.

E) All of the above.

Answer: C
Skill: Analysis

17) Eighty-one percent of the microorganisms in the soil are

A) Actinomycetes.

B) Algae

C) Bacteria.

D) Fungi.

E) Protozoa.

Answer: C
Skill: Recall

18) Most of the microorganisms in the soil are found at a depth

A) Between 3 and 8 cm.

B) Between 20 and 25 cm.

C) Between 35 and 40 cm.

D) Between 65 and 75 cm.

E) Evenly distributed from the surface to 76 cm.

Answer: A
Skill: Recall

19) $NO_3^- \rightarrow N_2$

    A) Ammonification

    B) Denitrification

    C) Nitrification

    D) Nitrogen fixation

    E) None of the above

Answer: B
Skill: Understanding

20) $N_2 \rightarrow NH_3$

    A) Ammonification

    B) Denitrification

    C) Nitrification

    D) Nitrogen fixation

    E) None of the above

Answer: D
Skill: Understanding

21) $NH_3 \rightarrow NO_2^- \rightarrow NO_3^-$

    A) Ammonification

    B) Denitrification

    C) Nitrification

    D) Nitrogen fixation

    E) None of the above

Answer: C
Skill: Understanding

22) Amino acids ($-NH_2$) $\rightarrow NH_3$

    A) Ammonification

    B) Denitrification

    C) Nitrification

    D) Nitrogen fixation

    E) None of the above

Answer: A
Skill: Understanding

23) *Nitrosomonas* and *Nitrobacter* are capable of performing this:

A) Ammonification

B) Denitrification

C) Nitrification

D) Nitrogen fixation

E) None of the above

Answer: C
Skill: Understanding

24) This process is done by *Rhizobium* and certain cyanobacteria:

A) Ammonification

B) Denitrification

C) Nitrification

D) Nitrogen fixation

E) None of the above

Answer: D
Skill: Understanding

25) Anaerobic respiration of *Thiobacillus denitrifivcans*:

A) Ammonification

B) Denitrification

C) Nitrification

D) Nitrogen fixation

E) None of the above

Answer: B
Skill: Understanding

26) Choose the correct wastewater treatment for removal of BOD.

A) Anaerobic sludge digester

B) Primary sewage treatment

C) Secondary sewage treatment

D) Tertiary sewage treatment

E) Water treatment

Answer: C
Skill: Understanding

27) Choose the correct wastewater treatment for BOD–containing effluent used for irrigation.

A) Anaerobic sludge digester

B) Primary sewage treatment

C) Secondary sewage treatment

D) Tertiary sewage treatment

E) Water treatment

Answer: C
Skill: Understanding

Choose the correct wastewater treatment for the following:

28) Residual chlorine is maintained.

A) Anaerobic sludge digester

B) Primary sewagae treatment

C) Secondary sewage treatment

D) Tertiary sewage treatment

E) Water treatment

Answer: E
Skill: Understanding

29) Sedimentation of sludge.

A) Anaerobic sludge digester

B) Primary sewage treatment

C) Secondary sewage treatment

D) Tertiary sewage treatment

E) Water treatment

Answer: B
Skill: Understanding

30) Product contains high BOD.

A) Anaerobic sludge digester

B) Primary sewage treatment

C) Secondary sewage treatment

D) Tertiary sewage treatment

E) Water treatment

Answer: B
Skill: Understanding

31) *Zoogloea* form flocculant masses.
   A) Anaerobic sludge digester
   B) Primary sewage treatment
   C) Secondary sewage treatment
   D) Tertiary sewage treatment
   E) Water treatment

   Answer: C
   Skill: Understanding

32) Uses rotating biological contactor.
   A) Anaerobic sludge digester
   B) Primary sewage treatment
   C) Secondary sewage treatment
   D) Tertiary sewage treatment
   E) Water treatment

   Answer: C
   Skill: Understanding

33) Aerobic respiration occurs here.
   A) Anaerobic sludge digester
   B) Primary sewage treatment
   C) Secondary sewage treatment
   D) Tertiary sewage treatment
   E) Water treatment

   Answer: C
   Skill: Understanding

34) Anaerobic respiration occurs here.
   A) Anaerobic sludge digester
   B) Primary sewage treatment
   C) Secondary sewage treatment
   D) Tertiary sewage treatment
   E) Water treatment

   Answer: A
   Skill: Understanding

35) Filtration to remove protozoa.

    A) Anaerobic sludge digester

    B) Primary sewage treatment

    C) Secondary sewage treatment

    D) Tertiary sewage treatment

    E) Water treatment

    Answer: E
    Skill: Understanding

36) Which of the following do *not* fix atmospheric nitrogen?

    A) Cyanobacteria

    B) Lichens

    C) Mycorrhizae

    D) *Frankia*

    E) *Azotobacter*

    Answer: C
    Skill: Recall

37) Which of the following pairs is mismatched?

    A) $CO_2 + 4H_2 \rightarrow CH_4 + 2H_2O$ — Methane–producing bacteria

    B) $Fe^{2+} \rightarrow Fe^{3+}$ — *Thiobacillus ferrooxidans*

    C) $6CO_2 + 6H_2O \rightarrow C_6H_{12}O_6 + 6O_2$ — *Clostridium*

    D) $N_2 + 6H^+ \rightarrow 2NH_3$ — *Beijerinckia*

    E) $SO_4^{2-} + 10H^+ \rightarrow H_2S + 4H_2O$ — *Desulfovibrio*

    Answer: C
    Skill: Understanding

38) Bacteria can increase the Earth's temperature by

    A) Generating a great deal of heat in metabolism.

    B) Producing CH_4, which is a greenhouse gas.

    C) Using the greenhouse gas CO_2.

    D) Providing nutrients for plant growth.

    E) Oxidizing CH_4.

    Answer: B
    Skill: Recall

39) The bacteria contributing most of the bacterial biomass to soil are
    A) Actinomycetes.
    B) Rhizobiaceae.
    C) Chemoautotrophs.
    D) Photoheterotrophs.
    E) Coliforms.

Answer: A
Skill: Analysis

40) In one hospital, *Pseudomonas aeruginosa* serotype 10 infected the biliary tract of 10% of 1300 patients who underwent gastrointestinal endoscopic procedures. After each use, endoscopes were washed with an automatic reprocessor that flushed detergent and glutaraldehyde through the endoscopes followed by a tap water rinse. *P. aeruginosa* 10 was not isolated from the detergent, glutaraldehyde, or tap water. What was the source of the infections?

    A) Bacterial cell walls in the water   B) A biofilm in the reprocessor
    C) Contaminated disinfectant   D) Fecal contamination of the bile ducts

Answer: B
Skill: Analysis

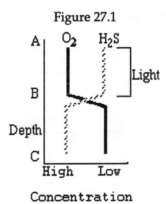

Figure 27.1

41) Where are photosynthetic bacteria most likely to be found in Figure 27.1?
    A) A   B) B   C) C   D) A and B   E) B and C

Answer: B
Skill: Understanding

42) Where are eucaryotic algae most likely to be found in Figure 27.1?
    A) A   B) B   C) C   D) A and B   E) B and C

Answer: A
Skill: Understanding

43) The bacteria that grow first in the microbial succession in a compost pile are

    A) Anaerobic mesophiles.

    B) Aerobic thermophiles.

    C) Anaerobic psychrophiles.

    D) Aerobic mesophiles.

    E) None of the above.

Answer: A
Skill: Analysis

44) The release of phosphate–containing detergents into a river would

    A) Kill algae.

    B) Increase algal growth.

    C) Kill bacteria.

    D) Increase oxygen in the water.

    E) None of the above.

Answer: B
Skill: Analysis

45) All of the following are true about releasing untreated sewage into a river *except*

    A) It is a health hazard.

    B) It increases the BOD.

    C) It decreases the dissolved oxygen.

    D) It kills bacteria.

    E) None of the above.

Answer: D
Skill: Understanding

ESSAY QUESTIONS

1) What is the effect of discharging primary-treated sewage on the BOD and dissolved O_2 (DO) of the receiving body of water? What is the effect of secondary-treated sewage on the BOD and DO?

Answer:

Skill:

2) A newspaper headline reported "Algal Bloom Kills Fish."
    a. What actually caused fish death since the algae were not toxic?
    b. What could have caused the algal bloom?

Answer:

Skill:

Figure 27.2
These results were obtained from a bioremediation experiment:

Number of colonies on				
Minimal salts agar inoculated with soil		Agar inoculated with soil		Uninoculated minimal salts agar
Without gasoline	With gasoline	Without gasoline	With gasoline	Without gasoline
3	33	0	0	0

3) Explain the data in Figure 27.2.

Answer:

Skill:

CHAPTER 28 Applied and Industrial Microbiology

OBJECTIVE QUESTIONS

1) Spoilage due to can leakage after processing:
   A) Thermophilic anaerobic spoilage
   B) Flat sour spoilage
   C) Spoilage by mesophilic bacteria
   D) Acid–tolerant fungi
   E) Putrefactive anaerobic spoilage

   Answer: C
   Skill: Recall

2) Spoilage of canned foods stored at high temperatures, accompanied by gas production:
   A) Thermophilic anaerobic spoilage
   B) Flat sour spoilage
   C) Spoilage by mesophilic bacteria
   D) Acid–tolerant fungi
   E) Putrefactive anaerobic spoilage

   Answer: A
   Skill: Recall

3) Spoilage of canned foods due to inadequate processing, not accompanied by gas production:
   A) Thermophilic anaerobic spoilage
   B) Flat sour spoilage
   C) Spoilage by mesophilic bacteria
   D) Acid–tolerant fungi
   E) Putrefactive anaerobic spoilage

   Answer: B
   Skill: Recall

4) Which of the following pairs is mismatched?
   A) *Propionibacterium* — Swiss cheese
   B) *Penicillium* — blue cheese
   C) *Streptococcus* — yogurt
   D) *Acetobacter* — vinegar
   E) *Bacillus* — hard cheese

   Answer: E
   Skill: Analysis

5) Which of the following is *not* an alternative fuel (energy source) produced by bacteria?

    A) Lactic Acid

    B) Methyl Alcohol

    C) Ethyl Alcohol

    D) Hydrogen

    E) Methane

Answer: A
Skill: Understanding

6) Which of the following food additives is *not* produced by microorganisms?

    A) Amylase

    B) Citric acid

    C) Sodium nitrate

    D) Glutamic acid

    E) Protease

Answer: C
Skill: Analysis

7) Wine is made from fruit juices by

    A) The addition of alcohol.

    B) Bacterial production of CO_2.

    C) The addition of sugar.

    D) Anaerobic fungal growth.

    E) Malting.

Answer: D
Skill: Understanding

8) Which of the following is an undesirable contaminant in wine–making?

    A) *Acetobacter*

    B) Lactic acid bacteria

    C) *Clostridium*

    D) *Bacillus*

    E) *Saccharomyces* (yeast)

Answer: A
Skill: Understanding

9) Commercial sterilization differs from true sterilization in that commercial sterilization
   A) Kills all microorganisms.
   B) Kills only bacteria.
   C) May result in the survival of thermophiles.
   D) Employs a higher temperature.
   E) May result in the survival of fungal spores.

Answer: C
Skill: Recall

Figure 28.1

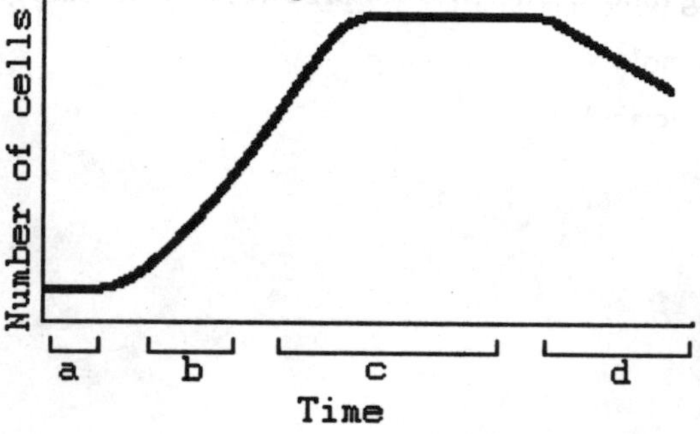

10) In Figure 28.1, assume that cells are your desired product. When would you harvest cells to maximize your yield?
   A) a   B) b   C) c   D) d

Answer: B
Skill: Understanding

11) In Figure 28.1, assume that a secondary metabolite is your desired product. When would you be able to obtain it?
   A) a   B) b   C) c   D) d

Answer: C
Skill: Understanding

12) In Figure 28.1, assume you want an enzyme that is not secreted. You would harvest at the end of what time?
   A) a   B) b   C) c   D) d

Answer: B
Skill: Understanding

13) Which reation leads to all the others?

   A) Sucrose ---> Ethanol

   B) Ethanol ---> Acetic acid

   C) Malic acid ---> Lactic acid

   D) Carbon dioxide ---> Sucrose

   E) None of the above

   Answer: D
   Skill: Analysis

14) Which reaction makes wine less acidic?

   A) Sucrose ---> Ethanol

   B) Ethanol ---> Acetic acid

   C) Malic acid ---> Lactic acid

   D) Carbon dioxide ---> Sucrose

   E) None of the above

   Answer: C
   Skill: Analysis

15) If the *lac* operon is used as a receptor for the lux operon, the presence of lactose would cause the cell to

   A) Die.

   B) Emit light.

   C) Produce lactose.

   D) Grow better.

   E) Use light energy.

   Answer: B
   Skill: Understanding

16) When compared to a flask culture, a bioreactor offers all of the following advantages *except*

   A) Larger culture volumes can be grown.

   B) Instrumentation for monitoring environmental conditions.

   C) Uniform aeration and mixing.

   D) Aseptic sampling.

   E) None of the above.

   Answer: E
   Skill: Recall

17) Microbial products can be improved by
   A) Isolating new strains.
   B) Mutating existing strains.
   C) Genetically engineering strains.
   D) Modifying culture conditions.
   E) All of the above.

   Answer: E
   Skill:

18) Cellulase attached to a membrane filter will
   A) Degrade cellulose.
   B) Degrade the membrane.
   C) Do nothing; it requires a cell.
   D) Degrade lactose.
   E) None of the above.

   Answer: A
   Skill: Analysis

Figure 28.2

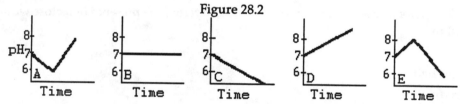

19) Which of the graphs in Figure 28.2 shows the pH in a culture flask as cells metabolize glucose and then protein?

   A) Graph A   B) Graph B   C) Graph C   D) Graph D   E) Graph E

   Answer: A
   Skill: Understanding

20) All of the following are industrial products producted by microbes *except*
   A) Amino acids in food supplements.
   B) Antibiotics.
   C) Industrial enzymes.
   D) Uranium.
   E) Vitamin B$_{12}$ and riboflavin.

   Answer: A
   Skill: Understanding

21) All of the following are produced using microbial fermentations *except*
   A) Aspartame.   B) Citric Acid.   C) MSG.   D) Riboflavin.   E) Saccharin.

   Answer: E
   Skill: Recall

22) All of the following are produced using microbial fermentations *except*

    A) Blue cheese.

    B) Ice cream.

    C) Sour Cream.

    D) Yogurt.

    E) None of the above.

    Answer: B
    Skill: Recall

23) Which of the following is an oxidation reaction that *Thiobacillus ferooxidans* might do?

    A) Fe^{3+} ---> Fe^{2+}

    B) U^{4+} ---> U^{6+}

    C) Cu^{2+} ---> Cu^{+}

    D) Cu^{+} ---> Cu^{0}

    E) Al^{3+} ---> Al^{2+}

    Answer: B
    Skill: Understanding

24) Methane made from biomass is produced by

    A) Anaerobic respiration.

    B) Fermentation.

    C) The Krebs cycle.

    D) Oxidation.

    E) None of the above.

    Answer: A
    Skill: Understanding

25) All of the following are industrial enzymes made by microbial fermentations *except*

    A) Glucose isomerase.

    B) Rennin.

    C) Proteases.

    D) Cellulases

    E) None of the above.

    Answer: E
    Skill: Recall

26) Ethonol for automobile fuel is produced from corn by
   A) Anaerobic repiration.
   B) Fermentation.
   C) The Krebs cycle.
   D) Oxidation.
   E) None of the above.

   Answer: B
   Skill: Understanding

27) The following steps are required for making chees. What is the second step?
   A) Enzymatic coagulation of milk
   B) Fermentation of curd
   C) Inoculate with lactic acid bacteria
   D) Press curd
   E) Separate curds and whey

   Answer: B
   Skill: Analysis

28) As cheese ages it gets
   A) More acidic.
   B) More whey.
   C) More protein.
   D) Saltier.
   E) None of the above.

   Answer: A
   Skill: Recall

29) Assume that you bought a semisoft cheese such as blue cheese and put it in your refrigerator. Three weeks later you cut the cheese and notice the center is now liquid because
   A) Lactic acid bacteria have grown in the center.
   B) Fungal enzymes have digested the curd.
   C) Fungal enzymes have made more curd.
   D) Fungi have accumulated in the center .
   E) None of the above.

   Answer: B
   Skill: Understanding

30) Your friend says he had stored a semisoft cheese (blue cheese) in his refrigerator for three weeks. He asks you why the outer "skin" of the cheese is so much thicker than it was when he originally purchased the cheese. You tell him that

A) Lactic acid bacteria have grown on the outside.

B) Fungal enzymes have digested the curd.

C) Fungal enzymes have made more curd.

D) Fungi have been growing.

E) None of the above.

Answer: D
Skill: Understanding

31) You are growing *Bacillus subtilis* in a bioreactor and notice that the growth rate has slowed and the pH has decreased. You suspect the bacteria are

A) Using the Krebs cycle.

B) Fermenting.

C) Photosynthesizing.

D) Using proteins.

E) None of the above.

Answer: B
Skill: Understanding

32) You are growing *Bacillus subtilis* in a bioreactor and notice that the growth rate has slowed and the pH has decreased. What could you do?

A) Add glucose.

B) Add lactose.

C) Add peptides.

D) Add oxygen.

E) None of the above.

Answer: D
Skill: Understanding

33) Radiation and canning are used for all of the following reasons *except* to

A) Prevent diseases.

B) Prolong the shelf life of foods.

C) Prevent spoilage of foods.

D) Sell more food.

E) All of the above.

Answer: D
Skill: Analysis

34) Radiation is used for all of the following *except*

    A) Foods that cannot be heated.

    B) Sterilizing food.

    C) Killing *Trichinella*.

    D) Killing insect eggs and larva.

    E) Preventing sprouting.

Answer: B
Skill: Analysis

35) Canning preserves food by

    A) Aseptic packaging.

    B) Chemicals.

    C) Heating.

    D) Radiation.

    E) All of the above.

Answer: C
Skill: Recall

36) Microorganisms themselves are commercial products. Which of the following microbes is available in retail stores?

    A) *Bacillus thuringiensis*

    B) *Lactobacillus*

    C) *Rhizobium*

    D) *Saccharomyces*

    E) All of the above

Answer: E
Skill: Understanding

Figure 28.3

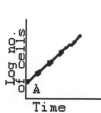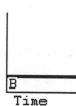

37) Figure 28.3 shows the growth curves of *Acetobacter* under different conditions. Which culture would have the highest amount of acetic acid?

    A) Graph A   B) Graph B   C) Graph C   D) Graph D   E) Graph E

Answer: A
Skill: Analysis

38) Canning works to preserve foods because of this:

A) Anaerobic environment

B) Anaerobic environment and heat

C) Heat

D) Mutations

E) pH

Answer: B
Skill: Analysis

39) Aseptic packaging works to preserve foods because of this:

A) Anaerobic environment

B) Anaerobic environment and heat

C) Heat

D) Mutations

E) pH

Answer: C
Skill: Analysis

40) Radiation works to preserve foods because of this:

A) Anaerobic environment

B) Anaerobic environment and heat

C) Heat

D) Mutations

E) pH

Answer: D
Skill: Analysis

41) Fermentation works to preserve foods because of this:

A) Anaerobic environment

B) Anaerobic environment and heat

C) Heat

D) Mutations

E) pH

Answer: E
Skill: Analysis

42) What will be produced if wine is aerated?
   A) Ethyl alcohol + CO_2
   B) $CO_2 + H_2O$
   C) CH_4
   D) Acetic acid
   E) None of the above

   Answer: D
   Skill: Understanding

43) Grape juice before fermentation is called must. What will be produced if must is aerated?
   A) Ethyl alcohol + CO_2
   B) $CO_2 + H_2O$
   C) CH_4
   D) Acetic acid
   E) None of the above

   Answer: B
   Skill: Understanding

44) A product made late in log phase and in stationary phase is called
   A) Fermentation.
   B) An idiotypic metabolite.
   C) A secondary metabolite.
   D) Undesirable.
   E) None of the above

   Answer: C
   Skill: Recall

Figure 28.3

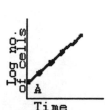 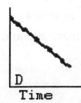 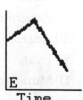

45) Which of the growth curves in Figure 28.3 will give the highest yield of a secondary metabolite?
   A) Graph A   B) Graph B   C) Graph C   D) Graph D   E) Graph E

   Answer: C
   Skill: Analysis

ESSAY QUESTIONS

1) Why does fermentation preserve foods?

   Answer:

   Skill:

2) Why would a farmer purchase *Rhizobium? Bacillus thuringiensis?*

   Answer:

   Skill:

3) Most of the world's population relies on wheat for food. Research is being conducted to produce wheat with a higher protein content. Design a biotechnological approach to improving the amino acid content of wheat.

   Answer:

   Skill: